Richard E. Prange   Steven M. Girvin
Editors

The Quantum Hall Effect
Second Edition

Foreword by Klaus von Klitzing

With Contributions by
Marvin E. Cage, Albert M. Chang, Steven M. Girvin,
F. Duncan M. Haldane, Robert B. Laughlin, Richard E. Prange,
Adrianus M.M. Pruisken, and David J. Thouless

With 116 Illustrations

Springer-Verlag
New York Berlin Heidelberg
London Paris Tokyo Hong Kong

Richard E. Prange
Department of Physics and Astronomy
University of Maryland
College Park, MD 20742
U.S.A.

Steven M. Girvin
Department of Physics
Swain Hall-West 117
Indiana University
Bloomington, IN 47405
U.S.A.

Series Editors

Joseph L. Birman
Department of Physics
The City College of the
City University of New York
New York, NY 10031
U.S.A.

H. Faissner
Physikalishes Institut
RWTH Aachen
5100 Aachen
Federal Republic of Germany

Jeffrey W. Lynn
Department of Physics and Astronomy
University of Maryland
College Park, MD 20742
U.S.A.

Library of Congress Cataloging-in-Publication Data
The Quantum Hall effect / edited by Richard E. Prange and Steven M.
  Girvin ; foreword by Klaus von Klitzing with contributions by Marvin
  E. Cage . . . [et al.]. — 2nd ed.
    p. cm. — (Graduate texts in contemporary physics)
  Includes bibliographical references.
  ISBN 0-387-97177-7
    1. Quantum Hall effect.   I. Prange, Richard E.   II. Girvin,
Steven M.   III. Series.
QC612.H3Q36   1990
537.6—dc20   89-21926
                                                              CIP

© 1987, 1990 by Springer-Verlag New York Inc.
All rights reserved. This work may not be translated or copied in whole or in part without the written permission of the publisher (Springer-Verlag, 175 Fifth Avenue, New York, New York 10010, USA), except for brief excerpts in connection with reviews or scholarly analysis. Use in connection with any form of information storage and retrieval, electronic adaptation, computer software, or by similar or dissimilar methodology now known or hereafter developed is forbidden.
The use of general descriptive names, trade names, trademarks, etc. in this publication, even if the former are not especially identified, is not to be taken as a sign that such names, as understood by the Trade Marks and Merchandise Marks Act, may accordingly be used freely by anyone.

Printed and bound by R.R. Donnelley & Sons, Harrisonburg, Virginia.
Printed in the United States of America.

9 8 7 6 5 4 3 2 1

ISBN 0-387-97177-7 Springer-Verlag New York Berlin Heidelberg
ISBN 3-540-97177-7 Springer-Verlag Berlin Heidelberg New York

Foreword: Some Remarks on the Discovery of the Quantum Hall Effect

It is a pleasure to accept the invitation of the Editors to write a foreword to this volume on the quantum Hall effect. I have been asked to briefly describe the circumstances surrounding the original discovery and to comment on the many subsequent developments. I shall begin with the former.

The first inversion layer Hall conductivity measurements in strong magnetic fields were done by S. Kawaji and his colleagues in 1975. Using a somewhat different experimental arrangement which measured the Hall voltage rather than the Hall current, Th. Englert and I had found flat Hall plateaus in 1978. In his diploma work, my student G. Ebert tried to analyze the deviation of the Hall conductivity σ_{xy} from the classical curve. No explanation of the experimental data was possible on the basis of published theories, which at the time did not include the effect of localization.

During the winter of 1980 I was working at the High-Field Magnet Laboratory of the Max Planck Institute in Grenoble investigating anomalies in the Hall resistance and magneto-resistance of inversion layers with the aim of studying how localization effects would modify the results of the self-consistent-Born-approximation theory of T. Ando, Y. Matsumoto, and Y. Uemura. I was initially using samples supplied by G. Dorda of Siemens AG. The experimental arrangement was such that two small magnets were operated during the daytime and the large Bitter magnet which I was using ran only at night.

On the night of 4–5 February at approximately 2:00 a.m. I made several runs which exhibited distinct Hall steps and I realized that while the magneto-resistance and the behavior of the Hall resistance at the edges of the steps was non-universal, the Hall resistance on the plateaus appeared to be unexpectedly constant and reproducible. Various experimental difficulties such as electrical noise from the Bitter magnet prevented precision measurements at that time. However, the birthday of the quantum Hall effect was on this night: It was just the idea to analyze the Hall effect in the plateau region relative to the fundamental value $h/e^2 i$ expected in the simple one-electron picture for integer filling factors of Landau levels. Subsequent work in my laboratory in Würzburg using a superconducting solenoid confirmed the constancy of the Hall resistance both in Dorda's samples and in samples supplied by M. Pepper of the Cavendish Laboratory. With technical assistance from the Physikalisch-Technische Bundesanstalt in Braunschweig, an absolute measurement of the Hall resistance confirmed the fundamental quantization relation $R_H = h/e^2 i$ to an accuracy of about *1* part in *10⁵*.

Recalling the practical applications of the Josephson effect, my initial thinking was oriented toward the idea of a resistance standard, but various groups at national laboratories which are involved in high precision measurements of fundamental constants pointed out that, in addition, the quantized Hall resistance yields a new fundamental measure of the fine structure constant α.

These then were the initial events which led to the remarkable surge of interest within both the metrology and condensed matter physics communities in quantum transport in inversion layer systems. Subsequent developments have been many and varied and are described in detail in this volume.

Extensive work at several national measurement laboratories has improved the experimental limit on the precision of the quantization to a few parts in *10⁸* and verified the absolute accuracy to a few parts in *10⁷*. It is quite astonishing that it is the total macroscopic conductance of the Hall device which is quantized rather than some idealized microscopic conductivity. The reasons for this have been laid out in a beautiful series of theoretical developments beginning with the original paper of R.B. Laughlin. The *sine qua non* is the presence of persistent currents due to the nearly complete freedom from dissipation. The Anderson localization effects which I was originally hoping to find have turned out to be essential to the unexpected quantization. A full understanding of localization in the presence of a magnetic field has not yet been achieved, however.

The other major development is of course the discovery in 1982 of the fractional quantum Hall effect by D.C. Tsui, H.L. Störmer, and A.C. Gossard. The existence of Hall steps with rational fractional quantum numbers has stimulated tremendous interest in the many-body physics of interacting electrons in fractionally filled Landau levels. Rapid theoretical development beginning with the original wave functions proposed by R.B. Laughlin has established a very good, though not yet perfect, understanding of this phenomenon and has connected it at a fundamental level with a variety of areas including plasma physics, superfluidity, and topological excitations in modern field theories.

One of the nice things about being a physicist is the many opportunities for friendship and cooperation with colleagues around the world and the chance to see that developments in one area of physics have connections with, and implications for, quite different areas of physics. It was a great pleasure and privilege for me to have the discovery of the quantum Hall effect recognized by the awarding of the 1985 Nobel Prize in Physics. I hope that this will bring further recognition of the importance of basic semiconductor research.

Stuttgart   Klaus von Klitzing

Preface to the Second Edition

During the three years since the publication of the first edition of this volume, the field of quantum Hall effect research has been quite active and several remarkable experimental developments have occurred. There have also been some very interesting new theoretical papers on the fractional effect and on possible connections between it and high-temperature superconductivity. A computer database survey reveals that about 1500 journal articles and abstracts for conferences have been published on the quantum Hall effect and related topics since the appearance of the first edition. This second edition contains a new Appendix whose purpose is to review some of the more important of these recent developments. Clearly we can only touch upon a few of the highlights in this rapidly expanding field. The reader may also wish to consult Chakraborty and Pietiläinen's detailed review of the FQHE (Chakraborty and Pietiläinen 1988).

Another feature of the second edition is a subject index. The editors are grateful to Carlo M. Canali for his valuable assistance in creating this index.

One of us (SMG) is also grateful to Diane Girvin, Michael Johnson, Allan MacDonald, Elena Fraboschi, and Geoffrey Canright for helpful comments and assistance in the production of the Appendix.

University of Maryland College Park	R.E. Prange
Indiana University Bloomington	S.M. Girvin

Preface

These notes resulted from a lecture course given at the University of Maryland in the Fall Semester, 1985. This course was the response to a request by Joe Redish, then chairman of the Department of Physics and Astronomy at Maryland. Professor Redish had conceived and arranged with Springer-Verlag for the publication of a new *Graduate Texts in Contemporary Physics* series and asked the Maryland faculty for exciting topics and lectures suitable for the Series. The quantum Hall effect certainly qualifies as such an exciting topic, and as of this writing, it has not generated even a comprehensive review article. [In organizing the course we learned that two such articles are in preparation.]

It was also realized that the University is situated fairly close to the laboratories where a good part of the important experimental and theoretical work is being carried out. Among these labs is the National Bureau of Standards in Gaithersburg, Maryland. Thus a natural collaboration between Steven Girvin of that institution and Richard Prange of the University of Maryland came about.

To take advantage of the proximity of so many of the major laboratories in the field a number of the leading workers were invited to a lecture and to provide the notes for a chapter. The intention was to provide a lecture notes volume which is more informal and pedagogical [and thus less complete] than a review paper or monograph. It is nevertheless intended to be authoritative and to cover the major aspects of the subject. In fact, several of the authors worked harder than we had anticipated, and have provided what is really a full-fledged review. In particular, Chapters 2 and 6 together give a very thorough coverage of the experimental situation.

By and large, the material in the lectures is already in the literature, but some new or previously unpublished material is to be found, particularly in Chapters 5, 6, 8, and 10. Most of the chapters should be accessible to students who have had an introductory graduate course in quantum mechanics. An exception is Chapter 5, which uses field-theoretic methods. The material in the remainder of the volume can be understood without mastering this chapter, however. One advantage of Chapter 5 is that it may explain the whole subject to those field theorists who have forgotten (or don't want to know) about theories of the solid state. In short, this book aims to be the place that the graduate student or non-expert physicist would go to learn the major aspects of the subject.

To further this aim, the chapters have been rather freely edited in order to avoid needless redundancy and unnecessary changes of notation. Because it was felt that prompt publication was important, these editorial improvements have been only partially achieved. The editors must take responsibilty for any errors, omissions, or misattributions which may have crept in.

In any case, it should be realized that each chapter is not meant to cover only or all of the particular author's work. It should also be realized that, although we have attempted to be complete in citing the literature, a main purpose is to provide the student with an easy access to that literature which would not be well served if all the papers in the field were indiscriminately cited. We regret it if important papers have been overlooked.

As one of the first publications in the *Graduate Texts in Contemporary Physics* series, we suffered the usual vicissitudes of such ventures, and there may be some relatively minor (we hope) typographical errors and inconsistencies in the final product. We trust that these do not detract seriously from the usefulness of the work.

We would particularly like to thank Professor Dr. Klaus von Klitzing for kindly consenting to write the foreword despite the demands being placed on his time at this especially busy juncture in his life.

The editors wish to acknowledge the help and forbearance of their colleagues and of the staffs of their respective institutions. In particular we thank Rachel Olexa, one of the world's fastest typists, and Sankar Das Sarma and Craig Van Degrift, who in addition to providing scientific advice made timely loans of computers. We also much appreciate the interest and technical help of Springer-Verlag New York. Finally we thank our families who were very understanding of two physicists who were learning that it's not so easy to be an editor.

University of Maryland
College Park

R.E. Prange

National Bureau of Standards
Gaithersburg

S.M. Girvin

Notation

A remark on the labelling of equations is in order. These are labelled by chapter, section, and number. When referenced within the same chapter, however, the chapter designation is usually dropped. A similar remark holds for figures and tables.

Acknowledgments

We thank the authors and journals in the table below for granting permission to reproduce their figures in this book. Reference to the article may be found in the figure caption.

Author	Journal	Figure
T Ando	J Phys Soc Japan	3.7
		3.8
		3.9
JP Eisenstein	Applied Phys Letters	6.4.1
	Phys Rev Letters	6.4.2
DR Hofstadter	Phys Rev	4.1
MA Paalanen	Phys Rev	6.4.11
HL Störmer	Physica	6.2.2
DC Tsui	Applied Phys Letters	6.2.1
		6.2.3

Contents

Foreword: Some Remarks on the Discovery of the Quantum Hall Effect
Klaus von Klitzing v
Preface to the Second Edition vii
Preface ix
Notation xi
Acknowledgements xii
Contributors xvii

CHAPTER 1
Introduction *Richard E. Prange* 1
   1.1. Introduction 1
   1.2. Preview of Coming Attractions 4
   1.3. The Ordinary Hall Effect 6
   1.4. Measuring the Conductance 7
   1.5. Introduction to the Quantum Case 9
   1.6. Impurity Effects 12
   1.7. Gauge Arguments 17
   1.8. Inversion Layers 22
   1.9. Acknowledgements 30
   1.10. Notes 30
   1.11. Problems 34

Part A The Integer Effect

CHAPTER 2
Experimental Aspects and Metrological Applications *Marvin E. Cage* 37
   2.1. Basic Principles 37
   2.2. The Devices 39

2.3.	The Basic Experiment	42
2.4.	Initial Experiments	43
2.5.	Precision Measurement Techniques	44
2.6.	Quantum Hall Resistors	46
2.7.	An Absolute Resistance Standard	50
2.8.	The Fine-Structure Constant	51
2.9.	Temperature Dependence of ρ_{xx}	51
2.10.	Temperature Dependence of ρ_{xy}	53
2.11.	Current Distribution and Edge Effects	56
2.12.	Current Dependence and Breakdown	58
2.13.	Hall Step Widths and Shapes	62
2.14.	Thermomagnetic Transport	65
2.15.	Magnetic Moment	66
2.16.	Magnetocapacitance	66
2.17.	Magneto-Photoresponse	66
2.18.	Conclusions	67
2.19.	Acknowledgements	67
2.20.	Notes	67

CHAPTER 3
Effects of Imperfections and Disorder *Richard E. Prange* 69

3.1.	Introduction	69
3.2.	The Impurity Potential	70
3.3.	Weak Potential	73
3.4.	Scattering Potential	74
3.5.	Smooth Potential	80
3.6.	Continuum Percolation Model	83
3.7.	General Potential	88
3.8.	Numerical Studies	89
3.9.	Acknowledgements	93
3.10.	Notes	93
3.11.	Problems	98

CHAPTER 4
Topological Considerations *David J. Thouless* 101

4.1.	Topological Quantum Numbers	101
4.2.	Quantum Hall Effect in a Periodic Potential	104
4.3.	Generalization of the Topological Interpretation	110
4.4.	Notes	115

CHAPTER 5
Field Theory, Scaling and the Localization Problem
Adrianus M.M. Pruisken 117

5.1.	Introduction	117
5.2.	Basic Notions	119
5.3.	The Search for the Principle	131
5.4.	Structure of the Effective Field Theory	146
5.5.	Instantons and Scaling	158
5.6.	Notes	171

Contents

Part B: The Fractional Effect

CHAPTER 6
Experimental Aspects   *Albert M. Chang* 175
6.1. Introduction 175
6.2. Conditions for the Observation of the FQHE and the
     Choice of Semiconductor Systems 177
6.3. The Fractional Quantum Hall Effect 183
6.4. Magneto-Transport in Two Dimensions 211
6.5. Future Experiments 228
6.6. Acknowledgements 231
6.7. Notes 231

CHAPTER 7
Elementary Theory: The Incompressible Quantum Fluid
Robert B. Laughlin 233
7.1. Introduction 233
7.2. Quantized Motion of Small Numbers of Electrons 234
7.3. Variational Ground State 244
7.4. Computational Methods 254
7.5. Fractionally Charged Quasiparticles 260
7.6. Exactness of the Fractional Quantum Hall Effect 276
7.7. More Than One Quasiparticle 280
7.8. Conclusions 301
7.9. Acknowledgements 301

CHAPTER 8
The Hierarchy of Fractional States and Numerical Studies
F. Duncan M. Haldane 303
8.1. Introduction 303
8.2. Pseudopotential Description of Interacting Particles
     with the Same Landau Index 305
8.3. The Principle Incompressible States: A Non-Variational Derivation ... 310
8.4. Quasiparticle and Quasihole Excitations 317
8.5. The Hierarchy of Quasiparticle–Quasihole Fluids 320
8.6. Translationally Invariant Geometries for Numerical Studies 325
8.7. Ground-State Energy Studies 329
8.8. Pair Correlations and the Structure Factors 333
8.9. Quasiparticle Excitations 341
8.10. Collective Excitations 343
8.11. Acknowledgements 352

CHAPTER 9
Collective Excitations   *Steven M. Girvin* 353
9.1. Introduction 353
9.2. Density Waves as Elementary Excitations 355
9.3. Collective Modes in the FQHE 357
9.4. Magnetophonons and Magnetorotons 360
9.5. Further Superfluid Analogies: Vortices 367

9.6.	Quasi-Excitons	373
9.7.	Magnetoplasmons and Cyclotron Resonance	374
9.8.	Other Collective Modes	376
9.9.	Role of Disorder	377
9.10.	Acknowledgements	378
9.11.	Notes	379
9.12.	Problems	380

Part C: The Quantum Hall Effect

CHAPTER 10
Summary, Omissions and Unanswered Questions *Steven M. Girvin* ... 381

10.1.	Integer Effect at Zero Temperature	381
10.2.	IQHE at Finite Temperatures	382
10.3.	Metrology	384
10.4.	Basic Picture of the Fractional Effect	385
10.5.	Remarks on the Neglect of Landau Level Mixing	388
10.6.	Wanted: More Experiments	388
10.7.	Towards a Landau–Ginsburg Theory of the FQHE	389
10.8.	Acknowledgements	398
10.9.	Notes	398
10.10.	Problems	399

APPENDIX
Recent Developments *Steven M. Girvin* ... 401

A.1.	Introduction	401
A.2.	Off-Diagonal Long-Range Order	403
A.3.	Ring Exchange Theories	415
A.4.	Chiral Spin Liquids, High-T_c, and the FQHE	417
A.5.	Even-Denominator and Spin-Reversed States	430
A.6.	Spin-Reversed Hierarchy	436
A.7.	Summary and Conclusions	438

References	439
Index	463

Contributors

Marvin E. Cage
Electricity Division
Center for Basic Standards
National Institute of Standards
and Technology
Gaithersburg, MD 20899 U.S.A.

Albert M. Chang
AT&T Bell Laboratories
Holmdel, NJ 07733 U.S.A.

Steven M. Girvin
Department of Physics
Swain Hall-West 117
Indiana University
Bloomington, IN 47405 U.S.A.

F. Duncan M. Haldane
Department of Physics
University of California,
San Diego
La Jolla, CA 92093 U.S.A.

Richard E. Prange
Department of Physics
and Astronomy
University of Maryland
College Park, MD 20742 U.S.A.

Robert B. Laughlin
Lawrence Livermore
National Laboratory
Livermore CA 94550 U.S.A.
and
Department of Physics
Stanford University
Stanford, CA 94305 U.S.A.

Adrianus M.M. Pruisken
Department of Physics
Columbia University
New York, NY 10027 U.S.A.

David J. Thouless
Department of Physics
University of Washington
Seattle, WA 98195 U.S.A.
and
Cavendish Laboratory
Cambridge CB3 OHE England

Klaus von Klitzing
Max-Planck-Institut
für Festkörperforschung
D-7000 Stuttgart 80
Federal Republic of Germany

CHAPTER 1

Introduction

RICHARD E. PRANGE

1.1 Introduction

The quantum Hall effect was discovered on about the hundredth anniversary of Hall's original work, and the finding was announced in 1980 by von Klitzing, Dorda and Pepper. Klaus von Klitzing was awarded the 1985 Nobel prize in physics for this discovery. In brief, it is found that under certain conditions in an effectively two-dimensional system of electrons subjected to a strong magnetic field, the conductivity tensor σ [Note 1] takes the form

$$\sigma = \begin{bmatrix} 0 & -ie^2/h \\ ie^2/h & 0 \end{bmatrix}. \qquad (1.1.1)$$

Here h is Planck's constant, e is the charge of the electron and i is a small integer. In other words, the current density $\boldsymbol{j}$ is directed precisely perpendicular to the electric field $\boldsymbol{E}$ according to

$$j_i = \sum_j \sigma_{ij} E_j$$

and it has the quantized magnitude

$$j/E = \sigma_{xy} \equiv \sigma_H = ie^2/h \ . \qquad (1.1.2)$$

The off-diagonal conductivity is thus given by a combination of fundamental constants.

The diagonal conductivity vanishes, so the system is dissipationless and thus is related to superconductivity and superfluidity. This is developed further in Chap. IX. Further, the vanishing of the dissipation is the essential ingredient in understanding the quantization of the Hall conductivity σ_H.

It should be emphasized that until recently, the conductivity was always found to be a *material dependent* quantity, which moreover depends on temperature, magnetic field, frequency and other variable conditions of the material or measurement. What is more, the directly measured quantities, the resistance or conductance, have a further dependence on the sample size and geometry and the nature of the contacts. Indeed, the fact that the combination of fundamental constants e^2/h has the dimension of a conductance or two-dimensional conductivity was not accorded much significance until recently. However, this combination of constants began to play a role in localization theory [Chap. V] as well as in earlier work on two-dimensional conductivity in a magnetic field. Still, the idea that the conductivity could be quantized was totally unanticipated.

There is a prehistory of the quantum Hall effect (Ando, Matsumoto and Uemura 1975, Kawaji, *et al.* 1975) which is discussed a bit in Chapter II. However, until the work of von Klitzing, *et al.* there was no inkling that Eq. 1.1 is essentially exact. It is now known to be correct to at least three parts in ten million. This is quite astounding, and has important implications in metrology. Notably it provides (since the speed of light c is an exactly defined quantity) a very accurate measure of the fine structure constant $e^2/\hbar c$, ($\approx 1/137$), surely one of the most important fundamental dimensionless constants of nature. It is very important as well in providing a new and convenient standard resistance. These matters will be discussed in Chapter II by Marvin Cage.

Introduction

The first theoretical papers were based on the model of *independent electrons* in the presence of *imperfections* and finally succeeded in understanding the main features of the earlier experiments. In the first half of this course we shall review that work.

However, the experimentalists (Tsui, Störmer and Gossard 1982b) had another great surprise in store. Namely, it was found that in some nearly ideal samples at very low temperatures, the value of i above could be replaced by one of a hierarchy of rational fractional values, f. In other words, the Hall conductivity can take on rational fractional values in units e^2/h, not just integer values. In particular, the denominator of the fraction $f = p/q$ is necessarily odd. The leading members of the hierarchy were found to be 1/3, 2/3, 2/5, etc. with 1/2 and 1/4 conspicuously absent. Daniel Tsui, Horst Störmer and Arthur Gossard won the 1984 Buckley Prize for this discovery. This completely unforeseen development is called the fractional or anomalous quantum Hall effect. Theoretically, electron Coulomb interactions are of the essence in this case. The theory of and experiments on the many body problem posed by fractional effect is the primary concern of the second half of this volume.

The present theory of the subject is quite convincing. In this course we discuss the main development of the material which is certainly close to the truth. However, there are still a few experts who think that some significant change will have to be made in our picture of the interacting system, particularly in the fractional case. For example, there are suggestions that the ground state of the ideal fractional Hall effect system is some kind of highly correlated Wigner lattice rather than being an incompressible liquid, the model proposed by Laughlin and adopted here. We will not attempt to discuss these alternatives exhaustively, although mention of them will be made where appropriate. In any case, there is no presently known alternate to the incompressible liquid picture which is at all complete or satisfactory. The present theory is also incomplete. In particular, questions which depend in detail on both interactions and impurities are not well answered. Specific unsolved problems will be mentioned throughout the book, and a summary made in the final chapter.

A number of short reviews of the QHE have appeared. A brief list of the more recent ones includes von Klitzing (1983), von Klitzing et al. (1983, 1984), Halperin (1983), Kawaji (1983), Laughlin (1984d), Cage and Girvin (1983a,b), Hajdu (1983), Störmer and Tsui (1983), and Ando (1983a).

1.2 Preview of Coming Attractions

In this section, the material to be covered in the remaining lectures will be previewed. In the rest of the first chapter, the elementary basis of the theory will be presented. This begins with the theory of the ordinary Hall effect and the simple solvable cases of two-dimensional electrons in a magnetic field. The elegant gauge arguments of Laughlin (1981) and Halperin (1982) which provide a more general understanding of the integer QHE and which are important in the fractional effect as well will be given. (The abbreviation IQHE and FQHE for integer and fractional quantum Hall effect will be used from now on.) We shall also describe the physics of inversion layers and heterojunctions, which are the systems in which the QHE is actually observed.

In Chapter II Marvin Cage discusses the basic IQHE experiments. He goes on to describe the most precise determinations of e^2/h presently made, and the implications for metrology. He also discusses other experimental matters, e.g., finite temperature effects.

In Chapter III we discuss the different types of scatterers which could influence the IQHE. The aim is to display the physics of electrons in such scattering arrays in as elementary a fashion as possible.

David Thouless discusses the homotopy properties of the IQHE in Chapter IV. The original papers considered the QHE in a periodic lattice potential. There have been several generalizations and the subject has taken on some importance even outside the QHE.

In Chapter V Adrianus Pruisken will bring field theoretic methods to bear on the problem. This chapter makes contact with the important recent developments in the general theory of quantum conductivity, namely localization and scaling. This methodology has led in particular to an understanding of the scaling properties of the conductivity.

Introduction

The experiments on the fractional QHE are described in Chapter VI by Albert Chang. These experiments are difficult in that they must be carried out at extremely low temperature and in extremely strong magnetic fields. Also, there are not many samples with the perfection required to observe the FQHE. Only a few laboratories in the world have successfully carried out experiments of this type. In addition, some important work on the integer QHE is described.

Chapter VII is devoted to the simplest picture of the fractional QHE, (for $f = 1/3$) which is that developed by the author of the chapter, Robert Laughlin, who won the 1986 Buckley Prize for this contribution as well as for his fundamental contributions to the theory of the integer quantum Hall effect. This picture represents the wave function of the fractional QHE state by a function whose norm can be interpreted as the partition function of a fictitious two-dimensional classical gas of charged rods.

Laughlin's theory has been advanced in several ways. One of these led to an understanding of the hierarchy of f values. Another class of efforts was numerical in nature. It allowed a check of theoretical ideas against the exact numerical results for a few (e.g., eight) electrons. These matters are discussed in Chapter VIII by Duncan Haldane.

The question of the excitation spectrum, particularly in the fractional QHE, is the material of Chapter IX presented by Steven Girvin. In addition to the direct construction of excited states based on Laughlin's wave function, and the numerical work, there is an elegant approximation scheme which exploits the analogy of the QHE with superfluid liquid helium.

The final chapter is devoted to a summary of the material and to corrections to the main picture of the fractional QHE, or more precisely, why there aren't corrections. Steven Girvin discusses in this chapter certain approximations made in Laughlin's picture, which is much more accurate than might *a priori* be estimated. The question of a possible transition to a Wigner lattice is also discussed. Finally, an attempt will be made to assess the theoretical and experimental position and to guess the future trends of research in this area.

1.3 The Ordinary Hall Effect

The simplest theory of the conductivity is semiclassical with quantum mechanics entering very indirectly. It is based on the concept of mean free time, τ_0 and/or mean free path, ℓ_0. In this picture, an electron of charge $-e$ and (Fermi) velocity v_F goes an average distance $\ell_0 = v_F\tau_0$ before scattering to a new velocity whose average is zero. If the electron (mass m) is in a weak electric field E it will on the average pick up an incremental velocity $\Delta v = -eE\tau_0/m$ between each scattering. Adding up the contribution of all the electrons leads to a current density $j = \sigma_0 E$, with σ_0 given by the famous formula

$$\sigma_0 = ne^2\tau_0/m. \qquad (1.3.1)$$

The electron number density is denoted by n. Quantum effects enter through the band structure, which changes the electron mass m_e to an effective mass, m, (often called m^*) and through the value of τ_0. There is also the Pauli principle which requires that only the electrons very near the Fermi surface be active. However, in the independent electron approximation, this effect does not change Eq. 3.1.

In the presence of a magnetic field B, the electron's path is curved, due to the Lorentz force $-ev{\times}B/c$. Taking this into account, it is easily found that

$$j = \sigma_0 E - \sigma_0 j{\times}B/nec. \qquad (1.3.2)$$

From now on we specialize to the case of two dimensions, (xy) embedded in ordinary three-dimensional space, with B in the z-direction. Then the resistivity tensor can be easily read off from (1.3.2). It is

$$\rho = \begin{bmatrix} \rho_0 & B/nec \\ -B/nec & \rho_0 \end{bmatrix} \qquad (1.3.3)$$

with $\rho_0 = 1/\sigma_0$. The conductivity tensor σ is the matrix inverse of ρ with components given by

$$\sigma_{xx} = \frac{\sigma_0}{1 + \omega_c^2 \tau_0^2}, \quad \sigma_{xy} = \frac{nec}{B} + \frac{1}{\omega_c \tau_0} \sigma_{xx}. \quad (1.3.4)$$

The cyclotron frequency ω_c is given by $\omega_c = eB/mc$. These expressions have been extensively compared with experiment and are often good approximations.

Before continuing to the next section, note the very simple form for the off-diagonal, or Hall resistivity (sometimes denoted ρ_H). It depends on the sign of the charge carriers but is independent of the scattering parameters. This detail is a consequence of the simple form of the scattering assumed above. Generally, ρ_H will depend weakly on the scattering parameters.

1.4 Measuring the Conductance

Section 1 is an oversimplication in one very important respect. Namely, the conductivity tensor is *not* what is measured. Rather, what is directly observed are the conductances and/or resistances. The conductance and resistance are related to the conductivity and resistivity by factors involving the sample dimensions, even assuming that the sample is otherwise completely homogeneous. The unusual and remarkable fact is that for the QHE the experimental "ances" are more fundamental than the "ivities", reversing the standard situation.

One way to measure these quantities is as follows. Electrons are passed from a 'source' S to a 'drain' D along a sample of rectangular shape, giving a total current I. Connections are made on the sides of the sample, as in Fig. 1.1. The measured quantities are the current I, the longitudinal voltage difference V_L between contacts A and B, and the transverse or Hall voltage difference V_H between contacts A and C. The length L between AB and the width W between AC can be measured as well, but these quantities are not defined to part per million accuracy.

Let it now be assumed that the current density is uniform and parallel to the long edge of the sample. The electric field *E* is also uniform but has a component, the Hall field, perpendicular to the current.

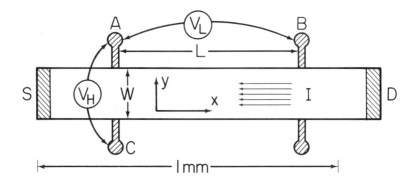

Figure 1.1

These approximations will be good if the length of the sample is large compared with the width W, and the connections are far from the ends. The current density is then $j = I/W$. The longitudinal electric field is $E_L = V_L/L$. Since I is fixed we compute the resistance. The longitudinal resistance $R_L = V_L/I = \rho_{xx}(L/W)$. Similarly the Hall resistance $R_H = V_H/I = \rho_H$. In this case (in two space dimensions) the dimensions of resistance are the same as those of resistivity, and indeed, for the Hall resistance, the size factors drop out, although if the geometry is less than ideal, the contacts are not points, etc., one might expect some modest geometric corrections. Note that the resistances R_L and R_H are defined independently of assumptions about the uniformity of the current distribution.

In Fig. 1.2 is a diagram (for the QHE case) of the normalized measured inverse Hall resistance $h/e^2 R_H$ as a function of normalized density nhc/eB. The longitudinal resistance is sketched as well. Rather than $1/R_H$ being the straight line of unit slope suggested by Eq. 3.3, it is found that there are a series of steps. The center of each step is very flat and very accurately given by the sequence of integers i. Where the Hall resistance is flat, the longitudinal resistance practically vanishes, so (1.1.1) holds. Between steps the longitudinal resistance is finite. The width of the flat part of the step varies according to the sample and temperature. In the best samples at very low temperatures the steps are wide and precise.

Introduction

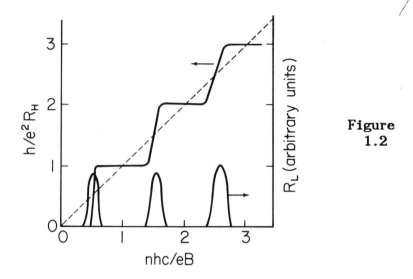

Figure 1.2

Note that, in some cases at least, it is possible to vary the density and magnetic field independently and to confirm that R_H depends only on their ratio.

It is evident from the remarkable accuracy of the data that the Hall resistance is free of geometrical corrections. A reason for this is the vanishing of the diagonal elements of (1.1.1), from which it follows that the separation and finite size of the electrodes (A,B,C in Fig. 1) do not matter [Probs. 1,2]. However, the experimental results are much deeper. For example, the current is not spatially uniform [Note 1.2, Chap. VI] so that if a conductivity is definable at all, it is not uniform as is (1.1.1).

1.5 Introduction to the Quantum Case

In this section we take up the simplest quantum problem, that of independent two-dimensional spinless [Note 3] electrons in a perpendicular magnetic field. This is, of course, old and standard stuff, but it will serve to establish the notation. To write down Schrödinger's equation explicitly it is necessary to choose a gauge. Two gauges which are particularly convenient are the Landau gauge, and the rotationally invariant symmetric gauge. The latter gauge is most useful in the study of interacting electrons. We shall give the results for it in Problem 3. In the Landau gauge, the vector potential **A** is

$$A_x = -yB, \quad A_y = 0. \tag{1.5.1}$$

Then Schrödinger's equation for the wave function ψ is

$$\frac{\hbar^2}{2m}\left\{\left[\frac{1}{i}\frac{\partial}{\partial x} - \frac{eB}{\hbar c}y\right]^2 - \frac{\partial^2}{\partial y^2}\right\}\psi = \mathcal{H}_0\psi = E\psi. \tag{1.5.2}$$

We take $\psi \propto e^{ikx}\varphi(y)$. Then

$$\frac{\hbar\omega_c}{2}\left\{-\ell^2\frac{\partial^2}{\partial y^2} + \left[\frac{y}{\ell} - \ell k\right]^2\right\}\varphi = E\varphi. \tag{1.5.3}$$

Thus φ satisfies a shifted harmonic oscillator equation. We have introduced the magnetic length

$$\ell \equiv (\hbar c/eB)^{1/2}$$

which is a fundamental scale of the problem. It is in the range 50-100 Å and is comparable with several other lengths of the system. It is worth noting that ℓ is independent of material parameters.

The different oscillator levels are labelled by $n = 0,1,2,...$, and define the Landau levels. These solutions are

$$\varphi_{nk}(y) \propto H_n(y/\ell - \ell k)\exp[-(y - \ell^2 k)^2/2\ell^2] \tag{1.5.4}$$

where H_n is an Hermite polynomial. The energy levels are

$$E_{nk} = \hbar\omega_c(n + \tfrac{1}{2}) \tag{1.5.5}$$

independent of k. The wave function in the y-direction is centered at $y = \ell^2 k$. If the y dimensions of the system are confined to $0 < y < W$ we see that $0 < k < W/\ell^2$. Let us impose periodic boundary conditions $\psi(x,y) = \psi(x + L,y)$. (A more careful treatment of the two boundary conditions is sometimes necessary [Note 4].) Then $k = 2\pi p/L$ with p an integer. Together with the condition on the range of k this implies that the total number of states N in a Landau level is $LW/2\pi\ell^2$,

Introduction

or that the number of states per unit area of a full Landau level is

$$n_B = 1/2\pi\ell^2 = eB/\hbar c. \qquad (1.5.6)$$

In this idealized model, then, the electronic energies lie in equally spaced but highly degenerate levels. The spacing and degeneracy are inversely proportional so that if the degenerate states were broadened into a uniform distribution, the density of states of the distribution would be $n_B/\hbar\omega_c = m/2\pi\hbar^2$, which is just the density of states of free electrons of mass m in two dimensions.

We take this opportunity to remark that the states ψ are extended in the x direction, but confined in the y direction. However, because of the massive degeneracy, linear combinations of states ψ exist which are confined in both directions: they are localized. [Problem 3 shows that in the rotationally invariant gauge, the natural eigenstates are localized.] Applied electric fields and/or impurity potentials lift the Landau level degeneracy and with it the freedom to choose between extended and localized states.

If a Landau level is full, the Fermi level must lie in the gap between occupied levels. It is plausible then that there is no scattering, i.e. τ_0 is infinite and ρ_0 vanishes. We denote the *filling factor* v by

$$v = n/n_B.$$

In this ideal case v is an integer i. Substitution of $n = in_B$ into (1.3.3) gives a resistivity tensor whose inverse is (1.1.1).

Another argument leading to the same result is as follows. The situation being considered is that of crossed electric and magnetic fields. According to standard relativistic transformation arguments (e.g., Jackson 1975), in a frame of reference moving with velocity $v = cE\times B/B^2$ with respect to the laboratory frame, the electric field vanishes. This gives the current density $j = -nev$ with the proper direction and magnitude if the Landau level is full.

This is a very powerful argument, which depends only on the translational invariance of the system. It is independent of electron-electron interaction. If such translational invariance is in force, it follows that $j = -nev$ no matter what the value of n. It is certainly not true that (1.1.1) is observed only for densities $n = in_B$. Rather as shown in Fig. 2, the conductivity is that of an ideal system whose density is in_B, as long as the density and magnetic field satisfy approximately the relation $v \approx i$.

To conclude this section, we introduce magnetic units, which are often convenient. The unit of length is taken to be ℓ, the unit of time $1/\omega_c$, and the unit of mass m, implying that the unit of action is $\hbar$. These units are used from time to time throughout these lectures to simplify the writing.

1.6 Impurity Effects

We have just seen that a breakdown of translational invariance is essential to the observed QHE. This breakdown is, in practice, provided by impurities or other imperfections. The traditional way of dealing with imperfections is through the concepts of mean free time and path, τ_0 and ℓ_0. We consider for now elastic scattering and discuss inelastic effects later. Then, for example, it might be guessed that each energy level should be given a complex part, i.e. $E_{nk} \to E_{nk} + i\hbar/\tau_0$. Thus, as illustrated in Fig. 3, each Landau level is broadened into a set of levels with a Lorentzian density of states [Note 5]. While the Lorentzian approximation has some merit when $\omega_c \tau_0$ is not too large, it fails at high enough fields. Indeed, the classical picture of the electronic motion in a magnetic field is that the electrons move in cir-

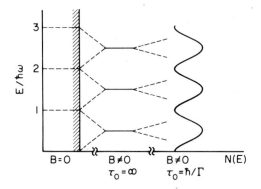

Figure 1.3. Schematic diagram of the density of states vs. energy for
a) zero field
b) zero scattering
c) finite field and scattering.

Introduction

cles. If the impurities are few and far between compared with the cyclotron radius, some electrons will circle forever in regions away from any impurities [and thus have no change in energy], while others might be constantly under the influence of an impurity potential [and thus suffer a shift in energy.]

This classical picture turns out to be the quantum picture as well, at least for some kinds of impurity systems, except that the cyclotron radius, v_F/ω_c, is replaced by the magnetic length ℓ. In addition, account must be made of the electric field which causes some of the electrons to drift through the array of impurities.

To get started, we solve the problem given by $\mathcal{H}_0$ of (1.5.2) with the addition of a single δ-function impurity potential, and an electric field in the y-direction, described by the potential eEy. The solution to this problem is sufficiently instructive that it enables one to guess correctly the results of more general situations (Prange 1981). Using magnetic units, the potential can be written as vy. [The drift velocity v ($= cE/B$) is small (10^{-4} in magnetic units) so perturbation theory (the Kubo formula) can be used. This method is used in Chapter IV. See also (Aoki and Ando 1981).] The eigenstates of $\mathcal{H}_v = \mathcal{H}_0 + vy$ are shifted harmonic oscillator states with φ_{nk} given in (1.5.4). Thus,

$$\psi_{nk}(x,y) = L^{-1/2} e^{ikx} \varphi_{nk}(y + v) \quad (1.6.1)$$

Again k has the form $k = 2\pi p/L$ with p an integer. The degeneracy of the Landau level is completely lifted, with

$$E_{nk} = n + vk \quad (1.6.2)$$

[We drop the zero point energy.] The k dependence of the energy is simple to understand. The kth state vanishes except near the line $y = k$ where the potential energy is vk. Note that the velocity in the x direction given by the standard group velocity formula

$$dE_{nk}/dk = v \quad (1.6.3)$$

agrees with the considerations of Section 5. Also note that the states are localized in one direction but are necessarily extended in the other direction because of the introduction of the electric field. In fact they extend along contour lines of constant potential energy.

Now add the potential energy of the scatterer $V_I = \lambda \delta(x - x_0) \delta(y - y_0)$. By expanding the eigenstates of $\mathcal{H}_v + V_I$ in terms of the states (1.6.1), it is easy to obtain the following relation determining the energy levels:

$$1 = \lambda \sum_{nk} \frac{|\psi_{nk}(x_0, y_0)|^2}{E - E_{nk}} . \qquad (1.6.4)$$

Without going through the details, (Prange 1981) we describe the results. Each E solving (1.6.4) lies between two unperturbed levels, (1.6.2) except possibly for one level lying below all unperturbed levels when $\lambda < 0$. There is one state completely localized for each Landau level. The energy of this state, for small v, lies well away ($\approx \lambda$) from the original Landau level position $n\hbar\omega_c$. An approximation to the lowest such state (at $x_0 = y_0 = 0$) is

$$\psi_{loc} \propto e^{-(x^2 + y^2)/4} e^{ixy/2} .$$

The energies of such states are shifted up or down from $n\hbar\omega_c$ according as λ is repulsive or attractive. [Repulsive potentials can bind electrons as easily as attractive ones, since there is a gap in the spectrum where the bound state can exist.] The states are centered on the position of the delta function. These localized states can carry no current, so the current does not depend on whether localized states are occupied by electrons.

All other energy levels solving (1.6.4) are constrained to lie very close to an unperturbed energy (1.6.2), (by an amount proportional to $1/L$). In the limit $E \to 0$ they are unshifted from $n\hbar\omega_c$. These states are extended and can be labelled by the integers n (Landau level) and $p = Lk/2\pi$ corresponding to the unperturbed state closest in energy below (above) it. There are $N-1$ extended states in each Landau level where N is given before Eq. 5.6. It may be shown

Introduction

[Chapter III] that the current carried by the extended states is

$$I = -\frac{e}{h}\sum_{np}(E_{np+1} - E_{np}) = -\frac{e}{h}\sum_{n}(E_{np_{max}} - E_{np_{min}}). \quad (1.6.6)$$

The sum is over occupied states. The state n, p_{max} is the last occupied state at the upper edge of the sample. The energy $E_{np_{max}}$ is $-e$ times the electrochemical potential at the upper edge of the sample and is independent of n. Correspondingly the energy $E_{np_{min}}$ is proportional to the chemical potential at the lower edge of the sample. Although the only states appearing in the final formula are edge states, current is carried throughout the bulk in general. Edge states are discussed further in the next section. It should be noted carefully that the current is proportional to the difference between maximal occupied energy levels at the edges of the sample, namely to the *electrochemical potential difference* between the edges $y = 0, W$. [It can happen in tricky cases that the experimentally meaningful electrochemical potential differs from the electric potential.]

The impurities reduce the number of states carrying the current but the current remains the same as long as all extended states are occupied. We shall see in Chapter III that electrons passing close to the impurity speed up and carry just enough more current to compensate for the current not carried by the localized state.

Further insight is gained by making the lowest Landau level approximation, valid if λ is small and for low enough density. This approximation restricts the basis to the states (1.5.4) with $n = 0$, for vanishing electric field [Note 6]. There are $N-1$ linear combinations of states which vanish at the point x_0, y_0, and thus have unshifted energies. The energy of the remaining state is shifted proportional to λ. If we generalize to the case where there are $M < N$ δ-function impurities, there will be $N-M$ states which are unshifted in energy, and M states which are shifted. The former are expected to be extended and the latter are expected to be completely localized. The density of localized states in energy depends on the distribution of λ's. For $M > N$ or other types of impurity po-

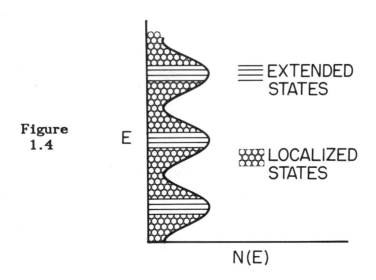

Figure 1.4

tential the singular δ-function in the density of states can disappear [Note 6]. However, it is still expected that some states will be localized and others extended, with the extended ones having energies close to the original $n\hbar\omega_c$. In Fig. 4 we illustrate this additional structure to the density of states.

A crucial point is that *only the extended states can carry current* at zero temperature. Therefore, if the occupation of the extended states does not change, neither will the current change (assuming the states themselves do not change). The flat part of the steps in Hall current are precisely those regions for which the occupation of extended states does not change.

Another crucial point is that the energy of the extended states in the center of the steps is *well away from the Fermi level*. Therefore, low-energy inelastic processes, e.g., phonon absorption at low but finite temperature, can have no influence on the occupation of these states. More precisely, the influence of inelastic processes will be small in proportion to $\exp(-W/k_B T)$ where W is the energy necessary to raise an extended-state electron to the Fermi level, k_B is Boltzmann's constant, and T is the temperature.

To summarize, it is expected that the density of states evolves from sharp Landau levels to a broader spectrum of levels as the impurity potential is turned on. Numerically it turns out that the resulting total density of states calculated in the earlier theories

Introduction

[Fig. 3.8] may not be a bad approximation (Ando 1983b). However, there are two kinds of levels, *localized* and *extended*, and it is expected that the extended states occupy a core near the original Landau level energy while the localized states are more spread out in energy. There are various questions about the width in energy of this core for realistic impurity potentials, etc. However, if it is assumed that the core of extended states (if any), in each Landau level, are separated from those of other Landau levels by an energy gap containing only localized states, then an elegant argument due to Laughlin and to Halperin shows that a) extended states indeed exist at the cores of the Landau levels, b) if these states are full, (i.e. the Fermi level is not in the core of extended states) then they carry exactly the right current to give (1.1.1). This argument is the subject of the next section.

We also remark that the existence of the localized states helps in another way. Namely, as the density is increased (or the magnetic field is decreased) the localized states gradually fill up without any change in occupation of the extended states, thus without any change in the Hall conductance. For these densities the Hall conductance is on a step of Fig. 2 and the longitudinal conductance vanishes (at zero temperature). It is only as the Fermi level passes through the core of extended states that the longitudinal conductance becomes appreciable and the Hall conductance makes its transition from one plateau step to the next. The localized states thus provide a sort of reservoir which allows the Fermi level to sit between Landau levels and away from the extended states [Note 7]. Finally, at finite temperature there is a small longitudinal conductivity due to hopping processes between localized states at the Fermi level [Chap. II]. Unlike the case of superconductivity, this longitudinal current is not 'shorted out' by the supercurrent, which is here perpendicular to it.

1.7 Gauge Arguments

A pretty argument (Laughlin 1981) will now be given in a more complete form due to Halperin (1982). This argument is based on gauge considerations, and thus has the possibility of being very general.

The first idea is that the QHE is basically a bulk effect. In other words, there is no evidence whatsoever that changes in size, shape, connectivity or edge conditions of a sample lead to any changes in the basic QHE results. Therefore, we can choose a convenient geometry for a *gedanken* experiment, and draw general conclusions.

The geometry of choice is one which is continuous but multiply connected, say a disk in the xy plane with a hole in it, Fig. 5. Experimental realizations of this geometry are called Corbino disks. The usual field B is present in the z direction, but there is also a variable flux Φ through and confined to the hole. This flux Φ can nevertheless affect those electrons in the disk whose wave functions actually circle the hole. This is directly related to the Aharonov-Bohm (1959) effect. The important feature of Fig. 5 is the connectivity of the geometry. It can also be imagined that the inner edge of the disk is maintained at a potential different from the outer edge, but this is not essential.

It is very convenient to follow Halperin and imagine that the impurity potentials are confined to a region of the disk GG' well away from both outer edge E and inner edge E'. Thus the sample consists of a ring of 'realistic' two-dimensional regions with imperfections 'guarded' by inner and outer rings which are free of imperfections. Fig. 5 also shows this. Now

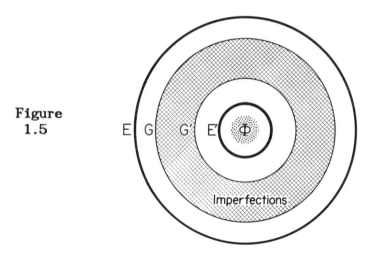

Figure 1.5

Introduction

we imagine that the total number of electrons is such that exactly i Landau levels are filled in the guard rings. The Fermi level thus lies in the gap between two Landau levels in the guard rings. By use of these guard rings we explicitly relate the conducting properties of the realistic system to that of the ideal system.

Next imagine adiabatically changing the flux Φ. There is then an electric field satisfying

$$\frac{1}{c}\frac{d\Phi}{dt} = \frac{1}{c}\int d\mathbf{A}\cdot\frac{\partial \mathbf{B}}{\partial t} = \oint_C d\boldsymbol{\ell}\cdot\mathbf{E} = \oint_C d\boldsymbol{\ell}\cdot\boldsymbol{\rho}\cdot\mathbf{j} \qquad (1.7.1)$$

where the line integral is along any contour C enclosing the hole. If there is a Hall current density perpendicular to the field there will be a charge transfer across the contour. Let the total change of flux be one flux quantum, $\Phi_0 = hc/e$. It is now almost obvious (but we will prove it below) that the total charge transferred across the loop C is ie where i is an integer.

Let the loop be in one of the guard rings, say EG. In these perfect regions $\rho_H = h/ie^2$ (by virtue of our assumption that exactly i Landau levels are filled) so the result follows. The electrons transported across EG must have come (say) from the edge E and been transferred into the impure region at G. At a physical edge, such as E, there is a quasi-continuum of edge states (for each Landau level). [Note 8]. The states of EG are illustrated in Fig. 6.

The edge states will be filled to the Fermi level. The way the current flows across the guard ring as the flux Φ changes, is that the state labelled p will adiabatically be transformed into the state $p+1$ [an explicit calculation will be given in Chapter 3.4]. After a change of one flux quantum, the states have their original wave functions and energies since an integer number of flux quantum can be removed by a gauge transformation [Chapter 3.4]. However, there has been one electron carried through the region, so that one fewer state at the Fermi level will be filled at E. This results in an infinitesimal change of energy, which is required since the flux change is adiabatic.

Now assume that there is no dissipation in the impurity region GG'. This can happen if the Fermi level

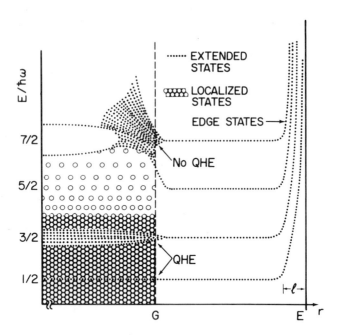

Figure 1.6. Energy levels of the ring system of Fig. 5 vs. radius. The levels (at least in EG) are extended around the ring but localized at a definite radius. Such states are indicated by points. States localized both radially and azimuthally are indicated by small circles.

is in a mobility gap, that is, if the states at the Fermi level are localized or nonexistent. This is also shown in Fig. 6. In other words, we have just assumed that any current is normal to the electric field, or that the diagonal conductivity vanishes.

Then there are two possibilities. The first is that the electrons delivered to the impurity region are transferred through it and there is no build up of charge in GG' or at its edges. This is the QHE case. It immediately follows that the conducting behavior of the impurity region is identical with the behavior of the perfect system with integrally filled Landau levels, namely the ideal current Hall voltage relationship is satisfied. The bulk conductivity of the impurity region, which can be defined by appropriate macroscopic averages if the impurity region is large enough, is therefore given by (1.1.1).

Introduction

The second possibility is that no electrons are transferred (or that additional electrons are transferred) through GG'. Obviously the electrons must go into or come from states in the interior of GG' or at the boundary G between the impurity region and the guard rings. This must happen if a strong insulator (e.g., the vacuum) is put in the 'impurity' region. These interface or impurity states must be at the Fermi level if an adiabatic process, the flux change, can transfer electrons into or out of them. This case is illustrated in the upper part of Fig. 6.

We emphasize again that the only states which are affected by the flux Φ are those whose amplitude is appreciable on a path circling the hole. These are exactly the extended states. Localized states are not affected by Φ and consequently their occupation or nature cannot change [Note 9]. This immediately implies (Halperin 1982) that if there are no extended states at the Fermi level, either in the bulk of the impurity region or at its boundaries, (the QHE case,) then there are extended states throughout the impurity region with energies away from the Fermi level. Further, any impurity or crystalline potential sufficiently strong to eliminate the extended states will also suffice to create extended states at its edge with energies near the Fermi surface. Halperin (1982) has also shown that there is a continuum of edge states at a physical boundary, such as E, even if impurities and surface imperfections are taken into account.

The original argument (Laughlin 1981) was very slightly different. He assumed that a static electrochemical potential difference V_H existed between inner and outer edges. As a result of the flux increase, i electrons are transported from one edge to the other causing a change in the total energy $\Delta U = eiV_H$. It may be shown that the total azimuthal current is $I = cdU/d\Phi$. In principle $dU/d\Phi$ is not necessarily constant but could depend periodically on Φ (Imry 1982). In practice, however, it suffices to use $I = c\Delta U/\Delta\Phi = (h/ie^2)V_H$.

Let us restate the result of this section. It has been shown that if the Fermi level lies in a mobility gap, so that the diagonal conductivity vanishes, then the off-diagonal conductivity has the form (1.1.1) with i integer. The 'trivial' case $i = 0$ is that of an ordinary insulator. If the impurity or crystalline potential is weak enough that the impurity region connects smoothly with an impurity free region, then the nontrivial case can be realized.

The above argument has been couched in the language of single electron states. Once the independent electron approximation is made that language is exact and it is quite clear that i is integer. Indeed, we can only construct guard rings with the Fermi level in a gap for integral i. The most convenient formulation in the interacting electron situation is not obvious. Nevertheless, it is clear that much of the above argument does not really rely on the independent electron approximation. The crucial point is that the nature of the states in the guard rings depends in a deep way on the electron Coulomb interaction, and it is possible that in the interacting case a fractional number of electrons are transferred when Φ is changed by an integer number of flux quanta. That is, in the interacting case, it will be possible that a fractionally filled Landau level possesses a gap. Thus once the nature of the states in the impurity free case is understood, general arguments are available to show that weak enough impurities do not have an effect on the conductance.

1.8 Inversion Layers

In this section we turn to the physical realization of two-dimensional electron systems, namely inversion layers. A great deal of physics and technology revolves about this subject, not only the QHE. We can give only the elementary features essential for the understanding of the material in this lectures. Some additional information is found in Chapter II.

Inversion layers are formed at the interface between a semiconductor and an insulator or between two semiconductors, with one of them acting as an insulator. The system in which the QHE was discovered has Si for the semiconductor, SiO_2 for the insulator. A good review of this system is (Ando, et al. 1982). Recently, the semiconductor-semiconductor system GaAs - $Al_xGa_{1-x}As$, the latter playing the role of the insulator, has achieved outstanding parameters. Notably the mobility can be made extremely large, that is, the effect of impurities and imperfections can be made

Introduction

very small. The parameter $x \approx 0.8$. We refer to AlAs for short. Both systems are extremely important technologically, with the silicon system already in wide use. The gallium arsenide system holds great promise for important increases in the speed of computer systems and in electro-optical devices.

The principle of the inversion layer is quite simple. It is arranged that an electric field perpendicular to the interface attracts electrons from the semiconductor to it. These electrons sit in a quantum well created by this field and the interface. The motion perpendicular to the interface is quantized and thus has a fundamental rigidity which freezes out motional degrees of freedom in this direction. The result is a two-dimensional system of electrons. Further, the wavelengths of these electrons are long so that an effective mass approximation with parabolic bands is quite good.

The source of the attracting electric field differs in the two systems. In the silicon case the insulating oxide layer is made relatively thick (e.g., 5000 Å) and a metallic electrode (Al) plated over it. This electrode is positively charged by application of an external 'gate voltage'. The resulting device is called a MOSFET — metal oxide semiconductor field effect transistor. This method of applying the electric field has the advantage of providing easy control over

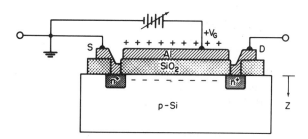

Figure 1.7. Schematic side view of a silicon MOSFET showing the aluminum gate, the SiO_2 insulator and the p-type Si crystal substrate. The source S and drain D contacts are heavily doped $n+$ regions with aluminum caps. Not shown are the potential contacts on the sides of the device; they are used to measure the Hall voltage V_H and the longitudinal voltage drop V_L. A top view of a Hall bridge is shown in Fig. 1.

the density of the sheet of electrons attracted to the interface.

The total self-consistent potential seen by the electrons is conveniently described by the picture of 'band bending'. That is to say, the periodic lattice potential gives rise to energy bands, and the slowly varying electric potential then is regarded as bending these bands. Fig. 8 gives the schematic diagram for this process.

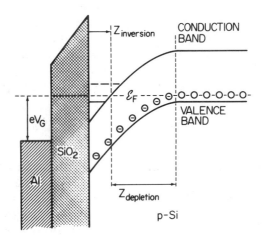

Figure 1.8. Electron energy level diagrams for a silicon MOSFET with a positive voltage applied to the aluminum gate. When the conduction band dips below the Fermi energy level, E_F, a 2D electron inversion layer forms in the Si crystal $\approx$ 25 Å from the SiO_2 interface. The crystal discontinuity at the interface and trapped impurities in the SiO_2 layer are major sources of scattering which limit the zero-field mobility μ to $\approx$ 40,000 cm^2/Vs or less.

In this figure, it is assumed that the semiconductor is doped p-type, i.e., some electrons of the valence band have become bound to acceptor dopant impurities leaving empty states, holes. The lowest energy holes are at the top of the valence band. This means that the Fermi level is close to the top of that band. The electrons attracted to the surface first fill up these hole states leaving a net negative charge of three-dimensional density equal to the acceptor density. However, if there is sufficient gate voltage, the

Introduction

bottom of the conduction band will become lower than the Fermi level, and if there is a way for them to get there, for example by an electrical connection to the surface region, electrons will occupy states in the part of the conduction band below the Fermi level. Obviously, the conduction band bends below the Fermi level near the insulator-semiconductor interface. This is the inversion layer, with the bottom of the conduction band below the top of the valence band, inverting the normal order. It is also possible to use an n-type semiconductor and reverse the sign of the gate voltage to achieve inverted hole states.

In the independent electron approximation, and assuming translational invariance along the interface, the electronic states [Note 10] are of the form $\psi(x,y) Z_n(z)$ where Z_n satisfies a Schrödinger equation

$$\left[-\frac{\hbar^2}{2m_z} \frac{\partial^2}{\partial z^2} + V_T(z) \right] Z_n(z) = \epsilon Z_n(z). \qquad (1.8.1)$$

Here $V_T(z)$ is approximately a triangular potential with a high, effectively infinite barrier at the interface representing the large band gap of the insulator, as shown in Fig. 9. [The mass m_z may not be the same as the transverse effective mass m. Note 11.]

There results a sequence of energy levels ϵ_n with corresponding eigenstates $Z_n(z)$. The accuracy of present measurements of this spectrum of states is such that the potential V_T must be obtained and solved self consistently to obtain agreement with experiment.

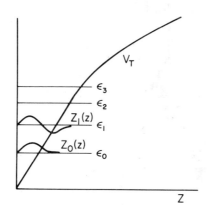

Figure 1.9

Nevertheless, the variational approximation for the lowest state Z_0 given by

$$Z_0(z) = 2(2b)^{-3/2} z\, e^{-z/2b} \qquad (1.8.2)$$

is useful and not misleading. The parameter b giving the scale depth of the wave function is typically $b \approx$ 30 - 50 Å. This is comparable with the magnetic length. The region for which all the acceptor levels are filled (the depletion layer) is much thicker, of order 500 Å in silicon. The depletion layer has no free carriers and is thus an insulator. The energy difference $\epsilon_1 - \epsilon_0$ is typically 20 meV (200 K), much greater than the temperature at which the QHE is observed.

The total energy of an electron is the sum of the transverse energy and ϵ_n. It is then said that there are surface subbands, one for each eigenstate Z_n. The transverse energetics in a magnetic field was estimated in Section 5 and constitutes the whole subject of the QHE. The ideal situation for the QHE is that the parameters are such that only the subband Z_0 is occupied. If the Fermi level is far below the bottom of the first excited subband, it will be a good approximation to neglect it [Note 12].

In the absence of a magnetic field it is usually a fair approximation to neglect the electron-electron Coulomb correlations. The usual dimensionless measure of the Coulomb interaction is $r_s \equiv (\pi a_B^2 n)^{-1/2}$, and $a_B \equiv (\kappa m_e/m)(\hbar^2/m_e e^2)$ is the effective Bohr radius. Since the dielectric constant $\kappa \approx 10$ and $m/m_e \approx 0.1$, the effective Bohr radius is 50 Å. Thus at areal densities above about $10^{12}/cm^2$, $r_s < 1$. This is a fairly high density, in practice, but correlation effects are not qualitatively important until r_s becomes rather large. Some quantitative correlation effects are observable however in the zero-field case. The strong magnetic field changes the picture dramatically and this is the central theme of the second half of the course. Electron-phonon effects are also interesting in the zero-field case but we shall for the most part ignore them.

The gallium arsenide system is quite similar to silicon. In this case, the GaAs is made as pure as possible but typically remains weakly p-type. It acts

Introduction

as the semiconductor. The AlAs has a wider band gap. It acts as an insulator. It has been found possible, using molecular beam epitaxy, to grow atomically sharp interfaces of these two materials called heterojunctions [Note 13]. Thus a certain controlled number of layers of GaAs is followed by an almost perfectly matched sequence of layers of AlAs. The two materials have nearly the same lattice and dielectric constants. Before taking into account certain electronic processes, the bands would appear as in Fig. 10.

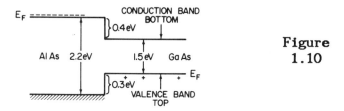

Figure 1.10

The AlAs is deliberately doped n-type, which puts mobile electrons into its conduction band. These electrons will migrate to fill the few holes in the top of the GaAs valence band but most of them will end up in states near the bottom of the GaAs conduction band. However, there is a positive charge left on the donor impurities which attracts these electrons to the interface and bends the bands in the process. This is the source of the electric field in this system. The transfer of electrons from AlAs to GaAs will continue until the dipole layer formed from the positive donors and the negative inversion layer is sufficiently strong. This dipole layer gives rise to a potential 'discontinuity' which finally makes the Fermi level of the GaAs equal to that of the AlAs. The density of electrons residing in the inversion layer is thus determined by the dopant density and is fixed for each sample. [One can put a 'backgate' on the AlAs, similar to Fig. 7, and gain some control over the electron density.] Fig. 1.11 shows this situation.

The major advantage of this system is that it lends itself so well to the molecular beam epitaxy techniques, resulting in the lowest impurity density, and the lowest density of interface defects presently achievable. A further extremely important technique is to implant the necessary donors physically away

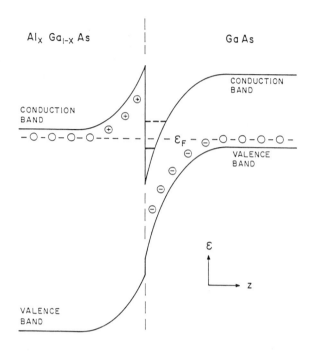

Figure 1.11. Electron energy-level diagram of a GaAs–AlGaAs heterostructure device. Donor electrons tunnel through the potential barrier and occupy the first conduction sub-band of the potential well. Not shown is an undoped AlGaAs layer which separates the donors from the interface. The 2D electron sheet is confined within ≈ 100 Å of the interface. This interface is smooth to within an atom and part of a single crystal; μ can be greater than 10^6 cm^2/Vs.

from the interface of the AlAs. This is known as modulation doping, and results in almost all of the impurity dopant atoms residing several hundred Angstroms from the inversion layer. In consequence, extremely high mobilities of the electrons in the inversion layer are achieved. What scattering there is, is largely near the forward direction, so the ratio of transport to scattering lifetimes in these devices is as much as two orders of magnitude more than in conventional situations [Das Sarma and Stern 1985].

A secondary advantage of GaAs is that its effective mass ($m = 0.07\, m_e$) is somewhat smaller than that in silicon ($m = 0.2\, m_e$). The dielectric constants are $\kappa_{GaAs} = 10.9$, $\kappa_{Si} = 11.7$, $\kappa_{SiO2} = 3.9$. GaAs is also somewhat simpler than Si in that there is only one set

Introduction

of electrons in the conduction band. In Si there are six pockets of electrons centered at symmetrical points in the Brillouin zone [Note 14].

The major defects causing departure from translational invariance are impurities and interface defects, such as steps. The effective potential seen by the two-dimensional electron gas due to an impurity of charge Q situated at distance Z from the interface (see Fig. 12) is given by

$$V_{imp}(x,y) = -\frac{Qe}{\kappa}\int_0^\infty dz \frac{|Z_0(z)|^2}{[(Z-z)^2 + (x-x_0)^2 + (y-y_0)^2]^{1/2}} \quad (1.8.3)$$

Here we have used the fact that the perpendicular wave function Z_0 does not change in the low-energy electronic processes of the QHE. It is clear that an impurity far from the layer will give a very smooth potential. Even one in the layer will have a potential smoothed on the scale of the parameter b of Eq. 1.8.2 which is comparable with ℓ. Similar considerations apply to the inter-electron Coulomb interaction, effectively smoothing and reducing this interaction as compared with the Coulomb interaction between electrons in an ideal truly two-dimensional layer [Note 15]. The effect of interface imperfections is smoothed and weakened in the same way. Eq. 8.3 leaves out

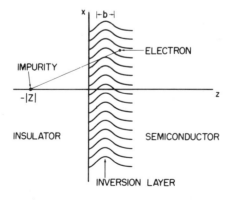

Figure 1.12. An impurity atom at position $(x_0, y_0, -|Z|)$ near the 2D electron system.

one effect, namely the image force resulting from the different dielectric constant on the two sides of the interface. This effect is rather small in the GaAs system, but is of some significance in silicon.

We emphasize that formula (1.8.3) does not include screening by the electrons in the inversion layer. These screening effects can be quite important in weak magnetic fields. They are imperfectly understood in the high field case at the present time. It is generally expected that screening will be reduced in the presence of a magnetic field, especially in cases were the Fermi level lies in a gap. In the fractional QHE where interaction effects are of the essence, all correlation effects, including screening, are in principle taken into account by the theory and it would not be correct to include screening in the impurity or interaction potentials.

1.9 Acknowledgments

This research was supported by NSF grant DMR-82-13768.

1.10 Notes

Note 1. The conductivity σ and resistivity ρ are tensors in the general case with $\sigma_{ij} = (\rho^{-1})_{ij}$. The general Onsager relation $\sigma_{ij}(B) = \sigma_{ji}(-B)$ holds (see for example Landau and Lifshitz 1960). This relation depends on the time reversal characteristics of the operators involved. It is clear intuitively that if the magnetic field deflects electrons from the x-direction towards the y-direction, then it will deflect them from the y-direction towards the negative x-direction.

Note 2. If σ has the off-diagonal form (1.1.1) then $R_L = 0$ and $R_H = h/ie^2$ independent of geometry. For example, the sample may have holes punched into it. Thus electrons cut out of the sample by the holes are compensated by extra currents generated elsewhere (Tsui and Allen 1981, Problem 1). Zheng et al. (1985b) have, by inserting probing electrodes into the central regions of samples, found nonuniform voltages which depend strongly and 'randomly' on sample and magnetic field. However these voltages are reproduc-

Introduction

ible. This means that the current density is far from uniform in the sample, presumably because of impurity effects and macroscopic inhomogeneities. Nevertheless, the quantization of I/V_H is unaffected by these interior current density fluctuations.

Note 3. For the integer QHE, spin plays a conceptually minor role. All that happens is that each energy level suffers a Zeeman splitting $\pm g\mu_B B$ so there are double the number of Landau levels, etc. These splittings are observed experimentally (Chapter II) and their magnitudes are understood by the standard theory of the g factor in solids. Even for the fractional QHE, theories assuming perfect spin polarization of Landau levels are successful, but the justification is far from trivial or complete (Halperin 1983, Girvin 1984b, Zhang and Chakraborty 1984, Kallin and Halperin 1984, Haldane Chap. VIII).

Note 4. Three possible boundary conditions are: Periodic boundary conditions in both x and y can be imposed for certain 'commensurate' magnetic field strengths. This is equivalent to a toroidal geometry [Chapters IV, VIII]. The upper and lower edges can be conveniently shrunk to points, a spherical geometry [Chapter VIII]. Physical confinement of electrons can be represented by potential walls, which give rise to edge states [Section 7].

Note 5. The Lorentzian approximation is an oversimplification. For one thing, it fails to take into account the fact that scattering cannot occur if there is no final state into which to scatter. This effect can be studied in a self-consistent single-scattering-site approximation. Although this type of approximation does not distinguish between localized and extended states, it does seem to give numerically a reasonable account of the total density of states in some cases. See Fig. 3.8 for some numerical results.

Note 6. The lowest Landau level approximation is often made in this field and seldom seems to be misleading. However, the justification offered is sometimes rather unsatisfactory. For example, the approximation is supposed to be good insofar as $\hbar\omega_c \gg E_c$ where E_c is a typical Coulomb or impurity energy. In practice, however, $E_c/\hbar\omega_c$ is seldom less than 0.5, while no correction bigger than 10^{-7} can be tolerated by experiment. The essential feature, as we shall see, is that numerical corrections to energy levels, etc. need not be terribly small as long as certain qualitative features, (namely mobility gaps) are maintained. See (Prange 1981) for a case where all Landau levels are kept. Chapter X discusses some numerical corrections to the lowest Landau level approximation which turn out to be much smaller than $E_c/\hbar\omega_c$.

Note 7. Baraff and Tsui (1981) suggest that there exist localized states not part of the Landau level but weakly connecting to it. These states reside on impurities of order 100 Å from the 2D QHE system. No doubt these states exist, [Section 1.8]. Their numerical importance in establishing the Fermi level probably varies from case to case although they are believed usually to make a relatively minor effect.

Note 8. An arbitrary potential depending only on y can be added to Eq. 6.1 and the solutions obtained. At a very large potential rise representing a boundary, there are edge states. These states possess a quasicontinuous spectrum which connects the bulk Landau level with the energy at the top of the potential wall. Such boundary states have been observed (in 3-dimensional cases) (Nee, Koch and Prange 1968). See Problem 3. Halperin (1982) has discussed the case of imperfect boundaries. There have been several important papers concerning edge states (e.g., Rammal et al. 1983). This emphasis may have led some to believe that the QHE cannot exist without edge states, or that all or most of the current is carried in edge states. Our point of view is that the QHE, like the Landau diamagnetism, is a bulk effect, with the main currents of the QHE in the bulk. Because of the integrability guaranteed by the gauge arguments, the results can be expressed as properties of edge states in the appropriate geometry. Indeed, this is a major theme of the field-theoretic approach discussed in Chap. V. However, we prefer to think that the bulk QHE guarantees the existence of the edge states, rather than the contrary. And in some geometries, for example on the torus, the QHE would presumably exist without edge states.

Note 9. In a finite ring, all states in principle extend around the ring, although the amplitude of what we call localized states is exponentially small in large parts of the ring. Strictly speaking the no-level-crossing theorem applies, although the crossings are 'avoided' by exponentially small gaps in the case

Introduction

of quasilocalized states. If this is taken literally, it requires that each state return to itself under adiabatic change of one quantum of flux. However, the gaps are of order $\exp(-L_1^2/\ell^2)$ where L_1 is some macroscopic length of magnitude the sample dimensions. Thus $L_1/\ell \approx 10^{3-4}$. Such gaps have to be neglected. Alternatively, we think of the rate of flux change $(1/\Phi_0)d\Phi/dt$ as being slow compared with ω_c but fast compared with $\omega_c \exp(-L_1^2/\ell^2)$. Of course, in numerical work on small systems it is not always easy to distinguish localized from extended states (Aoki 1983, Aers and MacDonald 1984).

Note 10. The wave functions under consideration are, strictly speaking, envelope wave functions. The true wave function $\Psi_{true}(xyz) = \psi(xy) \, Z_o(z) \, \Psi_{k_0}(xyz)$, where Ψ_{k_0} is the Bloch wave function at the band bottom (crystal momentum = k_0) and $\psi(x,y) \, Z_o(z)$ is the envelope wave function. Corrections to this formula renormalize the mass and other constants. The rapid variations of the true wave function near the ion cores are eliminated from the envelope functions.

Note 11. The effective mass approximation (and the parabolic band approximation) are not accurate to a part in 10^7. Again, corrections do not change qualitative aspects of the energy level structure. The techniques of Chapter IV can be used to show this explicitly (Středa 1982).

Note 12. Even if virtual transitions to higher subband states cannot strictly speaking be neglected, these transitions result only in a renormalization of some parameters. In cases that have been numerically investigated, this renormalization is quite small even when the Fermi level is rather close to the bottom of the empty subband.

Note 13. It is not completely understood why some materials are better candidates for molecular beam epitaxy (MBE) techniques than others. MBE is the designation for the art of growing crystals under high vacuum conditions by deposition of atoms or molecules from controlled beams. One important factor in making good interfaces is that the two materials have lattice constants which are well matched. Another is that the diffusion rate of atoms deposited on a surface be appropriately fast. (See Chang 1985).

Note 14. Silicon has six equivalent 'valleys' of electron states centered at different, but symmetrically equivalent points of the Brillouin zone. Depending on the orientation of the surface, this symmetry is broken but there is still a valley degeneracy which can be two, four, or six. For some reason not well understood, the observed valley degeneracy is always two, even for orientations which theoretically should have larger degeneracy. In the QHE context, the valley degeneracy acts like a spin which suffers no Zeeman splitting. If is of interest in the fractional QHE, since it will affect the correlation energy and give rise to a gapless Goldstone boson (Rasolt et al. 1985).

Note 15. It is customary in numerical calculations of the fractional QHE to use the ordinary Coulomb potential $e^2/\kappa R_{12}$. Much of the interest in these calculations lies in their comparison with one another and this is a convenient standard. Also, it avoids introducing a new length into the problem. In any case, the overall results must be insensitive to the exact potential used. Use of (1.8.3) and its interaction potential analog is essential for comparison with certain experiments, however. (Girvin et al. 1986b, Haldane and Rezayi 1985, Zhang and Das Sarma 1986).

1.11 Problems

Problem 1. Assume that σ is off-diagonal, as in (1.1). Show that the measured R_H and R_L are independent of sample geometry.

Problem 2. (Rendell and Girvin 1981). Assume $\sigma_{xx} = \sigma_{yy}$ is very small but not zero, while $\sigma_{xy} = -\sigma_{yx}$ is appreciable and constant in a rectangular two dimensional region, as in Fig. 1. (In S, D, A, B, and C σ_{xx}/σ_{xy} is large.) Assume a total current I, and a steady state. Show

a) $\partial E_x/\partial x + \partial E_y/\partial y = 0$.

b) The current flows into one corner of the sample and out the diagonally opposite corner.

c) The voltage drop from source S to drain D is equal to the Hall voltage drop, A to C.

Problem 3. Use the gauge $\mathbf{A} = -\mathbf{r} \times \mathbf{B}/2 = (-yB/2, xB/2)$ which is cylindrically symmetric. Show that the ground eigenstates of

$$\mathcal{H}_0 = \frac{1}{2}\left[\frac{1}{i}\frac{\partial}{\partial x} - \frac{y}{2}\right]^2 + \frac{1}{2}\left[\frac{1}{i}\frac{\partial}{\partial y} + \frac{x}{2}\right]^2 \quad (1.11.1)$$

(i.e. the lowest Landau level states) can be written

$$\psi(x,y) \propto z^m \exp(-|z|^2/4) \quad (1.11.2)$$

where $z = x - iy$ and m is the negative angular momentum about the B direction. [We shall later use the more customary $z = x + iy$, which is equivalent to taking B in the negative z direction and making some other sign changes.]

Problem 4. Add an infinite potential jump at $y = 0$ to the Hamiltonian (1.5.3). Find the lowest edge states, approximately, for k large and negative (Nee and Prange 1967). Show that their energies form a quasi-continuum. Remark: These states are very similar to the states $Z_n(z)$ of Eq. 8.1.

Part A

The Integer Effect

CHAPTER 2

Experimental Aspects and Metrological Applications

MARVIN E. CAGE

2.1 Basic Principles

The quantum Hall effect provides an unparalleled opportunity to study and utilize the physical properties of macroscopic, two-dimensional electron or hole systems in the presence of strong magnetic fields. These studies are revealing surprising results that are of particular interest to the disciplines of condensed matter physics and electrical metrology. This chapter discusses some experimental observations and metrological applications of the integer Hall effect.

Chapter I dealt initially with current densities, electric fields, resistivities and conductivities, but noted that the conductance and resistance are the fundamental quantities of interest both experimentally and theoretically. If it were the conductivity rather than the conductance which were quantized then precision measurements would be impossible since one would have to invoke assumptions of a homogeneous medium with a well-defined geometry in order to infer the microscopic conductivity from the macroscopic conductance. It is a remarkable feature of the QHE that this is not necessary.

When a current I is passed in the x direction through the longitudinal conductive channel in the presence of a magnetic field B along the z axis there is a Hall voltage V_H induced across the channel in the y direction. In an ideal homogeneous system with no disorder the Hall voltage is given (both classically and quantum mechanically) by

$$V_H = (B/ne) I \; , \qquad (2.1.1)$$

where n is the number of moving carriers per unit area with charge $-e$. For two-dimensional carrier systems in high magnetic fields each Landau level is spin-split, and in some semiconductors also valley-split. A split level is designated by the quantum integer i. We saw in Chap. I that, for the ith level, the maximum number n_B of allowed states per unit area is (in SI rather than cgs units)

$$n_B = eB/h \; , \qquad (2.1.2)$$

where h is the Planck constant.

At cold enough temperatures it should be possible in a perfectly made device for the n carriers per unit area to completely occupy all the in_B states per unit area of the filled levels at appropriate values of B --leaving all other levels empty because of an energy gap at the Fermi level. Under these ideal conditions expressions (2.1.1) and (2.1.2) may be combined to yield

$$R_H(i) = V_H/I = h/e^2 i \; , \qquad (2.1.3)$$

and the quantum Hall resistance R_H, defined as the ratio of V_H and I, would have a unique, universal value when the allowed states of the first i quantum levels were just-filled.

Eq. 1.3 was derived assuming a perfectly homogeneous medium with no imperfections or localization effects and no scattering processes of any kind. Under these ideal conditions $R_H = \rho_{xy}$, $\rho_{xx} = \sigma_{xx} = 0$, and $\sigma_{xy} = -1/\rho_{xy}$ as shown in Chap. I. What happens, however, in real devices? This is the subject of the present discussion.

2.2 The Devices

Two-dimensional sheets of conducting carriers required for quantum Hall effect studies have been realized in many types of devices, some of which are listed in Table 2.1, along with the effective masses m of the carriers, their zero magnetic field mobilities μ, and the average scattering time τ_0 in zero magnetic field. The basic structure of these devices was described in Sec. 1.8.

The effective masses are usually determined from cyclotron resonance experiments and the relationship $\omega_c = eB/m$ [see for example, Wagner et al. (1980)]. These authors measure absorption spectrum frequencies ω_c in a magnetic field using a far-infared (FIR) laser source. A small value of m is desirable for the QHE since a smaller, less expensive superconducting magnet is required to achieve the necessary quantization condition $\hbar\omega_c > k_B T$ because $\omega_c = eB/m$.

Zero magnetic field mobilities μ are determined at 4.2 K from 2D conductance (σ_{SD}) measurements at constant source-drain voltage V_{SD} using the relationship $\mu = \sigma_{SD}/ne$ and $\sigma_{SD} = 1/\rho_{SD} = (L/W)(I/V_{SD})$, where L and W are the length and width of the device. The mobility is the conventional figure of merit expressing the ability of the average electron to conduct. The density is conveniently found by the QHE relationship $n = ieB/h$ where B is the magnetic field necessary to occupy completely all states of the ith quantum level.

The mobility can also be expressed as $\mu = e\tau_0/m$, where τ_0 is the average time between collisions of the 2D gas at zero magnetic field. One wants to minimize collisions by maximizing τ_0, and to minimize the value of B required to reach a plateau. Therefore, devices in which τ_0 is largest and m is smallest [therefore μ largest] are the most desirable. By these criteria GaAs-AlGaAs is the best material available at present. We must note that τ_0 is the scattering time at zero magnetic field; τ and μ are many orders of magnitude larger in the quantized Hall regime where the devices become nearly dissipationless.

The most popular device types are Si MOSFETs (metal oxide-semiconductor field-effect transistors) and GaAs AlGaAs heterostructures. Si MOSFETs are fabricated by the usual silicon semiconductor technologies. Hetero-

Table 2.1

Interface Constituents	2D gas	m (m_e)	μ ($10^3 cm^2/Vs$)	τ_0 ($10^{-12}s$)
Si – SiO$_2$ [a]	e or (h)	0.19	10 [b] (2.5)	1.1
GaAs–Al$_{0.29}$Ga$_{0.71}$As [c]	e or (h)	0.068(0.38)	100 [d]	3.9
In$_{0.53}$Ga$_{0.47}$As–InP [e]	e	0.080	30	1.4
InAs–GaSb [f]	e and (h)	0.023(0.36)	170	2.2
InP–Al$_{0.48}$In$_{0.52}$As [g]	e	~0.08	10	0.5
Ga$_{0.25}$In$_{0.75}$As$_{0.50}$P$_{0.50}$–InP [h]	e	0.058	13 [i]	0.4
In$_{0.53}$Ga$_{0.47}$As–In$_{0.48}$Al$_{0.52}$As [j]	e	0.05	90	2.6
Hg$_{0.78}$Cd$_{0.22}$Te–HgCdTe oxides [k]	e	0.006	90	0.3

Table 2.1. Some device types in which inversion layers have been realized. The 2D gases are either electrons (e) or holes (h) and, except for the holes in GaSb, are formed in the crystal listed on the left-hand side of the interface constituents. All devices are III-V compound heterostructures except for the first, which is a silicon MOSFET, and the last, which is a HgCdTe MISFET (metal-insulator-semiconductor field-effect transistor with a II-VI HgCdTe compound semiconductor). The effective masses (m) are also listed along with typical zero magnetic field mobilities (μ) at 4.2 K and τ_0, the average scattering time in zero field. See also Table 6.2.1.

---

a) Fowler et al. (1966), Kawaji et al. (1975), von Klitzing, Dorda and Pepper (1980), Wagner et al. (1980) and Gusev et al. (1984).

b) Can be $4\times 10^4 cm^2/Vs$, but IQHE steps are too narrow (Pudalov and Semenchinskii 1984).

c) Tsui and Gossard (1981), Störmer et al. (1983b), Störmer et al. (1979) and Störmer and Tsang (1980).

d) Optimum mobility for integer step. Mobilities over $10^6\ cm^2/Vs$ are used for fractional steps.

e) Guldner et al. (1982) and Briggs et al. (1983).

f) Chang and Esaki (1980), Washburn et al. (1985) and Mendez et al. (1985).

g) Inoue and Nakajima (1984) Nakajima, Tanahashi and Akita (1982).

h) Razeghi, Duchemin and Portal (1985).

i) Mobility at 77 K.

j) Kastalsky et al. (1982).

k) Kirk et al. (1985) and Schiebel (1983).

structures can be grown by molecular beam epitaxy (MBE) (Dingle et al. 1978) or by metal-organic chemical vapor deposition (MOCVD) (Razeghi et al. 1981).

Typically, the devices are formed into rectangularly shaped Hall bridges (or bars) several millimeters long and several tenths of millimeters wide [e.g. Fig. 1.1]. They are used at constant source-drain currents to measure ρ_{xx} and ρ_{xy}. Corbino disks, having concentric ring geometries, are used with constant source-drain voltages applied between the inner disk and outer ring in order to measure σ_{xx}.

The two-dimensional character of the carriers in all the devices results from the quantization of the motion perpendicular to the interface, as shown in Figs. 1.8, 1.9 and 1.11. A nice feature of the silicon devices is that the 2D carrier density is continuously adjustable and is approximately linear in the gate voltage V_g. They are, however, very susceptible to damage by sparking because the electrostatic charges can generate electric fields on the order of 10^6 V/cm through the thin oxide layer. The carrier densities are fixed in heterostructures unless gates are included.

2.3 The Basic Experiment

Figure 1 shows chart recordings of the Hall voltage V_H and of V_x, the longitudinal voltage drop along the channel, [denoted V_L in Chap. I] vs. V_g for a Si MOSFET at constant current and fixed magnetic field. Eq. 1.1 predicts that under those conditions V_H would be proportional to $1/V_g$. The data do have this general tendency, but with additional features. Expression (2.1.3) implies that $R_H(i)$ would equal $h/e^2 i$ over only a very narrow range of V_g because as V_g is increased the extra induced, conducting electron would occupy a higher, unfilled quantum level and there would be no energy gap at the Fermi level; but clearly, V_H is independent of V_g over wide regions, forming flat plateaus or steps. As discussed in Chaps. I and III, these plateaus are the result of localization effects producing a mobility gap at the Fermi level.

Figure 2 shows plots of V_H and V_x vs. B for a GaAs heterostructure device. At small magnetic fields V_H is linear in B, as predicted by Eq. 1.1. Step-like

Experiments and Metrology

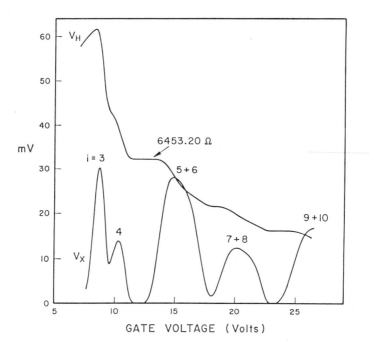

Figure 2.1. Plots of the Hall voltage V_H and the longitudinal voltage drop V_x vs. the gate voltage V_g for a Si MOSFET at 1.5 K and 13 T with a constant 5 µA source-drain current. The Shubnikov-de Haas (SbdH) oscillations in V_x are correlated with the plateaus or steps in V_H. The 6,453.20 Ω step corresponds to $R_H(4) = h/4e^2$. Data from Wagner et al. (1982).

features appear at higher fields. One goes from the classical ($i \to \infty$) to the quantum mechanical ($i \sim 1$) regimes in a single chart recording.

2.4 Initial Experiments

Kawaji et al. (1975), Igarashi et al. (1975), Kawaji (1978), and Wakabayashi and Kawaji (1978, 1980) performed the first experiments with Si MOSFET samples using Corbino disk and Hall bridge configurations. The diagonal conductivity σ_{xx} measured with the Corbino disks became very small at quantum level fillings as predicted by Ando, Matsumoto and Uemura (1975). The Hall current method used by Kawaji et al. to measure

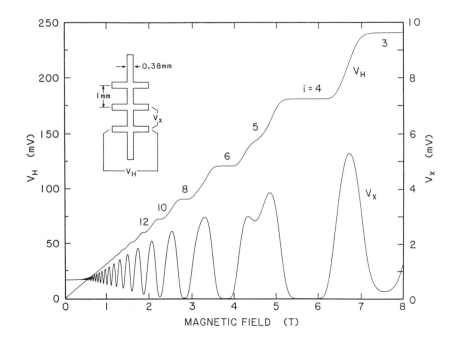

Figure 2.2. Chart recordings of V_H and V_x vs. B for a GaAs-AlGaAs heterostructure cooled to 1.2 K. The source-drain current is 25.5 μA and $n = 5.6 \times 10^{11}$ electrons/cm^2. Cage et al. (1985).

σ_{xy} was not suitable for precision measurements, and in any case sample quality had not yet reached the stage where finite width plateaus could be obtained.

K. von Klitzing, G. Dorda and M. Pepper (1980) made the first observation of quantized Hall plateaus. Using a Hall voltage method suitable for precision measurements they obtained good steps in high-mobility Si MOSFET devices and found that $R_H(i) = h/e^2 i$ to an accuracy of at least 5 parts-per-million (ppm), and thus clearly demonstrated the power and potential of the quantum Hall effect.

2.5 Precision Measurement Techniques

The work of von Klitzing, Dorda and Pepper (1980) was of immediate interest to the metrology community because of the possibility that the quantum Hall effect could be utilized as an absolute resistance standard

Experiments and Metrology

that depends only on fundamental constants of nature:

$$R_H(i) = h/e^2 i \approx 25\ 812.80\ \Omega/i \ . \tag{2.5.1}$$

These values of R_H happen to be in a very convenient range for resistance measurements. At least sixteen national laboratories around the world are currently developing such a standard. An additional point of interest is that, as discussed in Sec. 2.8, R_H can be used to determine the fine-structure constant.

The Hall resistance can easily be determined to one percent from chart recordings like those of Figs. 1 and 2 simply by dividing V_H by I. But to achieve maximum usefulness as a resistance standard, it must be measured to about 2 parts in 10^8 (0.02 ppm), the precision with which wire-wound resistance standards can be intercompared. One successful method of achieving the required factor of 10^6 increase in measurement sensitivity is shown in Fig. 3 (Cage, Dziuba and Field 1985). The Hall voltage V_H is potentiometrically compared with the voltage drop V_R across a series-connected room temperature wire-wound reference resistor that has been adjusted to have a resistance that is within a few ppm of $R_H(i)$ and is temperature controlled. Thus $V_R \approx V_H$. The potentiometer voltage is made almost equal to V_H and V_R, and the electronic detector amplifies the difference-voltage signal, thereby comparing R_H to R_R: $R_H = (V_H/V_R)R_R$. Another method employs a modified Wheatstone bridge (Field 1985) as shown in Fig. 4. Yet another method uses a cryogenic current comparator bridge to compare V_H with V_R and SQUIDs (superconducting quantum interference devices) as the detectors (Hartland 1984, Delahaye and Reymann 1985).

All of these methods are capable of providing random measurement uncertainties of 0.01 ppm. To obtain this resolution: (1) the current sources must be very constant; (2) the currents must be reversed to correct for thermally-induced voltages in the wires; and (3) electrical noise must be minimized. It is not too difficult to satisfy these conditions. The systematic measurement errors (and figuring out how to detect them) are major concerns, however. For example, in order to make a 0.01 ppm measurement with a 10 μA source-drain current one must assure that leakage cur-

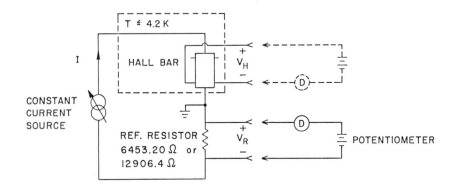

Figure 2.3. Simplified schematic of a high precision measurement circuit using the potentiometric method. The electronic detector **D** amplifies the difference - voltage signal. Leakage resistances must be greater than 10^{12} Ω for 0.02 ppm accuracies; so care is required to avoid fingerprints across connection terminals.

rents in the measurement system are less than 10^{-13} amperes. All three techniques can achieve 0.02 ppm total one standard deviation uncertainties.

2.6 Quantum Hall Resistors

The potentiometric measurement system of Fig. 3 was used to magnify by a factor of one million the $i=4$ (6,453.2 Ω) step of the GaAs device pictured in Fig. 2. The result is shown in Fig. 5. Each data point was obtained in one hour and typically had a ±0.01 ppm random measurement uncertainty. This Hall step is flat to within a least ±0.01 ppm over a magnetic field range that is 2% of the central value. Clearly this device has a Hall step that is suitable for use as a resistance standard.

High resolution measurement systems can be used to compare the Hall resistance R_H with that of wire-wound reference resistors. An example is displayed in Fig. 6 where $R_H(4)$ is compared with a 6,453.2 Ω reference resistor. The total uncertainty of each comparison is typically ±0.02 ppm. The value of this resistor is increasing at a rate of (0.052 ± 0.005) ppm/year relative to the quantized Hall resistance. This increase

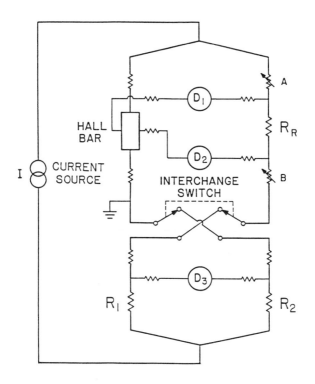

Figure 2.4. Simplified diagram of a modified Wheatstone bridge measurement circuit. Three electronic detectors **D1–D3** simultaneously amplify the difference voltage signals, making this method less sensitive to drifts of the constant current source or to changes in thermally-induced voltages. Digital voltmeters read the isolated detector outputs. Trim resistors **A** and **B** are adjusted to match the source and drain resistances. This method also is capable of achieving 0.02 ppm accuracies.

occurs even though the resistor is hermetically sealed, is continuously temperature controlled to within ±0.002 celsius, and has not been moved. Wire-wound resistors are intrinsically unstable. This is why metrologists want to use quantum Hall resistors as the primary standards.

The results shown in Fig. 6 demonstrate the power of this technique for calibrating resistors whose values just happen to be close to $25{,}812.80/i$ Ω. To be of general use however the Hall resistance $R_H(i)$ must be compared with the national standard, which in the

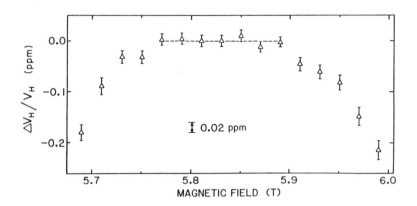

Figure 2.5. Digital mapping of a 6,453.20 Ω ($i=4$) QHE step of a GaAs heterostructure device cooled to 1.2 K. The current was 25.5 μA. Typical random uncertainties are ≈ 0.01 ppm. Cage et al. (1985).

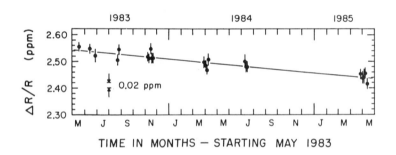

Figure 2.6. Relative comparisons as a function of time of a 6,453.20 Ω room temperature reference resistor with the $i=4$ steps of three different GaAs heterostructures. $\Delta R/R = (V_H - V_R)/V_R$. This resistor is increasing at a rate of (0.052 ± 0.005) ppm/year. Cage et al. (1985).

United States is legally defined in terms of the mean of five one ohm resistors maintained at the National Bureau of Standards and referred to as Ω_{NBS}.

The step-down from $R_H(i)$ to the ohm can be accomplished by either using cryogenic current comparator bridges (Hartland 1984, Delahaye and Reymann 1985) or Hamon-type series/parallel resistance networks (Hamon 1954, Small 1983, Cage et al. 1985). Figure 7 shows an example where a $R_H(4)$ step was first compared with a 6,453.2 Ω resistor, and then that resistance value R_R was stepped-down to the ohm via Hamon networks. It required about one month's work for each group of data points. The total uncertainties are ±0.05 ppm, mostly due to series/parallel connection uncertainties of the carefully matched terminal resistances. These uncertainties could be reduced to ±0.03 ppm by using cryogenic current comparator bridges.

The unit of resistance in the United States was thought to be stable, but these data provide the first direct evidence that Ω_{NBS} is drifting: its value is apparently decreasing at the rate of (0.05±0.02) ppm/year, which is consistent with a recent analysis of

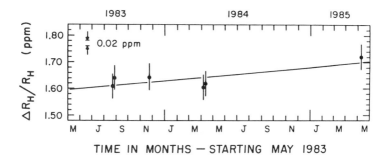

Figure 2.7. Monitoring as a function of time the value of $R_H(4)$, which is calibrated in terms of the national resistance unit Ω_{NBS}. $R_H(4)$ is expressed as a difference in ppm from a nominal value of 6,453.20 Ω_{NBS}. These preliminary data imply that the national unit Ω_{NBS} may be decreasing by (0.05 ± 0.02) ppm/year. These measurements will continue.

international resistance comparisons (Taylor 1985). The uncertainty of this drift rate is large but can be reduced by continued measurements. Quantum Hall effect devices will clearly become international resistance standards and will be used to monitor and maintain the working wire-wound resistor standards. This is analogous to the voltage unit, where the ac Josephson effect is used to define and maintain a working laboratory unit of voltage.

2.7 An Absolute Resistance Standard

The quantum Hall resistance can be expressed in terms of a national unit derived from wire-wound resistors, such as Ω_{NBS}. These resistors are artifacts that have no unique resistance value. Every country's unit is somewhat different, and this difference is difficult to determine since resistors often shift by as much as 0.2 ppm on transport, due to mechanical and thermal stresses in the wire.

It would be preferable to have an absolute standard based on fundamental constants of nature and expressed in SI (Système International) units. This can be done using a calculable capacitor (Clothier 1965) and a complicated sequence of capacitance and resistance bridges. In this method an as-maintained unit like Ω_{NBS} is expressed in terms of the change in length of the calculable capacitor. The most accurate measurement was done by Cutkosky (1974). He calibrated Ω_{NBS} in SI units to within ± 0.03 ppm. The experiment lasted fifteen years and required such feats as measuring a 0.5 picofarad change of the calculable capacitor to 1 part in 10^9. The Ω_{NBS} is apparently drifting so the experiment is being repeated. England and Australia (among other countries) have determined their resistance standards in SI units to 0.09 and 0.2 ppm (Jones and Kibble 1985 and Thompson 1968). Several other countries have measurements underway.

Once a national unit is calibrated in SI units then, in principle, the calculable capacitor experiment need not be repeated because the unit can be maintained via the quantum Hall effect. $R_H(i)$ then becomes an absolute resistance standard. Measurements to date find that the SI value of $R_H(i)$ is about $25,812.8063(77)$ Ω (± 0.3 ppm) (Tsui et al. 1982a, Bliek et al. 1983, Kinoshita et al. 1984, Hartland et al. 1985). This uncertainty should be reduced below 0.05 ppm in the United States when the calculable-capacitor experiment has been repeated. At sufficiently low temperatures R_H seems to be independent of the device and the quantum step number to within at least a few parts in 10^8.

2.8 The Fine-Structure Constant

The fine-structure constant α is a dimensionless measure of the coupling between matter and the electromagnetic field and lies at the heart of quantum electrodynamic (QED) theory. It can be expressed in terms of h/e^2 and therefore is directly related to R_H:

$$\alpha = \frac{\mu_0 c}{2} \frac{e^2}{h} = \frac{\mu_0 c}{2} \frac{1}{i R_H(i)} , \qquad (2.8.1)$$

where the permeability of vacuum μ_0 is by definition exactly $4\pi \times 10^{-7}$ H/m and the speed of light c has been recently defined to be exactly 299 792 458 m/s (BIPM 1983). [See Note 1 on the uncertainty of the meter.] Therefore one can use absolute quantum Hall effect measurements of R_H to determine α directly to the same uncertainty as for R_H. It is interesting that this fundamental coupling constant can be determined using a condensed matter system.

We have assumed that $R_H(i)$ is exactly $h/e^2 i$. Measurements of R_H to date agree within their 0.3 ppm uncertainties with the value obtained from the anomalous moment of the electron (Van Dyke, Schwinberg and Dehmelt 1981) and QED calculations (Kinoshita and Lindquist 1981): $\alpha^{-1} = 137.035\ 993(10)$ (± 0.07 ppm). Once the uncertainty in R_H is reduced below 0.05 ppm then the results could be interpreted either as a rigorous test of QED theory or as testing the exactness of expression (2.5.1).

2.9 Temperature Dependences of ρ_{xx}

Most theories of the quantum Hall effect assume that the temperature is at absolute zero. What happens at finite temperatures? We consider the temperature dependence effects of ρ_{xx} and σ_{xx} in this section, and then those of ρ_{xy} in the next section.

Referring to Figs. 1 and 2 we see that there are large Shubnikov-de Haas oscillations in the longitudinal component of the resistance $R_x = V_x/I$ and that this resistance is very small in the regions corresponding to Hall steps. Thus $\rho_{xx} = (W/L)R_x$ also becomes very small. [Note that ρ_{xx} is inferred from R_x measurements by assuming a homogeneous medium, a separation L between the V_x potential probes, and a channel width W. The medium is seldom homogeneous and often the potential probes extend beyond the channel (as in the inset of Fig. 2 where $W = 0.38$ mm) making a larger effective width. So, experimental values of ρ_{xx} should be treated with caution.]

The current theories for the minimum value of ρ_{xx} envision that at higher temperatures this dissipation is thermally activated with energy $E_g/2$ across a mobility gap and hence varies exponentially as $\exp(-E_g/2k_BT)$. [In this and the next section ρ_{xx} refers to this minimum value of the diagonal resistivity.] At lower temperatures, when that mechanism freezes out, variable range phonon-assisted hopping between localized states dominates. The variable range hopping leads to a resistivity which varies exponentially as $\exp[(-T_0/T)^\gamma]$ where γ is predicted to be 1/3 in the absence of a B field (Nicholas et al. 1977, Mott and Davis 1979) and 1/2 in the presence of a strong B field (Ono 1982, Wysokinski and Brenig 1983).

In Si MOSFETs the minimum ρ_{xx} still shows thermal activation in the 2-4 K temperature range where the Hall steps are beginning to form, but the activation starts to freeze out below 2 K (Englert and von Klitzing 1978, von Klitzing 1984, Tausendfreund and von Klitzing 1984). Surprisingly, E_g is 85% and 96% of the cyclotron resonance energy $\hbar\omega_c$ for $i=2$ and $i=4$ steps (von Klitzing et al. 1985). One would expect E_g to be substantially less because each Landau level is spin-split and valley-split into four levels for the (100) crystal plane orientation of Si. For GaAs heterostructures E_g is 20% larger than $\hbar\omega_c$ (von Klitzing et al. 1985) even though the Landau splittings greatly exceed the spin splittings (Englert et al. 1982). These results may indicate the importance of exchange and many-body interaction effects.

The thermal activation has frozen out well before the Hall steps are formed in GaAs devices (Tsui, Störmer and Gossard 1982c, von Klitzing 1984), in InP heterostructures (Briggs et al. 1983, Guldner et al. 1984), and in InAs-GaSb heterostructures (Washburn et al. 1985). The variable range hopping models work very well for heterostructures over the temperature range 10 mK to 4 K. There is a slight preference for $\gamma = 1/2$ over $\gamma = 1/3$ (Ebert et al. 1983b).

Cage et al. (1984) found the minimum ρ_{xx} to be extremely temperature dependent and obtained excellent empirical fits for GaAs devices using the function $\rho_{xx} = aT^7$ over the 1-4 K temperature range. The coefficient a was device dependent. The actual function is no doubt exponential.

At lower temperatures ρ_{xx} becomes so small that it is difficult to measure. Tsui, Störmer and Gossard (1982c) instead measured σ_{xx} at very low currents using Corbino disks of GaAs heterostructures cooled to 1.2 K. They then inverted the conductivity tensor to obtain ρ_{xx} and found that $\rho_{xx} < 10^{-7}$ Ω, which corresponds to a three-dimensional resistivity less than 5 × 10^{-13} Ω-cm. They point out that this resistivity is an order of magnitude smaller than that of any other nonsuperconducting material. Recently Haavasoja et al. (1984) reported a $\rho_{xx} < 10^{-10}$ Ω measurement for a GaAs device at 0.4 K - corresponding to a persistent current with a 300 second decay time. Hence ρ_{xx} does indeed approach zero when the temperature approaches absolute zero.

Plots of V_H and V_x vs. B for GaAs heterostructures, such as those in Fig. 2, become quite dramatic at dilution refrigerator temperatures below 0.1 K (Paalanen, Tsui and Gossard 1982, Ebert et al. 1982). The ρ_{xx} peaks are extremely narrow and vanish above 6 T, while the Hall steps are so wide that there is almost a stepwise transition between plateaus.

2.10 Temperature Dependences of ρ_{xy}

When the Hall steps are examined at high resolution it is found that as the temperature is increased there comes some point (typically between 2-4 K) where the steps begin to have a measurable slope. Tausendfreund and von Klitzing (1984) investigated the temperature dependence of these slopes for Si MOSFETs ($d\rho_{xy}/dV_g$) and GaAs heterostructures ($d\rho_{xy}/dB$). They found that the slopes were exponential functions: $\exp(-T_0/T)$, i.e. they were activated, even though ρ_{xx} for GaAs was not in the activated regime. For Si MOSFETs, where ρ_{xx} was activated, they found that the slopes were linear in ρ_{xx} over at least three orders of magnitude.

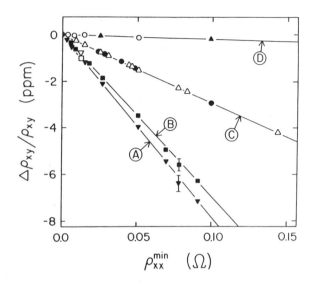

Figure 2.8. The error $\Delta\rho_{xy}/\rho_{xy}$ vs. minimum ρ_{xx} of several GaAs heterostructure devices when cooled to temperatures between 1.2 - 3.0 K. See also Fig. 9. Cage et al. (1984).

Yoshihiro et al. (1983) and Cage et al. (1984) measured ρ_{xy} at the values of V_g or B where ρ_{xx} was at a minimum for Si MOSFETs and GaAs heterostructures, respectively. They found the values of ρ_{xy} to be temperature dependent, being smaller at higher temperatures. The difference term $\Delta\rho_{xy} \equiv \rho_{xy}(T) - \rho_{xy}(0)$ satisfied the linear equation

$$\Delta\rho_{xy} = -s\, \rho_{xx} \qquad (2.10.1)$$

over at least four orders of magnitude change in ρ_{xx} (Cage et al. 1984) for the 1-4 K temperature range where the steps went from flat to sloped. An example of this linear dependence is shown in Figs. 8 and 9 for several GaAs devices. The values of s vary between 0.01 and 0.51 in those figures. In Si MOSFETs the values of s tend to cluster around 0.1, but have been between 0.06 and 0.15 (Yoshihiro et al. 1985a). Although the correction term $\Delta\rho_{xy}$ is small, it is strongly dependent upon ρ_{xx}: $-\Delta\rho_{xy} \lesssim 0.5\, \rho_{xx}$.

The mechanism for the linear dependence of $\Delta\rho_{xy}$ on ρ_{xx} is not known. It does not appear to be due to fin-

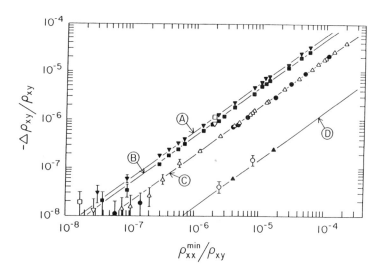

Figure 2.9. Log-log plots of $\Delta\rho_{xy}/\rho_{xy}$ vs. minimum ρ_{xx} over the temperature range 1.2 – 4.2 K for the same GaAs devices as shown in Fig. 8. The linear dependence of $\Delta\rho_{xy}$ holds over at least four orders of magnitude change in ρ_{xx}. Cage et al. (1984).

ite sample size effects which have been calculated by Rendell and Girvin (1981), Hall probe misalignment, variable range hopping conduction, or bulk leakage currents (Cage et al. 1984). van der Wel, Harmons and Mooij (1985) have recently found the slopes to be positive in Eq. 10.1 for GaAs heterostructures prepared by MOCVD. Therefore the device fabrication techniques may have something to do with this effect.

Cage et al. (1984) found the surprising result that even when the Hall steps were flat to within the 0.01 ppm instrumental resolution, the temperature-dependent error $\Delta\rho_{xy}$ could still be quite large. Thus, experimentally, there can be a correction to expression (2.5.1) because $\rho_{xy}(T)$ is not necessarily equal to $R_H(i)$.

All precision experiments to date have found that $\rho_{xy}(T)$ approaches a constant value $R_H(i)$ to within the instrumental resolution when the temperature is low enough. Presumably this value is $h/e^2 i$.

2.11 Current Distribution and Edge Effects

How is the current distributed in the 2D gas of a Hall bridge? There is considerable confusion about this question, due mainly to a series of theoretical papers dealing with current-carrying edge states (Halperin 1982, MacDonald and Středa 1984, MacDonald 1984 and Smrčka 1984). An impression could certainly be formed from these papers that all the current flows on the edges of the samples (see however Note 10, chapter I). For example Smrčka (1984) argues that these currents arise from "skipping orbits" in which electrons making cyclotron orbits elastically scatter from potential barriers at the edges. Pudalov and Semenchinskii (1984b) point out, however, that if the current were confined to the edges then the electron velocities would be two orders of magnitude greater than the speed of sound in the bulk material of GaAs for a 10 μA current.

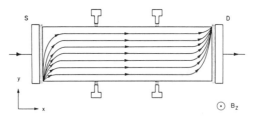

Figure 2.10. Top view of the 2D electron flow pattern at a quantum Hall step of an ideal device. The current is tilted from the x axis near the ends of the device due to the shorting of the Hall voltage by the source and drain contacts.

Figure 10 shows what the electron flow patterns might look like in an ideal device characterized by a uniform macroscopic conductance. In the steady state the Lorentz and Coulomb forces would be exactly opposed. This causes the electrons to enter and exit at diagonally opposite corners. The electrons would be executing tight spirals of radius $\ell \approx 80$ Å for both spin states of the first Landau level at 10 T with an angular frequency $\omega_c = eB/m = 2.6 \times 10^{13}$/second for a GaAs heterostructure device.

The source-drain resistance must be equal to or greater than R_H (Fang and Stiles 1983, Powell Dean and Pepper 1984), even if $\rho_{xx} = 0$. The resistance arises from current crowding in the corners. This "spraying resistance" could well be a source of hot electrons that are not in thermal equilibrium with the crystal lattice.

One can make multiple interconnections on the same device to obtain rational fractions of $h/e^2 i$ (Fang and Stiles 1984, Syphers, Fang and Stiles 1984, Powell, Dean and Pepper 1984). Rational factors of 3, 2, 3/2, 1, 2/3, 1/2 and 1/3 have been obtained by interconnections that act as series, series/parallel and parallel combinations. The results are easily explained by assuming that the interconnections act as additional sources and drains of individual quantum Hall resistors, with each resistor having a flow pattern similar to that of Fig. 10.

If $\rho_{xx} = 0$ then flow lines of Fig. 10 are also lines of constant potential. If $\rho_{xx} \neq 0$ then the isopotential lines tilt with respect to the current flow lines and intercept the sides of the device. The current then crosses these potential lines and the Hall angle $\theta_H = \arctan \rho_{xy}/\rho_{xx}$ becomes less than 90°.

Rendell and Girvin (1981) investigated the effects that such flow patterns would have on measurement of R_H. They predict a correction term that depends on ρ_{xx}, on the sample width and on the closest Hall probe set distance from the source or drain. It is not too difficult to fabricate devices having geometrical corrections below one part in 10^8.

A perfect device is hard to find. Most are inhomogeneous and infested with dissipative regions. These flaws make the percolation models of Iordansky (1982) [and several other authors whose work is discussed in Chap. 3.5, 3.6] attractive. They picture islands and continents of localized electrons, and lakes and seas of mobile 2D electrons. Source-drain conduction becomes likely when the topology is half land and half sea. The conduction patterns are complicated in highly disordered systems. They meander around, taking the paths of least resistance; but the condition $R_H(i) = h/e^2 i$ still holds. This concept is supported by the experiments of Syphers, Fang and Stiles (1985) who made nonconducting islands in the 2D electron layer of Si MOSFETs. They found the Hall voltage to be unaffected, as Prange (1981) had predicted.

There have recently been a group of experiments that studied the current distributions (Sichel, Sample and Salerno 1985, Ebert, von Klitzing and Weimann 1985, Zheng, Tsui and Chang 1985a). They mounted probes around the peripheries of the devices, as well as in the interiors, and monitored the voltages induced on the probes over the Hall step region where ρ_{xx} is small. The current is found to be generally distributed throughout the device. On rare occasions it is uniformly distributed across the sample for certain places on the Hall step. Often the current flow pattern moves around and sloshes from side to side as it avoids dissipative regions along the step. The flow pattern is very sensitive to inhomogeneities, such as a deliberately introduced 1% variation of the charge density n across the device by tilting a metal gate layer with respect to the 2D gas of GaAs heterostructures (Ebert et al. 1985). All the items mentioned in this section are further discussed in Chap. VI.

2.12 Current Dependence and Breakdown

The value of the minimum ρ_{xx} is only moderately current dependent in GaAs heterostructures until one reaches currents of hundreds of microamperes (Ebert et al. 1983a). Then there is a critical current I_c above which the dissipation along the channel ρ_{xx} suddenly rises many orders of magnitude (Ebert et al. 1983a, Cage et al. 1983, Kuchar et al. 1984). An example is shown in Fig. 11, where $I_c \approx 340$ μA. ρ_{xx} increases seven orders of magnitude between 25 μA and 370 μA in that device. The critical current density j_c at which this breakdown phenomenon occurs is estimated to be 0.6 - 0.9 A/m (Ebert et al. 1983a, Cage et al. 1983). Störmer et al. (1985) found that $j_c = 1.16$ A/m for the $i = 1$, 2 and 4 steps of a sample with a Corbino disk geometry. This geometry avoids the corner effects of Hall bridges. Kuchar et al. (1985) report that: (1) I_c is similar for devices having either Hall bridge or Corbino disk geometries; (2) I_c is independent of the current direction and of the perpendicular magnetic field directions $\pm B$; and (3) I_c remains unchanged when the 2D plane is tilted by angle θ and the field is increased to $B/\cos\theta$ so that B_z stays constant. von Klitzing et al. (1985) have observed breakdown in Si MOSFETS with $j_c \approx 0.2$ A/m.

Experiments and Metrology

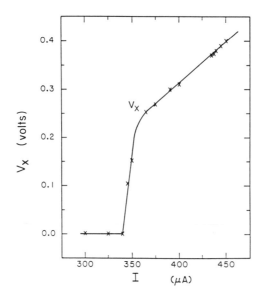

Figure 2.11. The minimum V_x over the critical current region of a GaAs heterostructure device at 1.2 K. $I_c \approx 340$ µA. Cage et al. (1983).

Figure 12 shows step-like features in V_x across the Hall plateau, indicating that breakdown occurs at lower currents as one moves away from the center of the Hall plateau. Cage et al. (1983) demonstrated that the breakdown is spatially localized into particular regions of the samples. They also observed a rich time dependent structure in ρ_{xx} - exhibited as transient switching and broadband noise. (See Figs. 13 and 14.) The transient switching occurred on a microsecond time scale among a numerous set of discrete, unstable dissipative states of current flow through the device. Along with this switching there was a strong background of structured broadband noise that spanned the entire spectrum from near dc to beyond 5 MHz, as in Fig. 14. This noise was much larger than that for electrons in thermal equilibrium with the crystal lattice. Kuchar et al. (1985) recently reported that GaAs devices initiate breakdown and reach a steady state condition within no more than 100 ns of a source-drain current pulse. Ricketts (1985) measured the noise over the 1-100 Hz frequency range and found it unlikely to limit the use of GaAs heterostructures as resistance standards.

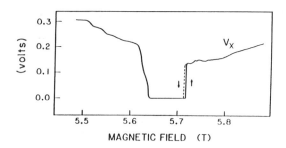

Figure 2.12. V_x over the Hall step region at 1.2 K for the same device as in Fig. 11. The 325 µA current is below the critical current. Arrows indicate the magnetic field sweep direction. Note the hysteresis in B. Cage et al. (1983).

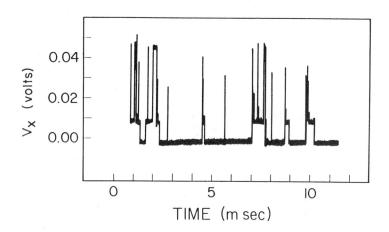

Figure 2.13. Oscilloscope trace of V_x^{min} taken at 325 µA and 1.2 K for the same GaAs device as in Figs. 11 and 12. Transient switching occurs between discrete dc voltage levels on microsecond time scales. The dc levels are characterized by structured broadband noise as shown in Fig. 14. Cage et al. (1983).

How do theorists explain all of this? Heinonen et al. (1984) and Středa and von Klitzing (1984) observe that the conduction electron drift velocities are near the speed of sound in GaAs at the critical current densities for breakdown. Spontaneous emission of phonons may occur in a manner analogous to the Cher-

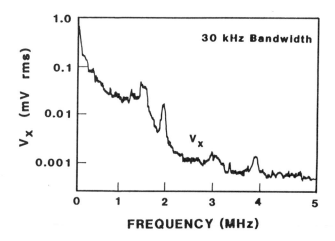

Figure 2.14. Frequency spectrum of the V_x^{min} broadband noise under the same conditions as Fig. 13. Cage et al. (1983).

enkov effect where charged particles that exceed the speed of light in a medium emit photons. These models are in qualitative agreement with experiment. Heinonen et al. (1984) predict that j_c scales as $B^{1/2}$, in agreement with Störmer et al. (1985) and von Klitzing et al. (1985). Their predicted values of j_c are an order of magnitude too large but Cage et al. (1983) showed that the breakdown regions are localized which would reduce j_c. The 1 µs phonon transit times are of the same order as the observed transient switching. They point out that GaAs is piezoelectric, so j_c should depend on the crystal orientation. These models look very promising. The critical current drift velocities of Si, however, are much smaller than the speed of sound (von Klitzing et al. 1985).

Another explanation for the origin of the breakdown is superheated electrons that are not in thermal equilibrium with the 2D electron gas (Gurewich and Mints 1984, Komiyama et al. 1985). Komiyama et al. (1985) measured ρ_{xx} and ρ_{xy} for GaAs devices and then calculated σ_{xx} from $\sigma_{xx} = \rho_{xx}/(\rho_{xx}^2 + \rho_{xy}^2)$. Plots of σ_{xx} vs. the Hall electric field E_y at different sample temperatures revealed breakdown at about 40 V/cm. Their superheated electron model provided excellent fits to these data. Yoshihiro et al. (1985b) proposed that hot electrons are produced in the vicinity of the source and drain electrodes. The electrons diffuse toward the central region of the channel by variable range hopping through localized states until they lose their excess energy. The calculated diffusion lengths are in reasonable agreement with their experimental configurations for both Si MOSFETs and GaAs heterostructures. von Klitzing et al. (1985) found that cyclotron radiation was emitted from a GaAs device and the intensity was substantially reduced when the opposite corners were masked by absorbers. Yoshihiro et al. (1985b) measured the frequency spectrum of ρ_{xx}. They found the noise spectrum to be proportional to $(\rho_{xx})^{1/2}$ and, like Cage et al. (1983), observed structure in the frequency spectrum. The noise was 4 orders of magnitude larger than the thermal equilibrium noise.

Yet another explanation of the breakdown is that the fraction of extended states for each quantum level i is dependent on the Hall electric field, varying approximately as E_y^2 (Trugman 1983, Störmer et al. 1985). The number of extended (mobile) states increases as the current increases, due to heating, moving the mobility edge closer to the Fermi energy E_F. Breakdown occurs when the mobility edge comes within $k_B T$ of E_F. This model also looks very promising

Zener tunnelling of electrons across the energy gap between quantum levels into the neighboring empty level is also a possibility (Tsui, Dolan and Gossard 1983b). This requires that $j_c \approx 6$ A/m, however, so the current would have to be squeezed into $\approx 1/50$ of the sample width. Certainly there is ample opportunity to experimentally test these breakdown models.

2.13 Hall Step Widths and Shapes

Hall step widths naturally depend on the measurement resolution. The standard approach is to choose the points where the plateaus first deviate by a fixed amount $\delta\rho_{xy}/\rho_{xy}$, such as ± 0.1 %. Most work has been done in Si MOSFETs. The widths *decrease* linearly with increasing current (Zavaritskiĭ and Anzin 1983, Pudalov and Semenchinskii 1984a,b), and with increasing temperature (Pudalov and Semenchinskii 1984b,c), [whereas Isihara (1983a,b) predicts that the widths should scale as $ln(1/T)$]. The widths also decrease linearly with increasing zero field mobility μ (Störmer, Tsui and Gossard 1982, Furneaux and Reinecke 1984a,b). They *increase* linearly with increasing magnetic field (Störmer, Tsui and Gossard 1982). There is a critical value of current, temperature or magnetic field at which the width disappears for a given sample. It also disappears at high mobilities. On the other hand, when the mobility is too small the steps once again shrink – except for the spin-split steps (Furneaux and Reinecke 1984a,b). Optimum mobilities for obtaining wide integer Hall steps with small values of ρ_{xx} seem to be around 10,000 cm^2/Vs for Si MOSFETs and 100,000 cm^2/Vs for GaAs heterostructures. The integer steps become quite narrow for μ > 1,000,000 cm^2/Vs, which are the values required for the fractional QHE studies.

The device mobilities in Hall step regions are many orders of magnitude larger than the zero-magnetic-field mobility μ. Although these quantities are perhaps not directly related, μ may be an indicator of the density of *immobile* states present in the Hall step regime (Störmer, Tsui and Gossard 1982). Clearly, μ is a good measure of the scattering at zero magnetic field. No data have been reported on the variation of ρ_{xx} with μ; but the widths of ρ_{xx} minima decrease with increasing μ (Furneaux and Reinecke 1984a,b), indicating that ρ_{xx} *increases* with increasing μ in the QHE regime.

This brings us back to the localization model of Aoki and Ando (1981). They propose that the step widths result from electrons occupying localized states. Electrons in these states are immobile and therefore do not contribute to the Hall voltage. Furneaux and Reinecke (1984a,b) have experimentally tested this concept. Sodium ions in the SiO$_2$ layer of Si MOSFETs induce localized states with comparable energies in the vicinity of the 2D electron gas. They therefore controllably diffused Na$^+$ ions in the oxide layer by heating the samples and applying forward or reverse bias gate voltages to move the ions towards or away from the 2D gas. The amount of interface charge Δn_{ox} could be determined from shifts in the threshold voltage ΔV_{th} and the capacitance per unit area: $\Delta n_{ox} = C\Delta V_{th}/e$. Their data support the localization picture. The plateau widths increase linearly with Δn_{ox} and decrease linearly with increasing μ.

Another indication of the importance of localization is provided in the experiments of Pepper and Wakabayashi (1982, 1983) and Long, Myron and Pepper (1984). They found the integer Hall plateaus of Si MOSFETs could be narrowed or removed by increasing the frequency of an ac source-drain current. These frequencies were in the kHz range. They propose that this is a delocalization process. If the electrons oscillate fast enough they no longer know they are localized, narrowing the steps.

Joynt (1985) believes that these frequency-dependent observations can be understood if semiclassical orbits are present in the sample. Using results from percolation theory (Trugman 1983), he predicts a frequency dependent conductance $\sigma_{xx}(\omega) \propto \omega^{-8/3} \exp(-\omega_0/\omega)$ where the threshold frequency ω_0 is 1 MHz or less. Experiments should be done to test this prediction.

There are two experiments which seem to conflict with the delocalization mechanism proposed by Pepper and Wakabayashi (1982, 1983). Iwasa, Kido and Miura (1983) used ac currents in Corbino disks and found that the σ_{xx} minimum broadened as the frequency increased. One would expect them to narrow if the Hall steps narrowed. Unfortunately they did not investigate the Hall steps. In the other experiment, Woltjer et al. (1985) found that the Hall step shapes were unaltered for current pulses of only 100 ns duration. These discrepancies need to be resolved.

The localization concept is disturbing. It implies that imperfections must be present in order to obtain good quantization, wide steps and accurate values of R_H, i.e. maybe you *can* make a silk purse from a pig's ear. There should not be too much imperfection, however, or the quantization is destroyed.

There is an equilibrium thermodynamic model by Heinonen and Taylor (1983) that predicts Hall plateaus in the absence of impurities. The steps do not occur exactly at integer values, however, and they treat the Coulomb interactions of the 2D electron gas only approximately.

Most Hall steps do not have "ideal" shapes. Instead they have structure on one or both sides of the step (Wagner et al. 1982). The structure can either be peaks or dips. An example of a dip on the high magnetic field side is shown in Fig. 5. This structure does not affect the width of the step at the

Experiments and Metrology

$\Delta\rho_{xy}/\rho_{xy} = \pm 1$ ppm resolution. The source of this structure is not known, but it is presumably due to inhomogeneities and impurities. Subtle changes can apparently cause dramatic effects since the structure has been observed to change from day-to-day even when the devices are stored in the dark at 4.2 K.

Pudalov and Semenchinskii (1983, 1984b,c) find that $\delta\rho_{xy}$ is linearly related to ρ_{xx} for ideally-shaped steps of Si MOSFETs: $|\delta\rho_{xy}| = (0.3 \pm 0.1)\rho_{xy}$ over the $10^{-5} < |\delta\rho_{xy}/\rho_{xy}| < 10^{-2}$ range. This is reminiscent of the $\Delta\rho_{xy} = -s\rho_{xx}$ relationship for the value of ρ_{xy}. They attribute the shapes of ρ_{xy} and ρ_{xx} over the Hall step region to changes in the carrier concentration. That assumption provides a fairly good account of their data.

2.14 Thermomagnetic Transport

There are other important properties associated with the quantum Hall effect, some of which are briefly indicated in these remaining sections - along with several relevant references. For a comprehensive review of the electronic properties of two-dimensional systems please refer to Ando, Fowler and Stern (1982).

Both theoretical and experimental investigations have been made on the thermal conductivity and thermopower. Girvin and Jonson (1982) predicted that: (1) in the absence of disorder, the thermopower is a universal function of $k_BT/\hbar\omega_c$; (2) its tensor components oscillate across the Hall step (Jonson and Girvin 1984); and (3) it is exponentially small (thermally activated) on the Hall steps.

The experimental work of Obloh, von Klitzing and Ploog (1984), Smith, Closs and Stiles (1984) and Davidson, Robinson and Dahlberg (1985) support these predictions. Other relevant theoretical studies are by Smrčka and Středa (1977), Středa (1983), Středa and Smrčka (1983), Götze and Hajdu (1978, 1979), Středa and Oji (1984), Oji (1984), Zawadzki and Lassnig (1984a,b) and Oji and Středa (1985).

Zawadzki and Lassnig (1984a,b) predicted that the specific heat of 2D electrons has an oscillatory character in the quantum Hall regime for a Gaussian density of states. Gornik et al. (1985a,b) performed experiments and obtained excellent agreement with theory using a Gaussian density of states distribution with constant background for the quantum levels. That distribution was preferred over a pure Gaussian or Lorentzian form.

2.15 Magnetic Moment

Another promising method of deducing the distribution of states within the quantum levels is via measurements of the magnetic moment μ of the 2D gas. Eisenstein et al. (1985a,b) performed such measurements by an ingenious torque method in GaAs heterostructures. This work is discussed in Chap. VI. A main result is that the density of states is relatively constant and does not become very small between Landau levels.

2.16 Magnetocapacitance

The capacitance between the gate and the 2D electron gas of a Si MOSFET also undergoes oscillations in the quantized Hall regime. For example, it goes to zero at the center of quantum level fillings in devices with Corbino disk geometries because σ_{xx} goes to zero (Kaplit and Zemel 1968). It was thought that this property might be a sensitive method of measuring the density of states distribution in the tail regions of the quantum levels; however, Goodall, Higgans and Harrang (1985) believe that these measurements sample the conductivity distributions and not the density of states.

Other pertinent references are Widom and Clark (1982), Smith and Stiles (1984), Smith, Heiblum and Stiles (1985b) and Smith, Goldberg and Stiles (1985a).

2.17 Magneto-Photoresponse

The values of V_H and V_x are sensitive to radiation directed on the samples – changing by ΔV_H and ΔV_x from the values in the dark. Lavine, Wagner and Tsui (1982) find that ΔV_x undergoes Shubnikov-de Haas oscillations when it is plotted against B for fixed photon wavelengths. They also find that the magnitude of V_H and V_x increase when the laser frequency is near an inver-

Experiments and Metrology

sion layer cyclotron resonance. They attribute this to heating, and believe it is the 2D gas which is heated, rather than the lattice, since the response is still unaltered for laser pulses of 3µs duration.

Some useful references are Shiraki (1977), Tsui et al. (1980), Stein, von Klitzing and Weimann (1983), Anzin et al. (1984), Sakaki et al. (1984), Wei, Tsui and Razeghi (1984) and Shah et al. (1985).

2.18 Conclusions

This chapter has outlined many of the experimental studies on the magnetotransport properties of the integer quantum Hall effect. Much has been accomplished in these studies; but many puzzles remain unsolved and additional surprises are certain to occur.

2.19 Acknowledgments

The author wishes to acknowledge the support of the United States Office of Naval Research, the Army Research Office and the Calibration Coordination Group of the Department of Defense, as well as the assistance of his colleagues: R S Allgaier, W J Bowers Jr., R F Dziuba, B F Field, S M Girvin, T E Kiess, B N Taylor and C T Van Degrift of the National Bureau of Standards, D C Tsui of Princeton University, A C Gossard of AT&T Bell Laboratories, and R J Wagner of the Naval Research Laboratory.

2.20 Notes

Note 1. The price one pays for defining the speed of light is that one must now determine the length of the meter by measuring how far light travels in $1/(299\ 792\ 458)$ seconds. For technical reasons this is an easier measurement than working the other way around. It is for this reason that the redefinition was done. The resulting uncertainty in the meter reappears in the absolute value of R_H because there is a length determination in the calculable capacitor. Hence we have not 'gained something for nothing'. At present the uncertainty in implementing the new definition of the meter is about 3×10^{-10} which is two orders of magnitude better than could be obtained prior to the redefinition.

It is amusing to recall the electrodynamic relation $\mu_0\epsilon_0=c^{-2}$ and note that since μ_0 is exactly defined, ϵ_0 must now be exactly defined. Since ϵ_0 is needed to determine the value of a capacitor, we appear to have gained something until we realize that we do not know exactly how far apart the capacitor plates are.

CHAPTER 3

Effects of Imperfections and Disorder

RICHARD E. PRANGE

3.1 Introduction

As was remarked in the first chapter, in order to have the QHE it is essential to break translational invariance. In practice this happens automatically because real material samples have some degree of imperfection. In this lecture, we deal with the effects of imperfections in some detail, although still pretty much at an elementary level. This will be done by introducing various models for impurities. A combination of these models is probably quite realistic. The aim is twofold: First, we want to appreciate better what kind of imperfection systems will give rise to a mobility gap at the Fermi level. This, we recall, is a situation in which there are no extended, current carrying states at the Fermi level. There may be, and in practice are, localized states there. A mobility gap is the essential condition for the QHE, according to the gauge argument of Laughlin given in Chap. 1.7. Secondly, we want to develop an understanding of what localized states are really like, and what extended states are like. We would also like some explicit, semirealistic models which exhibit the QHE.

The results are quite pictorial and vivid. However, they are obtained at the expense of generality, because it must be assumed that the magnetic field is strong enough that the system is well in the QHE regime. More powerful techniques are needed for certain deeper questions, as for example the question of how the QHE develops as one turns the magnetic field up. These techniques will be deployed in lectures four and five. This chapter summarizes the work of a number of authors, but will be organized along the lines of Joynt and Prange (1984).

3.2 The Impurity Potential

In the absence of a magnetic field, it is usually sufficient to characterize a potential by a mean free path ℓ_0, and the strength of the potential by $1/k_F\ell_0$, i.e. the ratio of the wavelength of the electron to the mean free path. Different potentials which give the same mean free path result in practically the same electronic behavior. In practice the mean free path is determined by fits to experiment. It is theoretically estimated by calculating the scattering rate, i.e. by solving a scattering problem. Then ℓ_0 is a more or less complicated construct depending on the potential and electronic parameters.

In a strong magnetic field the Fermi wave number has no meaning and neither does the mean free path. Rather, the characteristic length associated with electrons is the magnetic length ℓ, and the characteristic energy scale is $\hbar\omega_c$. These must somehow be compared with the potential. It has been found that a rather direct comparison is feasible.

The result is that the conditions for the QHE will be satisfied provided the impurity potential $V(r)$ can be written as the sum of three terms

$$V = V_W + V_{sc} + V_S . \qquad (3.2.1)$$

The first term, V_W, is characterized by the weakness condition

$$\max |V_W(r)| \equiv V^{max} \ll \hbar\omega_c . \qquad (3.2.2)$$

This potential can, however, be pervasive and have arbitrarily rapid variation. Thouless (1981) and Hal-

Imperfections and Disorder

perin (1982) were the first papers to exploit the simplicity of this case.

The second term V_{sc} is a sum of scattering potentials. A scattering potential is here defined to be a potential of arbitrary strength and variation which vanishes outside some region, the scattering region. A further condition is imposed on the sum of such potentials which requires that a potential-free region of minimum width d percolates throughout the system, where $d \gg \ell$. Scattering, or short range potentials are often considered to be generic in the field-free case. We set no limit on the size or extension of the scattering potentials as long as the potential-free region exists. (Prange and Joynt 1982, Joynt and Prange 1984 and Chalker 1984 worked out this case.)

The third potential V_S is required to be smooth in the sense that

$$|\nabla V_S| \ll \hbar\omega_c/\ell . \qquad (3.2.3)$$

In many ways, this is the most interesting and novel possibility which turns out to need percolation concepts. The smooth potential has new and qualitative experimental consequences which are observed [Chap. 2.11]. Such a potential does not exist in the usual low-field conductors as it is screened out. Iordansky (1982), Prange and Joynt (1982), Kazarinov and Luryi (1982) Luryi and Kazarinov (1983), Trugman (1983), Giuliani et al. (1983) and Joynt and Prange (1984), have made contributions in this case. A number of the results below were obtained by Tsukada (1976) before the QHE was discovered.

In words, the total potential of (3.2.1) will give rise to a structure containing mobility gaps between the Landau levels and thus to the QHE, if it locally varies considerably less than $\hbar\omega_c$. Further, isolated exceptions to this rule can be tolerated. This type of potential is illustrated in Fig. 1.

It is clear that in inversion layers all three components of the potential exist. The scattering potential comes from the few charged impurities which are in the middle of the inversion layer and from major interface defects. The weak potential comes from impurities at a distance of order ℓ from the layer, and the smooth potential comes from macroscopic

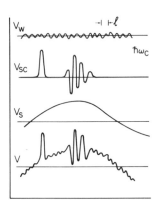

Figure 3.1 One dimensional schematic version of a potential satisfying the quantum Hall condition (3.2.1). The separate potentials making up the total potential are illustrated.

inhomogeneities and fluctuations in the impurity density at distances larger than ℓ from the interface.

Another approach is to start by making the lowest-Landau-level approximation (see Chap. 1.6) which results in considerable simplification. For example, a Gaussian white noise random potential, or a sum of δ-function potentials can be studied in this approximation either numerically (Aoki 1977, 1978, Ando 1983b, 1984a, Aers and MacDonald 1984) or by powerful field theoretic techniques (Wegner 1983, Brézin et al. 1984). The results are instructive although it is not always clear how to compare them to the 'true' impurity potential V. They do agree in a qualitative way with expectations based on (3.2.1). This indicates that if (3.2.1) holds, the lowest Landau level approximation is qualitatively pretty good.

The lowest Landau level approximation is only exact for nonlocal potentials. Therefore, the gauge arguments cannot be made with confidence once this approximation is made. There are some related difficulties which have to be avoided (Giuliani, et al. 1983). However Ando (1984b) has done numerical calculations keeping several Landau levels which show interesting quantitative effects but generally support the qualitative results of the other theories. Section 7 deals with the numerical results.

3.3 Weak Potential

The weak potential case is the easiest to treat, but at the same time the results are the least pictorial. One way to treat it (Thouless 1982) is by perturbation theory. For example, consider a propagator or Green's function evaluated at an energy midway between bare Landau levels. This propagator can be computed as a power series in V_W. It is possible to show that successive terms in the series decrease by factors smaller than $2V^{max}/\hbar\omega_c$. [The energy denominators are all greater than $\hbar\omega_c/2$.] The series will therefore converge under condition (3.2.2) and the resulting propagator is so smooth that there can be no state at that energy. In other words, (3.2.2) implies that there is a gap in the spectrum of the Hamiltonian $\mathcal{H}_0 + V_W = \mathcal{H}_W$ between each Landau level. If there is a gap, there is certainly a mobility gap.

The above result can be obtained in a rather trivial way by use of the Feynman-Hellmann theorem. Consider the Hamiltonian $\mathcal{H}_0 + \lambda V_W$ where the parameter λ will eventually be put equal to unity. Let $E(\lambda)$ be the energy of a state. Then the theorem asserts that

$$dE/d\lambda = \langle V_W \rangle , \qquad (3.3.1)$$

where the expectation value is in the state. It is easy to prove that $|\langle V_W \rangle| < V^{max}$. Integrating (3.1) over λ from zero to one gives the result that any exact energy level differs from an unperturbed level by at most V^{max}. The unperturbed levels are at $n\hbar\omega_c$ so the perturbed levels have gaps between them. While this argument suffices here, the perturbation version of the argument is needed in combining the several types of potential.

There is not much more to be said, but a few things are worth noticing. First, the above results together with the Laughlin-Halperin argument [Chap. 1.7] imply that *there exist some extended states of $\mathcal{H}_W$ in each level*. This is in direct contrast with the belief that in zero magnetic field, any random potential localizes all states in two dimensions. How many states are extended by the magnetic field is a question discussed further below and in Chaps. V and VI.

A second remark is that condition (3.2.2) is stronger than necessary. All that is required is $V^{max} < \hbar\omega_c/2$. However, in trying to combine the effects of V_W with V_S, one needs a somewhat stronger inequality, so to be on the safe side, we require (3.2.2).

3.4 Scattering Potential

In this section, we consider the case of a potential which is surrounded by a potential-free region of thickness $\delta \gg \ell$. This case is much more fun than that of the weak potential and is illustrated in Fig. 2 where the scattering regions are cross-hatched.

While it is not realistic to assume potential-free regions, it is necessary to show that isolated violent imperfections do not destroy the QHE. Such potentials would certainly cause localization at $B = 0$. Once the idealized scattering potential case is understood, it is easy to combine it with the other potentials and obtain the desired result.

Perhaps it is not too surprising that scattering potentials do not cause trouble for the QHE. Indeed, we saw in Prob. (1.1) that cutting holes in the system can do no harm, as long as it is possible to define a conductivity. However, the conductivity can not always be defined. [The conductivity involves a macroscopic average over a small region and is tricky to define in the presence of a potential V_S.]

The idea is to make a wave packet which impinges on the scattering region. It is straightforward to do this in the potential-free region, by superposition of the degenerate states ψ_{nk} of Eq. 1.5.4 [Problem 1]. The resulting wave packet will not move, however, so it can not impinge on the scattering center.

Therefore, we take the case in which the electric field is imposed, and form a superposition of states (1.6.1). The group velocity of this packet is then in the x direction and has magnitude v. Further, it suffers no dispersion, since the energy is linear in k, (Eq. 1.6.2). Although the packet is moving, it has no significant kinetic energy over and above the kinetic energy of localization in the magnetic field. It does however have a potential energy due to its localization along a line of constant potential energy. By making the packet long and narrow, this energy can be made as definite as one pleases.

Imperfections and Disorder

If the scattering center is in the positive x-direction from the packet, it will be struck by the packet and scatter it. Long after the scattering, the packet will be found in a state of the same energy as it started with, but outgoing from the center. There is only one such outgoing wave possible, namely, the packet must proceed in the positive x-direction with the same y-value (energy) as it had initially, but it must be on the positive x side of the scattering center. In short, only forward scattering is allowed for the simple reason that no other outgoing state has the right energy. This situation is depicted in Fig. 2.

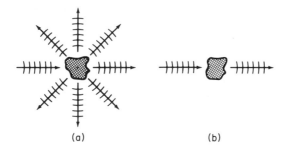

Figure 3.2. Two dimensional scattering in a) zero field, b) strong field.

In Fig. 3 we illustrate the path of a wave packet through a region of several scattering potentials. Roughly speaking, one expects the packet to skirt the edges of the scattering region where the potential

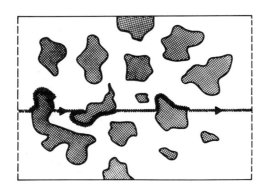

Figure 3.3.

becomes large. Whether it passes above or below this region depends on the sign of V_{sc}.

Once the simple and fundamental fact of this forward scattering is appreciated, it is not too difficult to see that the scattering center does not disturb the QHE. One could therefore, simply go on to the next case, but it is elementary and instructive to work things out in more detail.

First a few remarks: It is straightforward to prove that all scattering is in the forward direction in a formal way. We give the idea for this in Note 1. Second, this is doubtless the simplest quantum scattering problem conceivable, and should be in the textbooks. Ordinarily, time reversal invariance guarantees that even in one dimension backscattering (reflection) is possible.

Third, exactly the same situation arises in another context. Consider a thin film of fluid which supports vortices. The vortices will move perpendicular to applied forces, for the usual gyroscopic reasons. The moving vortices will scatter from scattering centers just as electrons do in the QHE system (Trugman and Doniach 1982). Fourth, if the scattering center is not completely isolated on all sides, but rather forms a barrier sticking out from one or both sample edges, then it is possible that the scattering will not be forward. Rather, there is another state of the same energy, an edge state, moving in the negative x direction into which the electron can scatter. Such an electron would eventually make an orbit closing on itself.

A final remark is that scattering theory studies the transition of a wave packet from an incoming initial state asymptotically far from the scatterer to the outgoing asymptotic state. In conventional scattering theory, 'asymptotically far' means many wavelengths and also far compared with the dimensions of the scattering region. In this case, asymptotically far means a long distance d in units of ℓ along the path of the particle from the scattering region. What is more, a distance d is asymptotic insofar as $\exp(-d^2/2\ell^2) \ll 1$. [Note 2]. This is a much weaker condition than for the zero B case and occurs because a fundamental length exists in the free electron Hamiltonian. Thus the length d separating scattering regions can be much smaller than the size of the typical region.

Imperfections and Disorder

We now make a phase-shift analysis of the scattering. It is clear that we may go to the limit of a wave packet whose energy is as well defined as is possible, so that the incident wave has the form ψ_{nk}. The result above is that the outgoing wave has the same form. Unitarity implies that it can differ from the incoming wave by at most a phase shift $\delta(k)$. We find, therefore, that

$$\psi_{nk}^{out} = \psi_{nk}^{in} e^{i\delta(k)}. \quad (3.4.1)$$

[We suppress the dependence of δ on n.] Since we know the form of the wave function on both sides of the scatterer, we can impose the condition of periodicity in x. We find, in the presence of the scattering center, that

$$kL + \delta(k) = 2\pi p. \quad (3.4.2)$$

For each integer p there is an allowed wave number k. The energy is obviously given by $E_{np} = vk + n$. The extended states are conveniently labelled by the integers p. There is thus a natural ordering of the states, which also coincides with the order in which they are localized in the y direction.

There will be a range of k for which $\delta(k)$ is not a multiple of 2π. This range is just that for which the scattering center lies across the line $y = k$. As k increases through this range, δ will necessarily be decremented by an integral number of 2π's, say by $2\pi M$. It is not difficult to count up the number of solutions of (3.4.2) and one finds that there are M fewer solutions if δ decreases by $2\pi M$ than if δ does not vary at all. The missing solutions are states bound to the scattering center since the total number of eigenstates is conserved.

We have just proven Levinson's theorem for this case. The scattering states are extended states. Levinson's theorem, in general, relates the number of bound states to the total change of the scattering phase shift (Seager and Pike 1974). It expresses a fundamental relationship between properties of scattering states and bound states in the same potential. Joynt and Prange (1984) give a more complete argument. The localized states are obviously confined to within a distance ℓ of the scattering region. It is not possible to discuss the bound states in more detail without making further assumptions on the nature of the scattering potential. If it is assumed to be smooth however, a rather complete analysis can be given [Problem 2].

Since a negative number of bound states does not make sense, it follows that the phase shift makes a net change which is negative or zero as the energy increases through the range in which scattering occurs. This is independent of the sign of the scattering potential. Also, if there is more than one scattering center which the electron must get by, the total phase shift is the sum of the individual phase shifts. The multiple scattering analysis is thus trivial.

An interesting quantity is the time delay, $d\delta/dE$. Recall that in the construction of a wave packet near energy E_0, we need $\int dE$ of

$$e^{i(\delta(E)-Et)} \cong$$

$$e^{i(\delta(E_0)-E_0 t)} e^{-i(E-E_0)(t-d\delta/dE)}.$$

Since in the present case $d\delta/dE$ is negative if there are bound states, there is a time *advance*. This explains in an intuitive way how the QHE works. If there are bound states, there will be fewer states to carry the current. But, if there are bound states, a wave packet appears sooner on the far side of the scattering potential than it would do in the absence of bound states. This means that in the neighborhood of the scattering center, there is a lower density of extended states, but these are moving faster and carry more current. This is not so surprising if one considers that a scattering center whose potential is strong enough to bind an electron will also have effective electric fields at its edge greater than the applied external electric field. These effective fields speed up the drift velocity of the electron.

We now turn to the explicit calculation of the Hall conductivity in this system. Let us use the gauge argument applied directly to the states of the system (which is possible because the extended states have a natural order). In the present geometry (periodic boundary conditions) the vector potential correspond-

Imperfections and Disorder 79

ing to the Aharonov-Bohm flux can be written as a constant in the x direction, which we write as $eA_x/\hbar c = 2\pi a/L$. [Think of the rectangle wrapped into a cylinder.] An increase in the parameter a by unity corresponds to an increase in the flux by one flux quantum.

A constant vector potential of this type can often be eliminated by a gauge transformation, $\psi \to \exp(2\pi i a x/L) \psi$. However, if periodic boundary conditions are imposed on ψ, (corresponding to the Aharonov-Bohm geometry), the gauge transformed state will not satisfy the boundary conditions, (unless a is an integer) and so the vector potential cannot be gauged away. So for extended states with significant amplitude extending from $x = 0$ to $x = L$, there is a dependence of the state and energy on a.

Localized states satisfy the boundary conditions trivially, since they vanish at both $x = 0$ and L and are willy-nilly periodic no matter what phase factor multiplies them. Their energies cannot depend on a therefore.

The total current operator is $I_{op} = (-e/\hbar) d\mathcal{H}_0/da$, by direct construction. Taking the expectation value of this operator and using the Feynman-Hellmann theorem gives

$$I = \langle I_{op} \rangle = -\frac{e}{h} \frac{d}{da} \sum_{np} E_{np}(a) , \qquad (3.4.3)$$

where the sum is over the filled levels. The only levels which depend on a are the extended ones whose energies are related to the phase shifts by the construction (3.4.2) with p replaced by $p+a$. This immediately implies $E_{np}(a=1) = E_{n,p+1}(a=0)$. Replacing dE_{np}/da by its average over the interval $0 < a < 1$, namely $E_{n,p+1} - E_{np}$ gives the result Eq. 1.6.6.

Another way to prove that the QHE exists in the presence of scattering potentials is to note that all extended states are at the original Landau level energies. Obviously, therefore, there is a mobility gap and the general gauge argument applies.

One may also consider the field ionization of bound impurity states. An isolated scattering center with zero electric field will succeed in binding all but the edge states to itself. The reason is that a wave packet, which to good approximation is in the field-free region, will have small tails overlapping the scattering center. These tails cause the wave packet to move very slowly around the center binding it. However, such effects are exponentially small. Even a weak electric field will ionize the weakest bound eigenstates. We may conclude that in zero electric field there are no extended states, (except at the edges) although there are numerous states of very small binding energy which break free at the first perturbation. A similar effect occurs for smooth potentials.

It might be concluded from this that the conductivity is a strongly nonlinear function of electric field. This is not the case as long as σ_{xx} remains small. Even though states are ionized as the field increases, and there are more extended states to carry the current, each state carries a smaller proportion of the current and the current remains proportional to the Hall voltage. Thus, although one might worry that there are some eigenstates which cannot be treated in perturbation theory in the electric field E, it is still permissible to use the Kubo formula for the conductivity, which is based on linearity in E.

It is amusing to study the propagation of a wave packet as it impinges on a potential. Generally, it speeds up, meanwhile distorting and eventually appearing beyond the scatterer. Some cases have been worked out (Joynt 1982, 1984) and we reproduce some of the results in Note 2.

3.5 Smooth Potential

A slowly varying potential provides a striking contrast with the conventional situation. The elementary aspects are again rather simple. A smooth potential can be considered as a local electric field, and so the electrons will drift locally perpendicular to it. It is clear that the wave functions will be confined to a distance ℓ from equipotential lines in first approximation so it is necessary to consider the nature of the system of equipotential lines. The situation is reminiscent of the approximation used in plasma physics in which a slow 'guiding center' drift motion is superimposed on a rapid, tight circular cyclotron motion (Northrup 1963).

Since the potential is smooth, it must be finite. It will therefore have maxima and minima which are

Imperfections and Disorder

well separated. [If this is not so, we can remove a potential of type V_W to achieve this situation.] The equipotential lines for high energies will be closed curves around the tops of the potential 'hills', while for low energies the closed curves will be around the bottom of the 'valleys'. In general, these closed curves will be rather long in units of ℓ so we may use semiclassical considerations to quantize them. These states obviously are localized.

Let ζ be the coordinate along the closed curve as illustrated in Fig. 4. Then the speed $d\zeta/dt$ is $c|\nabla V_S/eB|$. The Bohr frequency condition gives the increment of energy between two neighboring levels as $\Delta E = E_{p+1} - E_p = h/T$, where

$$T = \oint d\zeta/\dot{\zeta} . \qquad (3.5.1)$$

The integral is over the curve C of Fig. 4.

We can express the quantization condition as the quantization of area between the orbits. The area between successive orbits is

$$\Delta S = \oint \frac{\Delta E}{|\nabla V_S|} d\zeta = \frac{h}{T} \oint \frac{c}{eB} \frac{dt}{d\zeta} d\zeta = 2\pi \ell^2 . \qquad (3.5.2)$$

Thus each state occupies the same area. All this is an approximation, strictly valid only in the smooth potential ($B \to \infty$) limit. However, corrections can be made as a series in higher gradients of the potential,

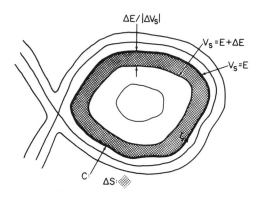

Figure 3.4

which we assume are small so that the series converges. [The condition (3.2.3) is thus symbolic and higher gradients are assumed small as well. However, given (3.2.3) it is usually possible to put all large higher gradient terms in V_W or V_{sc}.] Thus the true states will not quite follow the equipotentials, and they will not occupy exactly the area $2\pi\ell^2$ [Problem 3]. However, they will have qualitatively the properties of the zeroth approximation which is all that is necessary.

The states following closed equipotential lines are of course localized. As such, they cannot carry current. This is obviously quite a different type of localization than the Anderson localization encountered in the $B=0$ case. For example, even states of different energy are physically localized in different regions. There can be more than one closed equipotential curve with close to the same energy. If these curves are physically far apart there will be practically no communication between them and resonance between them can be ignored [Note 4].

The localized states in this case are thus quite simple. Of considerably more interest are the extended states. However, we don't have to know very much about them to conclude that the conditions for the QHE are satisfied in this case. Namely, it is clear that the equipotential lines, $V_S(x,y)=\epsilon$ will be closed for small ϵ and for large ϵ. They can be extended throughout an infinite sample for at most one energy ϵ. The gauge argument then implies that there will be Fermi energies for which there is a mobility gap and therefore that the QHE will exist. This in turn implies that there are extended states, and thus an energy at which the equipotential line extends throughout the system. Of course, it is known on other grounds that the extended equipotential exists.

Let us remark that the uniqueness of the extended state energy is a consequence of the tacitly assumed symmetry between x and y in the random function V_S. Thus, if the equipotential extends in the x direction, it must also extend in the y direction. Since equipotential lines of different energy obviously can't cross, we see that there is a unique energy for the extended state. However, the presence of an electric field removes this symmetry, and increases the number of extended equipotential lines. Thus in this case

Imperfections and Disorder

also, an electric field ionizes the most loosely bound states. Again, the current is still linearly proportional to the Hall voltage, even though the number of extended states is changing with voltage in some possibly nonlinear way.

If we use a finite disk (Corbino) geometry, as in Chap. 1.6 then the extended states are along equipotentials which circle the hole in the disk, while localized states correspond to equipotentials which do not do so. Kazarinov and Luryi (1982) explicitly (but approximately) calculate the Hall conductance without relying on Laughlin's gauge argument. Fig. 5 illustrates the states in this case. From this figure, we see that there is a net voltage change only across extended states. One can directly calculate the current carried by these states. Alternatively, one can note that the quasiclassical quantization of these states is given by a rule like (3.4.2). Thus the extended states circling the disk can be ordered just as the states in the scattering potential are ordered. Under adiabatic change of a threading flux, state p will transform into state $p+1$. Therefore a smooth potential will exhibit the QHE.

Figure 3.5
Equipotential lines for a Corbino disk. The voltage drop is $V_{SS} = V_{SA} + V_{AS}$ since $V_{AA} = 0$.

3.6. Continuum Percolation Model

It is of considerable interest to work out further aspects of the extended states in the smooth potential. To do this, we must make some further assumptions and imagine that the smooth potential V is somehow 'typical', for example, that V_S is not at the same time a scattering potential. This section is basically an encapsulation of the results of Trugman (1983).

Let the probability that $V_S(x,y) = \epsilon$ be given by a distribution function $\rho(\epsilon)$, which is continuous, bounded and independent of x,y. Let the correlation range b of the potential be finite, so that the continuum percolation problem defined by V_S is in the universality class of the uncorrelated lattice percolation problem. That is, we assume that $\langle V_S(r)V_S(r+r')\rangle - \langle V_S(r)^2\rangle$ rapidly vanishes for $|r'| > b$. We also assume that V_S is slowly varying on the length scale ℓ.

For large B, the density of states is just $\rho(\epsilon)$. Let us use the construction of coloring those areas for which $V_S < E$ (see Fig. 6). Then states of energy E are localized on the edges of the colored regions.

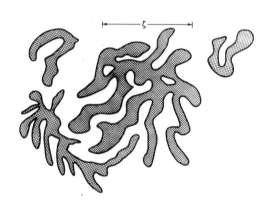

Figure 3.6

Percolation theory (see Stauffer 1979 and Essam 1980 for reviews) then implies the following: There is a unique energy E_c at which there is first a connected colored region of infinite extent. By instead coloring regions for which $V_S > E$ we see that E_c is the unique energy of extended states. There is a length

$$\xi(E) \propto |E-E_c|^{-\nu} \quad \text{(for } E \to E_c) \quad (3.6.1)$$

which describes the scale of the maximal connected region of energy E. In two dimensions, $\nu = 4/3$ (Riedel 1981).

The localization length ξ diverges at E_c as a power in this model. It can be defined as the rms radius of the connected colored region. This differs from the result of Ono (1982) who finds $\xi \propto \exp(\Gamma^2/|E-E_c|^2)$ where $\Gamma = \hbar/\tau_0$ is the zero magnetic field scattering rate. Ono's result is based on keeping certain graphs that give a good account of Anderson localization in the zero-field case (Vollhardt and Wölfle 1980), as well as use of a many δ-function scattering potential and the lowest-Landau-level approximation for the evaluation of certain quantities appearing in the formula. In any case Ono's treatment agrees with the results in other models that only one state is extended in zero field. It disagrees however with numerical results. However, the question is not settled conclusively.

Let us introduce $p = p(E)$ as a variable to replace E, where

$$p = \int_{-\infty}^{E} \rho(\hat{E}) d\hat{E}. \qquad (3.6.2)$$

[The variable p corresponds to the site occupation probability in the lattice percolation problem.] It is uniformly distributed from zero to one. Scaling hypotheses are then made, just as in the lattice case. Specifically, it is assumed that the number density n_s of clusters of large area s is given by $n_s \propto s^{-\tau} g[(p_c-p)s^\sigma]$ where $g(z)$ is a smooth scaling function with $g(0)=1$ and $g \to 0$ for $z \gg 1$. By cluster, we mean colored islands of large area. The eigenstate lives on the shoreline of this island. Since $p_c-p \propto E-E_c$ (because $p(E)$ is smooth) it follows that the relation between the perimeter and diameter of the largest clusters is $t \propto \xi^{1/\sigma v}$.

We also assume that the perimeter t of such an island increases in proportion to the area, i.e., $\sigma \propto t^{p'}$ with $p'=1$. This says that the large area islands are very convoluted. This is called the ramification hypothesis. Finally, because the potential is smooth, we assume that the perimeter is smooth, so that as $p \to p+dp$, a typical island of area s and perimeter t will increase in area by $ds \propto t dp$.

It follows from this that the fraction of states f with perimeter larger than t decreases proportionate to $t^{2-\tau-\sigma}$ for large t. Then $f(\xi) \propto \xi^{-p_1}$ where $p_1 = (\tau-2+\sigma)/\sigma v$. From known scaling relations among the exponents (Riedel 1981) we conclude $p_1 = 41/48$. Thus the fraction of states of extent larger than ξ vanishes for large ξ as a power. This is quite different from Anderson localization where $f(\xi)$ would vanish exponentially. There are some subtleties to this result which are mentioned by Trugman (1983).

This discussion has been for the case of zero electric field. Turning on the field will ionize some of the localized states and lead to additional extended states. Let us assume that max $|\nabla V_S| = M'$. We quote without proof a result of Trugman. Namely, (in the infinite B limit so that we can deal with contour lines of the potential energy),

$$f(E) \geq eE/(M'+eE), \qquad (3.6.3)$$

where $f(E)$ is the fraction of states of infinite extent as a function of electric field E. Further, all states are extended when $eE > M'$. This result is for any potential V_S, not necessarily a random one.

Trugman further conjectures that $f(E)$ increases more rapidly than the linear variation for small E of Eq. 6.2, in fact by a power law

$$f(E) \propto E^{p_2}, \qquad (3.6.4)$$

where $p_2 = (\tau+\sigma-2)/[\sigma(1+\upsilon)] = 41/84$. We will not reproduce his argument. However, there is an experimental consequence of this result which is worth discussion (see Chap. 2.12-13). The picture for the Hall effect is that, at zero temperature the ideal Hall result (1.1.1) is observed when all the extended states are full. Therefore, the Fermi level will lie in a region of localized states. At zero electric field or current, the extended states in this model are at a single energy, E_c. Then, the discontinuities in the Hall conductivity would be infinitely sharp and the plateaus as wide as possible. As the temperature rises, the plateaus will narrow and the jumps will broaden, because of hopping processes.

At finite electric fields, the region of extended states also increases, according to (3.6.2). Thus the width of the discontinuities will increase proportional to $(V_H)^{p_2}$. This effect may have been observed, but it is difficult to exclude other explanations of the data (Chap. 2.12).

This argument does suggest that the observed conductance at finite fields is not correctly given by the Kubo formula for all values of the Fermi energy or magnetic field. In other words, perturbation theory in the electric field may be adequate in finding small changes in localized states or in extended states, but it may not be able to account for the ionization process. If the Fermi level is such that the states at it are ionized by the electric field then the Kubo formula may break down. However, this does not detract from the use of the Kubo formula to study other properties. In fact, Niu and Thouless (1984b) show that when the Fermi level is in a gap, there is no correction to the Hall conductance expression to all orders in the electric field.

Another measurement which might display the percolation nature of the states is the frequency dependent conductivity. This has been discussed by Joynt (1985). The basic idea is to suppose that the Fermi level is in a region where the perimeters t of the extended states are long compared with the correlation range b of the potential so percolation arguments apply. However, the Fermi energy is supposed to be far from the percolation threshold.

At finite frequency, there is a contribution to the conductivity from localized states. This contribution arises from transitions between states differing in energy by $\hbar\omega$ and for which there is a significant dipole matrix element. For low frequencies, the main contribution will be between states whose wave functions live near the same closed equipotential line. For example, transitions between the states illustrated in Fig. 4 above will have large dipole matrix elements. This is because the states overlap strongly, the distance between orbits being much smaller than ℓ. The dipole matrix element is of order of the radius of the orbit of the state. Joynt (1985) was able to estimate that the ac conductivity $\sigma_{xx}(\omega)$ goes as

$$\sigma_{xx}(\omega) \propto \omega^{-4\rho} e^{-\omega_0/\omega} . \qquad (3.6.5)$$

The exponential factor is interpreted as the statement that states at the Fermi level with very long perimeters are exponentially rare. The parameter ω_0 is estimated to be in the megahertz range providing the correlation range b of the potential is of order 10^5Å. The prefactor contains the exponent $\rho = 2/3$ (Stauffer 1979) relating the rms radius of the state ξ with its perimeter t according to $\xi \propto t^\rho$. This is for the case well away from percolation threshold. The increase of the prefactor for small ω reflects the increase of the dipole moment of the state for large perimeter.

The necessity that the correlation range of the potential be quite long means that all samples cannot be expected to exhibit this effect. It is not believed to be necessary to measure σ_{xx} directly. Rather, if an ac measurement is made, σ_{xx} becomes finite and the QHE breaks down. Thus a measurement of the ac breakdown of the QHE effect is sufficient to test this theory. Chap. 2.13 mentions the experiments which bear on this theory.

3.7 General Potential

In this section we give briefly the arguments which allow one to conclude that the potential of Eq. 2.1 satisfies the conditions for the quantum Hall effect (Joynt and Prange 1984). First, let us combine the scattering potential V_{sc} with the smooth potential V_S. It is clear that there is no problem in this case, since wave packets in the asymptotic regions of the scattering potential still move smoothly. They may curve a bit through space and be slightly modified but this has no effect on their fundamental nature. Thus the only difficult task is to combine the weak potential V_W with the other types of potential.

In order to maintain a mobility gap, what we must show is that there are not unwanted extended states generated at all energies, in particular not halfway between the Landau level cores. First, let us combine V_W with V_{sc}. Consider the Green's function $G(r,r',E)$ at an energy E halfway between Landau levels. This is the propagator for a specific potential configuration, not the propagator averaged over random potentials. Then if $G(r,r',E)$ vanishes exponentially for all large $R = |r-r'|$ we may conclude that there is no extended state at this energy. Expand G in powers of V_W. This gives

$$G(r,r',E) = G_{sc}(r,r',E) +$$
$$+ \int dr_1 G_{sc}(r,r_1) \, V_W(r_1) \, G_{sc}(r_1,r',E) + \cdots ,$$
$$+ G_{sc} \, V_W \, G_{sc} \, V_W \cdots V_W \, G_{sc} + \cdots , \qquad (3.7.1)$$

Imperfections and Disorder

where G_{sc} is the propagator for the scattering potential. Since V_{sc} is arbitrary inside the scattering regions we may take V_W to vanish there. Now G_{sc} vanishes exponentially with R unless $\mathbf{r},\mathbf{r}'$ are in the same scattering region. In particular, for neither $\mathbf{r},\mathbf{r}'$ in a scattering region, $G_{sc} \propto \exp[-R^2/4]/(\hbar\omega_c/2)$.

Therefore, G of Eq. 7.1 can only be appreciable for large R if there is a term with a chain of intermediate points $\mathbf{r},\mathbf{r}_1,\mathbf{r}_2 \cdots \mathbf{r}_n,\mathbf{r}'$ such that $|\mathbf{r}_i-\mathbf{r}_{i+1}| \approx \ell$ for each i. The maximum contribution of such terms (and the sum of all higher order terms) can be shown to be $[V^{max}/(\hbar\omega_c/2)]^n$. The exponent n is clearly at least D/ℓ where D is the minimal distance in scattering free regions required to traverse the sample from $x=0$ to $x=L$. On the assumption that D increases with L we have exponential decay of G at this energy and therefore no extended state.

A similar argument can be used for the smooth potential. To generate extended states midway between Landau levels it is required that V_W connect localized states across the regions of unfavorable energy, in particular across the regions of the percolating extended states. The width of these regions is large compared with ℓ, and only in a high power of perturbation theory can this take place.

The above argument is in fact rather crude, and there is no reason to believe that extended states will be automatically generated at energies of the order of V^{max} from the 'unperturbed' extended states. On the contrary, what evidence there is suggests that there will still be only one extended state energy. How the inclusion of V_W changes the nature of the states is less clear however. In particular, whether the percolation picture retains some validity is open to doubt. There is also some evidence reported in the next section that V_W changes the localization from Gaussian to exponential, but this is not conclusive.

3.8 Numerical Studies

There has been important numerical work which gives insight into the nature of two-dimensional electronic eigenstates in a strong magnetic field. We shall briefly summarize that work in this section, particularly stressing the contributions of Ando (1983b, 1984a,b).

For the most part, this work has been carried out within the lowest Landau level approximation. A random potential is chosen, usually consisting of a sum of δ scatterers with a high concentration ($n_i = 5/2\pi\ell^2$). A finite sample then has a finite number of basis states N and the resulting Hamiltonian matrix can be diagonalized. Values of N of order 400 were used.

A number of results can then be obtained. Among them are the density of states and the localization length. The latter is obtained by consideration of the Thouless number (Edwards and Thouless 1972, Licciardello and Thouless 1978). This theory was developed for the $B=0$ case and is not expected to be applicable in its entirety in the high field limit. The Thouless number $g(L)$ is defined as the ratio (shift in the energy levels due to a change in boundary conditions from periodic to anti-periodic)/(mean level separation $[L^2 N(E)]^{-1}$) where $N(E)$ is the density of states per unit area. For localized states at $B=0$ the Thouless number is expected to decrease exponentially with sample length L. The formula used,

$$g(L) = g(0)\exp[-\alpha(E)L], \qquad (3.8.1)$$

defines the inverse localization length $\alpha(E)$. There are some indications that (8.1) should also hold for strong field, but these indications are from theories which are essentially transcriptions of the zero-field results to the high-field case in the lowest Landau level approximation. Certainly, for states localized in potentials of type V_{sc} and V_S, $\ln g(L) \propto -L^2$.

The result obtained by Ando is exponential localization with $\alpha(E)$ vanishing at E_c roughly as $(E-E_c)^2$. We can conclude from these results that all states of the system of δ-function impurities in the one Landau level approximation are localized except probably at the midpoint energy. Detailed exponents do not agree with other theories, (Ono 1982, Trugman 1983) however, but there are a number of reasons for this discrepancy, including the smallness of the numerical sample size.

Ando (1984a) has also studied a potential consisting of a sum of Gaussian scatterers with characteristic length scale 2ℓ. In this case the states with energies well away from the percolation energy have

Imperfections and Disorder

very small Thouless number, and it has been checked that they have Gaussian localization. Near the percolation energy there is apparent exponential localization with $\alpha(E)$ vanishing at E_c. Again the power is roughly two. Probably the system size is not large enough to see percolation effects of the type predicted by Trugman (1983). However, the potential is certainly not in the smooth limit either, and it may be that there is an effect of V_W which modifies the percolation result.

In Fig. 7 we reproduce some wave functions for this case which were obtained by Ando. The interest is in comparing them with the contour lines of constant potential. It is apparent that there is a strong correlation of eigenstate with potential but that one is still quite far from the high-field limit.

Ando (1984a) also calculated the dynamical conductivity, $\sigma_{xx}(\omega)$. He finds results in qualitative agreement with those of Joynt in the sense that there is a maximum in σ_{xx} at some value of the frequency. This frequency is probably too high to agree with Joynt's (1985) estimate, and the suppression of the low frequency conductivity is much weaker than Joynt would predict. A numerical comparison is not warranted however, as Ando's finite system is not in the same range of parameters as is Joynt's model. However, the turnover at low frequencies is a localization effect in both cases.

A study of the mixing effects of higher Landau levels is carried out in (Ando 1984b). Three Landau levels were retained in the calculation. Again, a high concentration of δ impurities was used. The energy parameter Γ of the single-site self-consistent Born approximation (Ando and Uemura 1974) [SCBA] continues to be appropriate for this system. It is given by

$$\Gamma^2 = 4 \, V_0^2 \, n_i / 2\pi \ell^2 \, . \tag{3.8.2}$$

Fig. 8 gives the density of states as found numerically for several values of the ratio $\Gamma/\hbar\omega_c$. The numerical density of states is narrower but with wider tails than the SCBA. There is no gap, even when the SCBA predicts one. Notice also the asymmetry induced by the virtual transitions to the higher Landau levels. Nevertheless, the SCBA is a reasonable guide to the results.

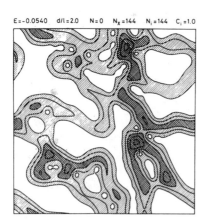
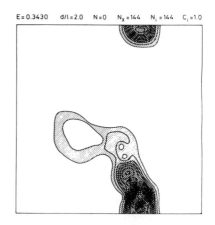

Figure 3.7 An example of equipotential lines (upper) and wave functions for a potential formed from a random sum of Gaussians with scale length 2ℓ and rms energy width Γ_0. Pluses and minuses give the position and sign of the 'impurity' potentials. There are 144 Gaussians and 144 basis states in the lowest Landau level. An extended state on the lower left is near the level center at $E = -0.0540\Gamma_0$ and a localized state away from the center at $E = 0.3430\Gamma_0$ is shown on the lower right. Electron probability is greatest in the hatched regions. (Ando 1984a.)

The localization was also calculated (see Fig. 9). It was found that the states between Landau levels were localized for $\hbar\omega_c > 1.25\Gamma$, although by this degree of disorder, the localization length was relatively long. At $\hbar\omega_c = \Gamma$ however, there was no evidence of

Imperfections and Disorder

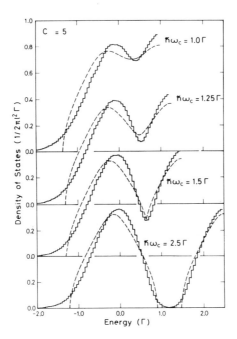

Figure 3.8. Histograms of the density of states when mixings among different Landau levels are taken into account. The dashed lines represent the results calculated in SCBA. The energy is normalized by the broadening Γ and its origin is chosen at the position of the lowest Landau level. (Ando 1984b).

localization between the lowest two levels. This might therefore be an example of the breakdown of the conditions for the QHE.

3.9 Acknowledgments

This research was supported by NSF Grant DMR-82-13768.

3.10 Notes

Note 1. (Joynt and Prange 1984). The scattered wave function can be written

$$\psi_{scat}(r,E) = \int d^2r' G_0(r,r',E) V_{sc}(r')\psi(r',E) , \quad (3.10.1)$$

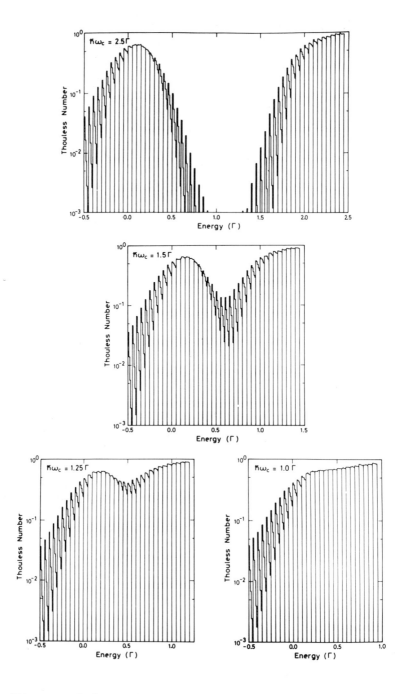

Figure 3.9. Calculated Thouless number $g(L)$. Within each energy interval, $g(L)$ is plotted as a function of L. (Ando 1984b).

Imperfections and Disorder

where ψ is the total wave function. Clearly r' is in the scattering region. The expression for the free Green's function (in an electric field) is

$$G_0(r,r',E) = \sum_{n=0}^{\infty} \frac{L}{2\pi} \int dk \, \frac{\psi_{nk}(r) \, \psi_{nk}^*(r')}{E - E_{nk} + i\delta} . \qquad (3.10.2)$$

From this the asymptotic form of G_0 can be readily extracted for large $|r - r'|$. This can be done, for example by moving the k contour into the upper or lower half plane to the saddle point of the integration. The saddle point contribution is exponentially small, and only the explicit pole will contribute. Further, this pole contributes only if $x > x'$ so that the contour is moved into the upper half k-plane. If $E = m + k_0 v$, the result is

$$G_0(r,r',E) \propto \theta(x-x') \, \psi_{mk_0}(r) \, \psi_{mk_0}(r') \qquad (3.10.3)$$

which proves the assertion of the text.

Note 2. It may be thought that in order to have the Hall conductance equal to the ideal value to 0.1 ppm it is necessary that the thickness d of the asymptotic region satisfy $\exp(-d^2/2\ell^2) < 10^{-7}$. However, it is not necessary that the scattered state be just like the free one to tremendous accuracy. It is sufficient that the scattered state be qualitatively like the free one and that will be the case under less stringent conditions.

Note 3. (Joynt 1982, 1984). Some numerical results on the scattering of wave packets from strong scattering centers are presented. These results were obtained by representing the wave function in the basis ψ_{nk} for a finite number of k values. A direct numerical integration of the real space equations is not feasible. In Fig. 10 the propagation of a wave packet, initially a Gaussian, is shown as it scatters from a positive Gaussian potential. The central circle shows the position of the half maximum ($3\hbar\omega_c$) of the potential. The density contour lines at various times are shown. The × marks the position of a packet propagating at constant speed v. This potential supports three bound states and is so strong ($6\hbar\omega_c$) that the packet does not penetrate it significantly.

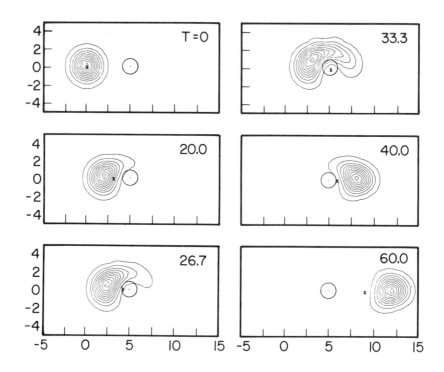

Figure 3.10

Fig. 11 shows a similar set of plots for scattering from a square well potential. This well is 'square' in two senses, namely the scattering region is a square of side 2ℓ and the potential jumps discontinuously from a value 0 to $-2\hbar\omega_c$ at the edge of the square. The packet develops two maxima near the center. In this case, the main part of the packet passes under the center, because the potential is attractive. This potential supports two bound states. Its phase shift versus k is plotted in Fig. 12.

Note 4. At a given energy there are many separate closed equipotential lines in general. It might be thought that the approximate eigenstates of nearly the same energy which live on separated lines would couple quantum mechanically. That is, maybe the approximate eigenstates obtained by quasiclassical quantization on individual closed contours resonate together and the true states are extended. This is something like the problem of resonating states in the usual Anderson localization problem. The energy separation between levels, calculated on one closed contour is propor-

Imperfections and Disorder 97

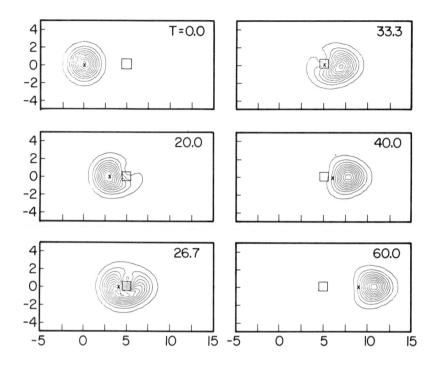

Figure 3.11

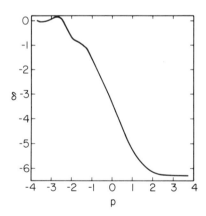

Figure 3.12

tional to the inverse of the length C of the level's orbit. The coupling between orbits on distinct contours is proportional to $\exp[-(D/\ell)^2]$ where D is the distance between distinct orbits of practically the same energy. In most cases D is of the order of the circumference, and there is little chance of an accidental degeneracy which couples the two levels. Orbits along equipotentials near a saddle in the potential could couple appreciably. However, even if there are a few such resonating pairs of orbits, the result will be new pairs of localized states, and nothing much will change in the final results. It would take an infinite sequence of resonances to convert bound states into extended ones, and that, it is safe to say, cannot happen for most energies. Simple arguments cannot rule out that extended states could be formed near the percolation energy, which might modify the percolation results of Section 6. However, the basic picture would be unchanged.

Note 5. There is a misperception about the nature of the percolating orbit which has appeared in the early literature of the subject. It seems at first obvious that an orbit which ramifies throughout space in both x and y directions must cross itself many times. That is, there is a suggestion that the extended orbit at E_c will have a finite density of saddle points where the contour lines cross. However, the probability that a smooth random potential will have a saddle point at any given energy is zero and the actual percolating contour will not cross itself.

Note 6. The parameter Γ introduced into the self-consistent Born approximation should not be confused with a scattering rate, even though it gives a good measure of the energy width of a Landau level. Rather, it is related to the quantity V^{max}. A distribution of scatterers of the form $\pm V_0 \delta(x-x_0) \, \delta(y-y_0)$ is assumed. We want to estimate the fluctuations in the potential on the scale ℓ. A measure of this is the rms potential averaged over a region of area $2\pi\ell^2$. The average potential is assumed to be zero. Γ is the rms average.

3.11 Problems

Problem 1. Construct a wave packet $\psi(x,y)$ by finding a weight function $C(k)$. That is

Imperfections and Disorder

$$\psi(x,y) = (2\pi)^{-3/2} \int dk \; C(k) e^{ikx} e^{-(y-k)^2/2} e^{-iE_k t}.$$

(3.11.1)

Find $C(k)$ such that

$$|\psi(x,y)| \propto \exp\left[-\frac{(x - x' - vt)^2}{2(1 + \kappa^{-2})} - \frac{(y - y')^2}{2(1 + \kappa^2)} \right]$$

(3.11.2)

where κ is an arbitrary parameter.

Problem 2. Let V_{sc} be a smooth potential, zero outside some region of area A. Let the electric field gradient be small compared with the gradients of V_{sc}. Find the time advance as a function of incident energy. Show that the integral of the time advance over energy is 2π times the number of bound states supported by the potential, in other words that the number of bound states is $A/2\pi\ell^2$. Show that this result is independent of the sign of the potential. Hint: Compare the paths shown below with and without a potential.

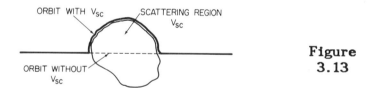

Figure 3.13

Problem 3. Consider the Hamiltonian $\mathcal{H}_0 + vy + wy^2$. Show a) the density of a full Landau level is not given by eB/hc, b) formula (1.6.6) still holds, implying that the exact QHE effect is still correct, but c) the exact QHE relation must be expressed in terms of the electrochemical potential difference, $(E_{pmax}-E_{pmin})/e$, and not the electric potential difference $EW = vW$. The Kubo formula fails in this case as it is perturbation in the electric potential. It gives $\sigma_H = nec/B$ with the 'wrong' n, i.e. the actual density of the level, not the density of the ideal level. This result also explains why the density of bound states in a smooth potential is only approximately eB/hc.

CHAPTER 4

Topological Considerations

DAVID J. THOULESS

4.1 Topological Quantum Numbers

The integer or rational quantum numbers that occur in quantum theory seem to be of two different sorts. There are those related to symmetry, and those that are determined by topology. The distinction is not necessarily an unambiguous one, and it may be possible to look at the same quantum number in more than one way, but there are certainly two quite different approaches to the problem of quantization.

Angular momentum is a famous example of a case in which integer quantum numbers occur as a result of symmetry. In classical mechanics rotation invariance of the Lagrangian leads to the existence of angular momentum as an integral of the equations of motion. If there is a slight departure of the Lagrangian from spherical symmetry, angular momentum is no longer a strict constant of the motion but varies slowly with time. In quantum mechanics there is more sensitivity to the detailed structure of the rotation group - the fact that it is compact leads to discrete quantum numbers. The degeneracy of levels can be deduced from the algebra of the group generators. In classical mechanics the three-dimensional rotation group and the two-dimensional unitary group are equivalent to one another, but in quantum theory the topological differences between them lead to important differences in the quantum numbers – integer in the first case, integer or half integer for the unitary group. Departures from ideal spherical symmetry lead to a breaking of the degeneracy, and the original quantum numbers can no longer be associated with the energy levels in an unambiguous fashion. Most solid state physicists are aware, for example, of the fact that in the band theory of solids it is possible to go continuously through an energy band from a state which is s-like to a state which is p-like.

There are two closely related examples of topological quantum numbers in condensed matter theory. The first of these is the quantization of circulation in superfluid helium (Onsager 1949, Vinen 1961). From the single-valuedness of the superfluid wave function it is deduced that the integral of the superfluid velocity round any closed circuit is equal to an integer multiple of h/m, where m is just the mass of a helium atom. The integer quantum number is a "winding number", the number of times the phase of the superfluid wave function goes through a complete cycle on the closed path. It has not proved possible to test this with great precision, since no method has been devised for measuring the superfluid velocity with high accuracy, but it is generally thought that this quantization would hold with high precision if the measurement could be made despite impurities in the liquid, irregularities in the container, and thermal excitations. None of these effects alters the requirement that the superfluid wave function should be single-valued, and therefore its phase must change by an integer multiple of 2π on any closed path.

The quantization of flux in superconductors is very similar. On the basis of the existence of a single-valued condensate wave function London (1950) predicted that the flux (or rather the fluxoid, which is a combination of the flux contained in a loop and the current flowing round the loop) should be an integer multiple of h/e. Experimental measurements by Doll and Näbauer (1961) and by Deaver and Fairbank (1961) showed that the flux was actually quantized in multiples of $h/2e$, which was explained by Byers and Yang (1961) to be a result of the fact that the fundamental units of superconductivity theory are electron pairs, rather than single electrons as London supposed. In this case the quantization is known to hold with very high accuracy, and it forms the basis for most modern precision measurements of magnetic field variations.

The distinction between the two reasons for quantization is perhaps not as clear as a superficial discussion suggests. Although angular momentum conservation appears to be lost as soon as a small symmetry-breaking term is introduced, it has been known for a long time that for many classical systems the loss of such a symmetry does not cause the system to wander all over phase space, but it remains confined to a surface close to the one determined by the strict conservation law (see, for example, Arnol'd and Avez 1968). The quantum numbers may still be determined by the topology of these surfaces, even if the symmetry is broken.

If the quantization of Hall conductance has to be fitted into one of these two classes of quantum number, it is obviously more attractive to fit it into the class of topological quantum numbers. The Hall conductance makes no strong demands on the purity or symmetry of the system in which it is manifested, and it does not seem to be related in any obvious way to a symmetry group. One of the attractive features of Laughlin's (1981) theory of the quantum Hall effect was that it seemed to use a topological argument and to be independent of many of the details of the Hamiltonian. Laughlin argued that, if the Fermi energy lay in an energy gap, change of flux by one quantum unit through a ring of two-dimensional electrons would transport an integer number of electrons from one edge to the other. A few other assumptions have to be made, such as that the ratio of two discrete changes can be replaced by a derivative, but this argument leads to the conclusion that the conductance must be a multiple of e^2/h. Although this argument has topological aspects, it does not seem to me to show the topological basis for the quantization of Hall conductance. The work I am going to describe, which is primarily concerned with Hall conductances for strictly periodic systems, was not undertaken to provide such a topological basis, but I believe it does provide one.

4.2 Quantum Hall Effect in a Periodic Potential

Numerous studies have been made over the years of the quantum theory of electrons which are simultaneously in a magnetic field and in a periodic potential (Onsager 1952, Harper 1955, Zil'berman 1957a, Azbel' 1964, Zak 1964, Pippard 1964, Rauh et al. 1974, Hofstadter 1976, Wannier 1978). These are all essentially concerned with the two-dimensional problem, as the motion in the direction of the magnetic field is trivial if the field lies along a lattice vector. It was found that, if the number of flux quanta passing through a two-dimensional unit cell is a rational number p/q, each Landau level is broken into p subbands by the periodic potential. Equivalently, each energy band is split into q sub-bands by the magnetic field. This is a paradoxical result, since it implies that if the flux through a unit cell is an irrational multiple of h/e there are infinitely many energy levels. The paradox was made less startling by Zil'berman's (1957b) observation that these splittings become exponentially small. The process was shown very graphically by Hofstadter's (1976) calculations of these energy bands for all fractions with denominators up to about 50 which is shown in Fig. 4.1. An important feature to notice in this picture is that, while the energy bands are wildly discontinuous functions of the flux per unit cell, the gaps are continuous except at discrete points. In the limit of an irrational flux the spectrum shrinks to a Cantor set whose measure is zero, as Azbel' (1964) pointed out.

The properties of such a system can be studied in the simple approximation that leads to Harper's (1955) equation. In magnetic units, with $\hbar = eB/c = m = 1$, the wave function in the Landau gauge has a y-dependence of the form $\exp(iky)$ if the substrate potential is uniform. [Here it is convenient to deviate from Chap. I and take the vector potential in the y-direction.] If this system is now perturbed by a weak periodic potential of the form

$$V_1' \cos(2\pi x/a) + V_2' \cos(2\pi y/b),$$

the effect of the first term is to break the degeneracy of the Landau levels and give a sinusoidal modulation of the diagonal term, while the second gives a

Topological Considerations

Figure 4.1. Energy bands obtained by Hofstadter (1976) for a square lattice $vs.$ q/p, the inverse of the flux per unit cell (running horizontally from 0 to 1). Energy bands are plotted vertically and shown for all values of p up to 50. In principle, there are p bands but for even p the two central bands touch one another.

coupling between states whose values of k differ by $2\pi/b$. The resulting equation for the amplitudes C_n of these coupled Landau wave functions is of the form

$$V_2 \, C^{\alpha}_{n-1} + 2V_1 \cos(4\pi^2 n/ab + \gamma) \, C^{\alpha}_n + V_2 \, C^{\alpha}_{n+1} = \epsilon_\alpha C^{\alpha}_n, \quad (4.2.1)$$

which is Harper's equation. The parameters V_i are proportional to V_i' [Note 1]. The phase γ depends on the initial value of the parameter k which was taken, and it must be varied over the range $4\pi^2/ab$ to get the complete spectrum.

Eq. 2.1 was originally derived as the equation which gives the perturbation by a weak magnetic field of a tight-binding model energy band. The same form for the equation is obtained, but the flux per unit cell is replaced by its reciprocal, so the physical interpretation looks rather different. I will not discuss that weak magnetic field limit.

If the number of flux quanta per unit cell $ab/2\pi$ is a rational number p/q then Eq. 2.1 is periodic with period p, and so the level splits up into p sub-bands in which the levels can be parameterized by the Bloch wave number and by the phase γ. The problem has an important duality property which has been studied by Aubry and André (1980). In this context it can be seen most readily by repeating the derivation of Harper's equation with the vector potential in the x direction, in which case the same equation is derived with V_1 and V_2 interchanged. They showed that for V_1 greater than V_2 the eigenfunctions are localized in the x direction and extended in the y direction, so that the isotropic case $V_1 = V_2$ is a critical point. They also showed numerically that the sum of the widths of the sub-bands tends to $4|V_1 - V_2|$ as p tends to infinity, and I have since shown this analytically (Thouless 1983b). This suggests that the spectrum is singular continuous for an incommensurate flux density at the critical point, since the density is everywhere either zero or infinity, but there are no points with finite weight.

It is an odd result that a Landau level carrying one quantum unit of Hall current is split up into p sub-bands, each of which carries an integer Hall current, but it is an inescapable conclusion from

Topological Considerations

Laughlin's (1981) work. There is a nice paper by Středa (1982) which shows that the Hall conductance is $e(dn/dB)$, the rate of change of charge density when the magnetic field B is changed at fixed Fermi energy. This quantity is constant within each continuous energy gap. Since the ordinate in Fig. 4.1 is inversely proportional to B, it can be shown that the Hall conductance can be read off such a diagram by extrapolating the gap back to the axis at $1/B = 0$. The gaps meet this axis at an integral number of Landau levels, and the number is just the coefficient of e^2/h in the Hall conductance. For complicated fractions p/q the sub-bands carry rather large positive or negative Hall currents in such a way that the sum of all the sub-bands gives unity.

We (Thouless et al. 1982; see also Thouless 1984a) showed this result in a very different way which forms the basis for most of the rest of what I want to say. Harper's equation gives us something concrete to think about, but the derivation of these results is not subject to the special assumptions which lead to that equation. We assume that the electrons are noninteracting, and are subject to a uniform magnetic field and an ideal periodic potential $V(x,y)$. It is assumed that the flux is commensurate with the unit cell of the potential in the sense that p unit cells of the potential contain q quanta of flux, so that there is a magnetic unit cell p times the size of the potential unit cell. Under these conditions it is known (Zil'berman 1957a) that the eigenfunctions of the Schrödinger equation can be written in the form

$$u_{k_1 k_2}(x,y) \exp[i(k_1 x + k_2 y)],$$

where $u(x,y)$ is periodic in the y coordinate but has an extra gauge factor in its periodicity condition in the x direction. For the function $u(x,y)$ there is an effective k-dependent Hamiltonian

$$\mathcal{H}(k_1,k_2) = \frac{1}{2}\left[\left(-i\frac{\partial}{\partial x} + k_1\right)^2 + \left(-i\frac{\partial}{\partial y} + k_2 + x\right)^2\right] +$$
$$+ V(x,y). \qquad (4.2.2)$$

The velocity operators are given by the partial derivatives of this effective Hamiltonian with respect to k. The Kubo (1957) formula gives the conductivity in terms of the velocity-velocity correlation function, and this gives the transverse conductivity of a square of area A as

$$\sigma_{xy} = \frac{ie^2}{A} \times$$

$$\times \sum_{\epsilon_\alpha < E_F} \sum_{\epsilon_\beta > E_F} \frac{(\partial \mathcal{H}/\partial k_1)_{\alpha\beta}(\partial \mathcal{H}/\partial k_2)_{\beta\alpha} - (\partial \mathcal{H}/\partial k_2)_{\alpha\beta}(\partial \mathcal{H}/\partial k_1)_{\beta\alpha}}{(\epsilon_\alpha - \epsilon_\beta)^2}$$

(4.2.3)

This is a product of $\partial \mathcal{H}/\partial k_1$ times an energy denominator and $\partial \mathcal{H}/\partial k_2$ times an energy denominator, and it is reminiscent of perturbation theory for the changes of the wave function due to changes in k, and indeed that is just what it is. First order perturbation theory gives the change of the wave function u_α due to a change $\delta k_2 \partial \mathcal{H}/\partial k_2$ of the Hamiltonian as

$$\delta u_\alpha = \sum_\beta (\epsilon_\alpha - \epsilon_\beta)^{-1} \delta k_2 (\partial \mathcal{H}/\partial k_2)_{\beta\alpha} u_\beta, \qquad (4.2.4)$$

so the sum over β in Eq. 2.3 can be expressed in terms of the overlap $(\partial u_\alpha/\partial k_1, \partial u_\alpha/\partial k_2)$ between such perturbed wave functions. When the sum over α is replaced by an integral over the Brillouin zone it gives the sum over all sub-bands below the Fermi energy of

$$\sigma_{xy} = \frac{ie^2}{(2\pi)^2} \iint \left\{ \left[\frac{\partial u_\alpha}{\partial k_1}, \frac{\partial u_\alpha}{\partial k_2} \right] - \left[\frac{\partial u_\alpha}{\partial k_2}, \frac{\partial u_\alpha}{\partial k_1} \right] \right\} dk_1 dk_2 \quad (4.2.5)$$

The inner product, (u,v), is given by

$$\int u^*(x,y) v(x,y) \, dx dy.$$

It can be shown, either by elementary means, or by appealing to general results of differential geometry, that $h\sigma_{xy}/e^2$ as given by (2.5) is an integer, and so each filled sub-band contributes an integer to the Hall conductance of the two-dimensional electron gas. In general terms (Avron et al. 1983, Kohmoto 1985) it

Topological Considerations

is the topological invariant that defines the first Chern class of the mapping of the Brillouin zone (a two-dimensional torus) onto the complex projective space (because phase factors are arbitrary) of wave functions $u_\alpha(x,y)$. This can be shown in elementary terms (Thouless 1984b,c) [Note 2] by defining the phase factors of the wave functions $u_\alpha(x,y)$ to satisfy the conditions

$$\int u^*_{k_1 0} \frac{\partial u_{k_1 0}}{\partial k_1} \, dx dy = 0,$$

$$\int u^*_{k_1 k_2} \frac{\partial u_{k_1 k_2}}{\partial k_2} \, dx dy = 0 \quad (4.2.6)$$

where the integrals are taken over the unit cell which contains the origin. This defines a parallel translation of the wave functions u within the Brillouin zone, first along the k_1 axis, and then parallel to the k_2 axis. Since the equation for u is periodic in the Brillouin zone, u can only change by an overall phase factor when it is translated by a reciprocal lattice vector. The second of these equations shows that the phase due to displacement parallel to the k_1 axis is independent of k_2, while displacement parallel to the k_2 axis gives

$$u_{k_1 k_2 + K_2} = \exp[i\delta(k_1)] \, u_{k_1 k_2}, \quad (4.2.7)$$

where K_1, K_2 are the reciprocal lattice vectors. Since translation first in the k_1 direction and then in the k_2 direction must give the same wave function as translation in the opposite order, the phase must satisfy

$$\delta(k_1 + K_1) = \delta(k_1) + 2\pi t, \quad (4.2.8)$$

where t is an integer. Substitution of these equations in Eq. 2.5 shows immediately that te^2/h is the Hall conductance of the sub-band. It is, of course, independent of the particular choice of the phase of u made in Eq. 2.6.

This derivation assumes that the periodic potential is commensurate with the flux, but it is easy to alter the formulation so that an incommensurate situation can be handled (Thouless 1983a). If the variable k_2 is eliminated in favor of the energy $E(k_1,k_2)$, then the Hall conductance can be written as an integral over k_1 and E of the same form as Eq. 2.5, where the integral over E is taken to be round a contour in the complex plane which surrounds the energies of all occupied states. This then gives the Hall conductance of the occupied sub-bands as a topologically invariant integral over a torus in k_1,E space, and the only restriction is that the Fermi energy must lie in a band gap.

There is another important aspect of the result given in Eq. 2.5 which was derived in a rather more restricted form by Thouless et al. (1982), and then obtained as a general result on the basis of Lorentz invariance by Thouless (1983a), and, more straightforwardly, on the basis of the magnetic translation group by Dana et al. (1985). The conductance t of a single sub-band satisfies the Diophantine equation

$$pt + qs = 1 \qquad (4.2.9)$$

where p/q is the number of flux quanta per unit cell and s is another integer. A sum over the filled sub-bands gives an equation which is valid for arbitrary values of the magnetic flux density, which is

$$(B\sigma_H/e - n)ab = \text{integer}, \qquad (4.2.10)$$

where n is the electron density and ab is the area of the unit cell. This determines the Hall conductance uniquely for irrational values of the flux per unit cell, but for rational values it is only determined up to a multiple of q, and detailed examination of the structure of the energy gaps is necessary to resolve this ambiguity.

4.3 Generalization of the Topological Interpretation

The theory explained in Sec. 4.2 is confined to an infinitely extended periodic system of noninteracting electrons. If this theory is to be used as the basis of a more general interpretation of the Hall conductance as a topological quantum number, it is important

Topological Considerations

to understand how the restrictions implied by the last sentence can be removed. Some of them have already been successfully overcome in the work of Niu and Thouless (1984a) and Niu et al. (1985), and I will try to indicate how I think it may be possible to remove the remaining restrictions. [Other topological approaches have been proposed (Friedman et al. 1984a,b, Ishikawa 1984), which discuss the quantum Hall effect in terms of the chiral anomaly in 2+1 dimensions. This chiral anomaly is the same quantity that we have used.]

As a first step towards a more realistic situation we consider a cylinder or an annulus in a uniform magnetic field instead of considering an infinitely extended system. Such systems were considered by Laughlin (1981) and Halperin (1982) in their discussions of the quantum Hall effect. The y coordinate is taken to go around the cylinder, while the x coordinate is along the cylinder as shown in Fig. 4.2. An electromotive force is applied in the y direction, perhaps by making a steady change in the flux that passes through a solenoid along the axis of the cylinder; just as in the work of Laughlin (1981) and Halperin (1982), the uniform field normal to the surface of the cylinder is not supposed to be changed during this process. The current induced by this emf is along the axis of the

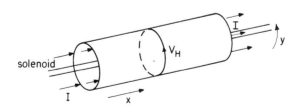

Figure 4.2. Hall conductance in a hypothetical cylindrical device. A current I is passed from one end of the cylinder to the other, in the x direction parallel to the axis. The Hall voltage V_H in the tangential (y) direction is supplied by a solenoid along the axis. A changing flux through the solenoid generates an electric field in the y direction without affecting the uniform magnetic field normal to the cylinder surface.

cylinder, and so electrons must be supplied at one end and removed from the other to carry the Hall current. In this geometry there is no edge current, and generalized periodic boundary conditions are satisfied in the y direction. This does not correspond to an experimentally desirable geometry, but it is one that can be realized in principle.

An important feature of the theory can be seen by rewriting the Kubo formula in terms of Green functions. The transverse current-current correlation function can be rewritten as

$$\sigma_{xy} = -\frac{e^2}{2\pi m^2 A} \sum_{\alpha\beta} \int \frac{dz}{(z-\epsilon_\alpha)(z-\epsilon_\beta)^2}$$

$$\times \int dx_1 dy_1 \psi_\alpha^* \frac{\partial \psi_\beta}{\partial x_1} \int dx_2 dy_2 \psi_\beta^* \left[\frac{\partial \psi_\alpha}{\partial y_2} + ix_2 \psi_\alpha\right] \quad (4.3.1)$$

where the contour of integration surrounds all the occupied energy bands. This is now in the form of the trace of the product of a Green function with the derivative of a Green function, where the argument z of the Green function is always away from the region of physical states. These Green functions are therefore exponentially small unless their arguments are within a range of the order of the magnetic length from one another. Furthermore the continuity of the current ensures that the current crossing a circle of constant x is the same for any value of x, and so the integration over one of the x variables can be restricted to a narrow range well away from the two edges of the cylinder. It should therefore be possible to replace the proper boundary conditions, which are generalized periodic in the y-direction (the phase of the matching condition depends on the instantaneous value of the flux through the solenoid) and of the source and drain type in the x direction, by generalized periodic boundary conditions in both directions. The current should be independent of the phases ξ, η of the matching conditions for the wave functions. Now it is possible to average over these phases, and the Hall conductance becomes

Topological Considerations

$$\sigma_{xy} = \frac{ie^2}{4\pi^2 A} \sum_{\epsilon_\alpha < E_F} \int_0^{2\pi} d\xi \int_0^{2\pi} d\eta \left\{ \left[\frac{\partial u_\alpha}{\partial \xi}, \frac{\partial u_\alpha}{\partial \eta} \right] - \left[\frac{\partial u_\alpha}{\partial \eta}, \frac{\partial u_\alpha}{\partial \xi} \right] \right\}.$$

(4.3.2)

It is the same topological invariant, but now the boundary conditions in the two directions have taken over the role of the two components of the reciprocal lattice vector in Eq. 2.5.

The argument is insensitive to small perturbations that do not change the gap structure in the interior of the system. Irregularities in the substrate, variations in the magnetic field and so on can all be dealt with. Also the existence of localized states in the mobility gaps do not undermine the argument, since they do not alter the exponential decay with distance of the Green function in the range of z integration. A many-body formulation of this theory is also possible, and it just involves replacing the sum over single-particle states in Eq. 3.2 with a single many-body ground state. It can then be argued that the Hall conductance is unchanged by electron-electron interactions, provided they are not sufficiently strong to close the gaps which exist for the noninteracting electron system.

If the energy gaps are due to the electron-electron interaction we can obviously no longer argue on the basis of small perturbations from the noninteracting electron system. It has been argued (Anderson 1983, Tao and Wu 1984, Niu et al. 1985) that the fractional Hall effect is evidence of a discrete broken symmetry when appropriate boundary conditions are used. I think this is probably correct, and it is certainly easy to see how a discrete broken symmetry can lead to fractional values for the topological invariant in Eq. 3.2, but I am not sure that the case for a broken symmetry is completely convincing.

We have shown how the Hall conductance can be exhibited as a topological invariant in a cylinder in which the electric field is tangential and the current axial, but how can we generalize this argument to the more usual situation in which the current is passed along a strip (or round a ring), and the potential difference between the two edges is measured? The problem is that in that situation a large fraction of the current may be carried on the edges (MacDonald et al. 1983, Thouless 1985), and on the edges our argument about the exponential localization of the Green function is of no use to us. However, in order to measure the potential it is necessary to introduce probes, as shown in Fig. 4.3, and the probes could be imagined as connected to one another to give the cylindrical geometry again, with a balancing emf round them to prevent current flowing in the probes. Of course that is not how the potential difference is actually measured, but the detailed way in which the voltage is produced should not be important. All that seems to be required for this argument to go through is that the probes be fairly wide compared with the magnetic length. This itself is an undesirably strong requirement, but it suggests that the Hall conductance can be understood as a topological quantum number in reasonably realistic situations, and we should not be surprised that measurements have shown the quantization to be so precise.

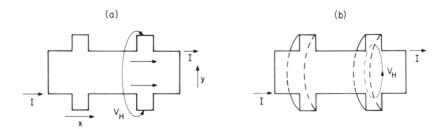

Figure 4.3. (a) A typical experimental Hall bar in which the current I is passed in the x direction. The Hall voltage V_H is measured across pairs of probes as shown. The current flows along equipotential lines, and its injection and emission are concentrated into the two corners. (b) If the voltage probes in (a) are connected together by more two-dimensional material around which an electromotive force V_H is supplied, the topology becomes similar to the cylindrical topology shown in Fig. 4.2.

4.4. Notes

Note 1. When these matrix elements between Landau wave functions are calculated, they give

$$V_1 = V_1' \exp(-\pi^2/4a^2), \quad V_2 = V_2' \exp(-\pi^2/4b^2). \quad (4.4.1)$$

Note 2. We supply a few more steps of the argument ending at Eq. 2.8. The integrand of Eq. 2.5 may be written $\mathrm{curl}_k \mathbf{A}$ where

$$\mathbf{A} = \int dx dy \, u_\alpha^* \, \boldsymbol{\nabla}_k \, u_\alpha = - \int dx dy \, u_\alpha \, \boldsymbol{\nabla}_k \, u_\alpha^* \quad (4.4.2)$$

since $(u_\alpha, u_\alpha) = 1$. Thus, by Stokes theorem

$$h\sigma_{xy}/e^2 = (i/2\pi) \int_c d\mathbf{k} \cdot \mathbf{A}. \quad (4.4.3)$$

The contour c goes around the Brillouin zone in four legs, (i) $k_1 : 0 \to K_1$, $k_2 = 0$, (ii) $k_2 : 0 \to K_2$, $k_1 = K_1$, (iii) $k_1 : K_1 \to 0$, $k_2 = K_2$, (iv) $k_2 : K_2 \to 0$, $k_1 = 0$. The choice indicated in Eq. 2.6 causes $\mathbf{A}$ to vanish except on leg (iii). Indeed, let

$$u = e^{i\beta(\alpha)} \, e^{i\psi(\alpha, xy)} \, w_\alpha(xy),$$

where w is real. The gradient in $\mathbf{A}$ acts, in effect, only on the phases of u. The space independent phase β is arbitrary. The choice

$$\beta(k_1 k_2) = - \int_0^{k_1} d\hat{k}_2 \, [w(k_1 \hat{k}_2)]^2 \, \frac{\partial \psi}{\partial k_2}$$

$$- \int_0^{k_1} d\hat{k}_1 [w(\hat{k}_1 0)]^2 \, \frac{\partial \psi}{\partial k_1} \quad (4.4.4)$$

leads to (4.2.6). On this leg, (4.2.7) may be used so

that

$$\int_C d\mathbf{k}\cdot\mathbf{A} = -i\int_0^{K_1} dk_1 \frac{d\delta}{dk_1} = -2\pi i t \qquad (4.4.5)$$

according to Eq. 2.8.

CHAPTER 5

Field Theory, Scaling and the Localization Problem

ADRIANUS M.M. PRUISKEN

5.1 Introduction

At roughly the same time as von Klitzing's discovery of the QHE, there was an astounding development in the theory of electronic disorder. With the discovery of "weak localization", Abrahams, Anderson, Licciardello, and Ramakrishnan (1979) were able to translate earlier scaling ideas on electronic transport into the framework of the renormalization group. Wegner (1979a) complemented the development with the formulation of a renormalizable field theory of interacting matrices. Considerable theoretical, numerical, and experimental work has been done since then in order to verify the one-parameter scaling ideas and extend the weak localization methodology in several ways. [See the articles in *Anderson Localization*, (Nagaoka and Fukuyama, eds. 1982).]

It became increasingly clear, however, that there existed a serious discrepancy between von Klitzing's discovery and localization theory. The quantization of the Hall conductance was interpreted as being a result of Anderson localization due to random impurities. On the other hand, delocalized states were needed to carry the Hall current. The one-parameter scaling ideas, however, always seem to predict localized wave functions in 2D.

On its simplest level, the problem resided in the fact that the scaling picture of electronic transport made no reference to the Hall conductance whatsoever. In brief, it was discovered in the field-theoretic derivation (Pruisken 1984) that the effective statistical average over impurity scattering involves phase factors of the type

$$\exp[2\pi i \sigma^0_{xy} q]$$

signaling the presence of the magnetic field through the parameter σ^0_{xy}, the Hall conductance (expressed in units of e^2/h) in the self-consistent Born approximation [SCBA].

The integer q is a *topological invariant*, associated with the fluctuating fields in the theory and invisible in diagrammatic analyses. This phase factor turned out to be an application of the θ-vacuum concept, which was discovered several years ago in four-dimensional gauge theories. In subsequent work (Levine, Libby and Pruisken, 1983, 1984a,b,c) the analogy with existing field-theoretic literature on θ-vacua was emphasized and exploited. Much effort was devoted to issues of σ_{xy} or θ-renormalization which did not appear in field-theoretic contexts before. The picture was finally made conceptually complete through the formulation of a two-parameter scaling theory on the basis of instanton considerations (Pruisken, 1985a,b).

More recently, Wei, Tsui, and Pruisken (1986) have given a detailed experimental study of the temperature dependence in the QHE. The measurements were done on InGaAs-InP heterostructures and the results are consistent with the theory. So, the necessary first agreement has been achieved. These experiments will be discussed in the next chapter [Sec. 6.4.3] as well as at the end of this one.

In this chapter, I will attempt to give a more or less complete report on these developments. Much of the material presented here can be found in (Pruisken, 1985b) and the original literature. However, emphasis will be given to such topics as edge currents and the physical meaning of topology. Moreover, I will report the results of a more extended analysis of the semiclassical method, which deals with instantons as nonperturbative objects in the weak coupling (localiza-

Field Theory, Scaling and Localization

tion) regime. This revises earlier statements on "replica symmetry breaking", as well as on the manner in which the instanton survives the replica limit.

5.2 Basic Notions

5.2.1 Introduction

We will be largely concerned with the evaluation of the advanced, (retarded) propagators $[E - \mathcal{H} \pm i0^+]^{-1}$, averaged over disorder, which contain all the information of physical interest. Here, E denotes the Fermi level and $\mathcal{H}$, the free electron Hamiltonian is

$$\mathcal{H} = \frac{1}{2m}(p - \frac{e}{c}A)^2 + V(r)$$

$$= \frac{1}{2m}\Pi^2 + V(r) = \mathcal{H}_0 + V(r) , \qquad (5.2.1)$$

where $V(r)$ represents the impurity potential and $A_\mu = (0, Bx)$. We recall from Chap. I that the energy spectrum for $V = 0$ consists of highly degenerate Landau levels with energies $(n+\frac{1}{2})\hbar\omega_c$. The degeneracy per Landau level per unit area is $eB/hc = 1/2\pi\ell^2$. In the following, we often use magnetic units, in which $\hbar = \ell = m = 1$, to make the equations less cumbersome, but since we shall be interested in scaling, we shall return to conventional units when necessary for clarity.

Generally speaking, the averaged single-particle Green's functions describe the system in thermodynamic equilibrium. The nonequilibrium properties, however, involve the simultaneous average of advanced and retarded Green's functions. Examples of the former are the density of states and magnetization; the conductance parameters, magnetic susceptibility and the participation ratios are examples of the latter. There is a profound distinction between these two aspects of the electronic system and this distinction is characteristic for problems with quantum diffusion.

One-particle propagators are generally found to be short-ranged, nonsingular objects which can be easily evaluated by elementary numerical means or well approximated by simple analytical means (e.g. the effective medium approximation, coherent potential approximation [CPA], and the SCBA). The averages of combined advanced and retarded propagators, on the other hand, involve a different length scale in the problem, which is the characteristic one for electronic transport. This involves the localization length, i.e. the spatial extent of the electronic wave functions near the Fermi energy. This length can be infinite (conducting system) or finite (insulating system) and the study of the singularities near the possible metal-insulator transition forms the actual subject of localization theory.

The developments in recent years have shown that the transport problem exhibits scaling behavior and is amenable to renormalization group computations. The theory is formally analogous to that of the Heisenberg ferromagnet. Such aspects as $D=2$ being the lower critical dimension below which the localization length must be finite, and ϵ [$= D-2$] expansions for critical exponents characterizing the metal-insulator transition, all fall within the scope of field theory and renormalization group calculational techniques (again, see Nagaoka and Fukuyama 1982). In what follows, I will give the definition and a brief description of the physical quantities involved, which is then supplemented by an elementary discussion of the IQHE.

5.2.2 Density of states

The averaged density of states at the Fermi level E can be expressed as

$$\rho(E) = \frac{i}{2\pi} \overline{\left[G^+(r,r) - G^-(r,r) \right]}^{Av} ;$$

$$G^{\pm}(r,r') = \langle r | (E - \mathcal{H} \pm i0^+)^{-1} | r' \rangle . \quad (5.2.2)$$

This quantity has been evaluated exactly for the lowest Landau level in the strong-magnetic-field limit in the case of a Gaussian white-noise potential (Wegner 1983). Brézin et al. (1984) generalized this result to the case of an arbitrary uncorrelated point-like scattering potential. A main achievement is that these results confirm the notion that ρ is a nonsingular function of energy and gives no indication of a metal-insulator transition. This was actually expect-

Field Theory, Scaling and Localization

ed long before on the basis of approximate methods (see Ando et al. 1982). This notion was furthermore used as a basic starting point in the discussion of the localization problem (Pruisken 1984).

The simplest approximation, which shows all of the essential ingredients and is appropriate to a white-noise potential, is given by the SCBA [or CPA] for the self-energy. This approximation covers the whole spectrum of weak and strong magnetic fields. [See Section 5.4.] The SCBA self-energy $\Sigma_0^\pm$ is given by

$$g^{-1}\Sigma_0^\pm + \langle r|(E - \mathcal{H}_0 + \Sigma_0^\pm)^{-1}|r'\rangle = 0 . \quad (5.2.3)$$

Here, g denotes the variance of the distribution P for the impurity potential $V(r)$

$$P[V] \propto \exp\left[-\frac{1}{2g}\int d^2r \, V^2(r)\right] . \quad (5.2.4)$$

Eq. 2.3 is simplified by representing it in terms of the eigenfunctions and energies of the Hamiltonian (Chap. I) and can be written

$$g^{-1}\Sigma_0^\pm + \sum_{n=0}^{M_0} (E - E_n + \Sigma_0^\pm)^{-1} = 0;$$

$$E_n = n + \frac{1}{2} . \quad (5.2.5)$$

This equation exhibits a logarithmic singularity which can be absorbed in a redefinition of Σ_0 and E through $\Sigma_0 \to \Sigma_0 - \ln M_0$, $E \to E + \ln M_0$ with M_0 the maximum of n. We remark that this singularity is unphysical and does not occur in a more realistic effective Hamiltonian for the electron gas, involving higher powers of the velocity operator Π_μ. We will limit ourselves to the simplest case, however.

For strong magnetic fields, only one term in the sum will dominate and Eq. 2.5 reduces to the quadratic equation ($E \cong E_n$)

$$(\Sigma_0^\pm)^2 + (E - E_n)\Sigma_0^\pm + g = 0 , \quad (5.2.6)$$

the solution of which can be written

$$\Sigma_0^\pm = -\frac{1}{2}(E - E_n) \pm \frac{i}{2}\left[4g - (E - E_n)^2\right]^{\frac{1}{2}} . \quad (5.2.7)$$

The density of states in SCBA, $\rho_0 \propto \text{Im } \Sigma_0^+$ reduces to the well-known semicircle around E_n. Fig. 1 shows Wegner's exact result and that of the SCBA [Note 1].

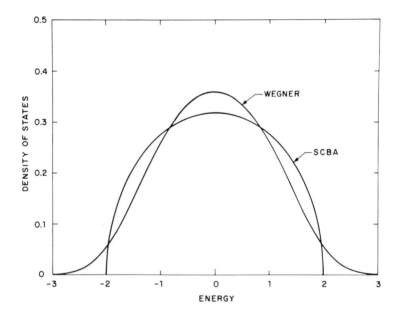

Figure 5.1. The exact (Wegner) and SCBA density of states in the lowest Landau level for a white-noise impurity potential.

5.2.3 Magnetization

The Fermi energy contribution to the magnetization can be expressed in terms of the velocity operator

$$\frac{\partial M}{\partial E} = \frac{-ie}{8\pi c \Omega} \text{Tr } \overline{r \times (\Pi^\dagger + \Pi)(G^+ - G^-)}^{Av} ,$$

$$\Pi = -i[r, \mathcal{H}] = \left(p - \frac{e}{c}A\right) . \quad (5.2.8)$$

The vector cross product above, and in what follows, is short for the z-component of the cross product, i.e.

$$a \times b \equiv \epsilon_{\mu\nu} a_\mu b_\nu ,$$

where $\epsilon_{\mu\nu}$ is the antisymmetric unit tensor, and we adopt a summation convention for spatial subscripts. The area of the system is $\Omega = L^2$. [We take a square geometry for notational convenience. The shape of the system does not matter, of course. With this convention it is unnecessary to distinguish between conductance and conductivity.]

Eq. 2.8 can be written in a somewhat more transparent fashion as

$$\frac{\partial M}{\partial E} = \frac{e}{2c\Omega} \int_\Omega d^2r \sum_\alpha \overline{r \times v^\alpha(r) \, \delta(E-E_\alpha)}^{Av} ;$$

$$v^\alpha = \tfrac{1}{2}\xi_\alpha^*(r)(\Pi^\dagger + \Pi)\xi_\alpha(r) . \qquad (5.2.9)$$

The E_α, $\xi_\alpha(r)$ denote the eigenenergies and functions for a single realization of the electronic disorder. This expression has a peculiar property in that only the edge states contribute. Due to translational and rotational invariance in the bulk, which is accomplished by the ensemble averaging, it is easy to show that the expectation value of the current density vanishes in the interior of the system. In fact, the expression is a realization of the well known result that the magnetization can be obtained from an edge current I_e

$$M = I_e/c, \qquad \partial M/\partial E = (\partial I_e/\partial E)/c . \qquad (5.2.10)$$

On the other hand, one can show that the magnetization and hence the edge current is entirely determined by the bulk. Namely, by operating on Eq. 2.8 with $\int dE' d/dE$ we obtain

$$\frac{\partial M}{\partial E} = \frac{-ie}{4\pi c\Omega} \int_{-\infty}^{E} dE' \mathrm{Tr}\left[\overline{G^+(E') \, r \times \Pi \, G^+(E')} - (+ \leftrightarrow -) \right]$$

$$(5.2.11)$$

where the notation Tr indicates a trace over both matrix *and* space indices [while the notation tr, to be used later, is the trace over matrix indices only]. Eq. 2.11 can be evaluated one step further by noting

$$\frac{\partial M}{\partial E} = \int_{-\infty}^{E} dE' \cdot \frac{\partial \rho(E')}{\partial B} = \left[\frac{\partial n}{\partial B}\right]_{E} , \qquad (5.2.12)$$

with n the electronic number density. So, the magnetization is given in terms of the bulk density of states. Eq. 2.12 can also be derived from the thermodynamic definition of magnetization and actually forms a Maxwell relation. Namely, the magnetization is defined by

$$M = -\frac{d}{dB}\langle E' \rangle_N = -\frac{d}{dB}\left[\int_{-\infty}^{E} E' \rho(E') dE'\right]_N , \qquad (5.2.13)$$

where the derivative with respect to magnetic field is to be performed for a fixed number of electrons. This means that one has to account for a variation in E with changing B

$$\frac{dn(E)}{dB} = \frac{\partial E}{\partial B} \rho(E) + \int_{-\infty}^{E} \frac{\partial \rho(E')}{\partial B} dE' = 0 , \qquad (5.2.14)$$

and the partial derivative under the integral indicates that E' is held constant. The expression for the magnetization can be written

$$M = -\frac{d}{dB}\int E \, dn = -\frac{d}{dB}\left[En(E) - \int_{-\infty}^{E} n(E') dE'\right] , \qquad (5.2.15)$$

which leads to the desired result

$$M = \int_{-\infty}^{E} \frac{\partial n(E')}{\partial B} dE' . \qquad (5.2.16)$$

Finally, using Eqs. 2.12, 16 and 10, we conclude that the equilibrium edge current can be understood as

$$\frac{\partial I_e}{\partial E} = c\frac{\partial n}{\partial B} \ . \qquad (5.2.17)$$

Hence, the amount of edge current near E is determined by the bulk density of levels below the Fermi energy. This relation is quite general, holds for any type of disorder and can also be microscopically derived for discretized versions (see Fig. 2) of the magnetic field problem [Notes 2,3].

A relation can be derived which is useful for checking Eq. 2.17 for approximate calculational schemes of one-particle Green's functions. Namely, by writing the exact averaged one-particle propagator in terms of a self-energy $\Sigma^{\pm}(\mathbf{r},\mathbf{r}')$

$$\overline{G^{\pm}(\mathbf{r},\mathbf{r}')}^{Av} = \langle \mathbf{r}|(E - \mathcal{H}_0 + \Sigma^{\pm})^{-1}|\mathbf{r}'\rangle \qquad (5.2.18)$$

we have, instead of Eq. 2.11

$$\frac{\partial M}{\partial E} = \frac{-e}{4\pi c\Omega} \int_{-\infty}^{E} dE' \ \mathrm{Tr}\Big\{ \overline{G^+(E')} \ \mathbf{r}\times\boldsymbol{\pi} \ \overline{G^+(E')} \Big[\mathbb{1} + \frac{\partial \Sigma^+}{\partial E'} \Big]$$
$$- (+ \leftrightarrow -) \Big\} \ , \qquad (5.2.19)$$

where $\mathbb{1}$ is the unit matrix. Eq. 2.19 can be evaluated as

$$-\frac{\partial M}{\partial E} + \frac{\partial n}{\partial B}$$

$$= \frac{i}{2\pi\Omega} \int_{-\infty}^{E} dE' \mathrm{Tr}\Big\{ \Big[\frac{\partial}{\partial E}\Sigma^+(E')\frac{\partial}{\partial B}\overline{G^+(E')} - \frac{\partial}{\partial B}\Sigma^+(E')\frac{\partial}{\partial E}\overline{G^+(E')} \Big]$$
$$- \Big[+ \leftrightarrow - \Big]\Big\} \ . \qquad (5.2.20)$$

Comparing Eqs. 2.20 and 2.12, we conclude that the following equality should hold

$$\frac{1}{\Omega}\sum_{p=\pm} s^p \ \mathrm{Tr}\Big[\frac{\partial \Sigma^p}{\partial E}\frac{\partial \overline{G^p}}{\partial B} - \frac{\partial \Sigma^p}{\partial B}\frac{\partial \overline{G^p}}{\partial E}\Big] = 0. \qquad (5.2.21)$$

Here s^p is the sign of p. By the arguments presented above, this equality guarantees the validity of Eq. 2.17. For instance the loop expansion for the one-particle propagators, to be discussed later, fulfills this relation order by order in perturbation theory (Pruisken and Wegner, unpublished). The zeroth order term is given by the SCBA, Eq. 2.5, and one can easily check that Eq. 2.21 holds within the SCBA.

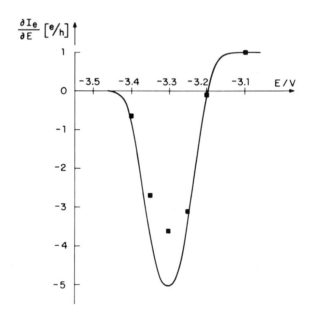

Figure 5.2. The edge current $[\partial I_e/\partial E]$ in the lowest Landau level as a function of energy for the disordered tight binding Hamiltonian. The data [■] are numerically obtained for an infinite strip 15 lattice constants wide. The solid curve is obtained from Eq. 2.17 using the bulk density of states for the infinite width system. There is one flux quantum per eight plaquettes. The ratio W/V of the width of square diagonal disorder to hopping strength is 1/2.

5.2.4 Transport parameters

According to linear response theory, the dc conductance at finite temperatures can be expressed as

$$\sigma_{\mu v}(T) = \int dE' \, \frac{\partial f(T)}{\partial E'} \, \sigma_{\mu v} \,, \qquad (5.2.22)$$

Field Theory, Scaling and Localization

with f denoting the Fermi-Dirac distribution and $\sigma_{\mu\nu}$ the zero-temperature conductance. [Note: The parameters $\sigma_{\mu\nu}$ will be considered as dimensionless numbers, measuring the conductance in units $[e^2/h]$]. For the latter we use the specific version obtained in the work Smrčka and Středa (1977)

$$\sigma_{xx} \times [\frac{e^2}{h}] = - \frac{\hbar e^2}{4\pi\Omega} \overline{\text{Tr } \Pi_x(G^+ - G^-)\Pi_x(G^+ - G^-)}^{Av} . \quad (5.2.23)$$

$$\sigma_{xy} = \sigma_{xy}^{I} + \sigma_{xy}^{II};$$

$$\sigma_{xy}^{I} \times [\frac{e^2}{h}] = - \frac{\hbar e^2}{4\pi\Omega} \overline{\text{Tr } \epsilon_{\mu\nu}\Pi_\mu G^+ \Pi_\nu G^-}^{Av} , \quad (5.2.24)$$

$$\sigma_{xy}^{II} \times [\frac{e^2}{h}] = \frac{ie^2}{4\pi\Omega} \overline{\text{Tr } \epsilon_{\mu\nu} r_\mu \Pi_\nu (G^+ - G^-)}^{Av} . \quad (5.2.25)$$

Whereas σ_{xx} is expressed in the familiar fashion as a current-current correlation function, the expression for the Hall conductance is remarkable in that it involves a single energy E, namely the Fermi energy. Such a result is of important conceptual value because the formulation of renormalization group ideas is heavily based upon this aspect of the free particle problem. Indeed, these formulae become particularly meaningful within the field-theoretic formalism, to be discussed later.

First we notice that the part σ_{xy}^{II} has been discussed before as the Fermi energy contribution to the magnetization. This emphasizes the role of edge currents in the localization problem, if one chooses to formulate the conductance as a process at the Fermi level. However, Eqs. 2.8, 11, 12 show that the edges can be "transformed away" and σ becomes a bulk quantity, to which now all the bulk states below the Fermi level contribute:

$$\sigma_{xy}^{II} [\frac{e^2}{h}] = - ec \int_{-\infty}^{E} dE' \frac{\partial}{\partial B} \rho(E') = -ec \frac{\partial n}{\partial B} . \quad (5.2.26)$$

As a first approximation, one can evaluate these expressions in the SCBA, which amounts to the insertion of the self-energy, Eq. 2.3, for the random potential. This leads to the result

$$(\sigma_{xy}^I)_0 = -\omega_c\tau(\sigma_{xx})_0 , \qquad (5.2.27)$$

where $\tau^{-1} = \text{Im } \Sigma_0^+$. This form was discussed first by Středa (1982) and reproduces the classical Drude-Zener result for the Hall conductance in the weak magnetic field limit, where the electron density becomes B-independent. For strong magnetic fields, the sum of both parts in the Hall conductance precisely reduces to Ando's version of the SCBA results (Pruisken, 1985b,c; see also Sec. 5.4.4).

The SCBA for the conductance parameters will serve as a starting point for scaling. Here, we will discuss a slightly different interpretation of the exact formulae in the opposite limit where scaling has been achieved and the states near the Fermi level are localized. In the end, this interpretation brings about some of the most essential theoretical aspects of the IQHE. It will enable us furthermore to understand better the formal field-theoretical apparatus and it is necessary to clarify what is physically going on.

To fix the thought, consider the macroscopic continuity equation

$$e\frac{\partial n}{\partial t} + \nabla \cdot \mathbf{j} = 0 . \qquad (5.2.28)$$

We are specifically interested in the current response to electric fields. For fields $E(r,t)$ which vary slowly in space time we may insert

$$j_\mu = \sigma_{xy} \times [\frac{e^2}{h}]\epsilon_{\mu\nu}E_\nu , \qquad (5.2.29)$$

so that

$$-ec\frac{\partial n}{\partial t} + [\frac{e^2}{h}]\sigma_{xy}\frac{\partial B}{\partial t} = 0 , \qquad (5.2.30)$$

where we have made use of Maxwell's equations. [Without the assumption of localization one has to include diffusive currents.] Integration over space-time leads to the desired result

Field Theory, Scaling and Localization   129

$$(\Delta N)_{ad} = \sigma_{xy}(\Delta \Phi)_{ad} , \qquad (5.2.31)$$

where N, Φ refer to the total number of electrons and the dimensionless flux respectively. [See Eq. 1.7.1.]

This result underlines the difference between the equilibrium quantity σ_{xy}^{II} and the transport parameter σ_{xy}. The former refers to the process in which the system is at any instant of time in thermodynamic equilibrium with the changing magnetic field, keeping the Fermi level fixed [one has to imagine here the presence of some relaxation mechanism such that electrons can be transported to and from the system]. The latter however refers to the system taken out of thermodynamic equilibrium by a switch-on process at $T=0$ for a changing magnetic field.

Going back to the expression for the Hall conductance, we conclude that σ_{xy}^{I} precisely makes up for the difference in the equilibrium and nonequilibrium processes. Although this quantity is formally expressed as a Fermi level contribution alone, its interpretation, like σ_{xy}^{II}, involves all electronic levels below E.

In order to discuss this interpretation, we remark that localized electrons cannot be transported to and from the system as a result of external perturbations. If the system were to have only localized wave functions, then all of the eigenstates would follow the changing magnetic field in a continuous fashion. The Fermi level would adjust itself such that the total number of electrons remains constant. Clearly, the N in Eq. 2.31 refers to the number of extended bulk electrons present; without these states there could not exist a QHE.

Our present understanding of the situation is illustrated by Fig. 3. In the strong field limit, such that the Landau levels are separated, there is a singular energy at the band center, characterized by a divergent localization length. The states near this energy are the ones capable of exchanging electrons with the exterior by passing through the extended edge states (Fig. 4).

Actually, one can push the discussion a little further and prove the existence of the IQHE. Namely, we imagine the situation where E is located between two Landau levels where, for large enough B, $\rho(E)$ becomes

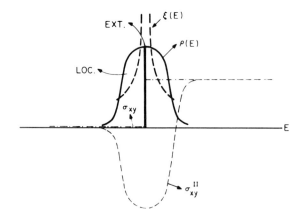

Figure 5.3. Hall conductance and nature of the electronic states in the lowest Landau level.

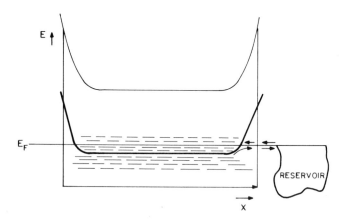

Figure 5.4. Exchange of charge with the exterior because of an adiabatic change in the magnetic flux.

arbitrarily small. Under these circumstances σ_{xy}^I vanishes, as well as σ_{xx}. Furthermore, the degeneracy per Landau level equals eB/hc so that the Hall conductance is quantized and equals $i(e^2/h)$ with i the total number of occupied Landau levels. [Notice that this is entirely a consequence of the density of states, an equilibrium quantity which we have taken to be harmless, and nonsingular.] In the present context, this means that an (adiabatic) change in flux by q quanta, i.e. $(\Delta\Phi)_{ad} = q(hc/e)$, will result in the transport of

Field Theory, Scaling and Localization

precisely iq electrons to (from) the system, namely q electrons per Landau level.

Next we imagine the Fermi level to move inside the Landau level, such that $\rho(E) \neq 0$. As long as E moves through a region of localized states, both the dissipative conductance and the Hall conductance can not alter their values by the very arguments presented above; i.e., the aforementioned charge transfer is solely a consequence of whatever extended bulk levels are present below E. The resulting picture for the Hall conductance at $T=0$ is illustrated in Fig. 3 for the lowest Landau level. Apparently, the quantity σ_{xy}^I acts in such a fashion as to bring the sum $\sigma_{xy}^I + \sigma_{xy}^{II}$ back to quantization. The way the electronic system manages to do this forms, of course, the actual subject of a microscopic theory.

5.3 The Search for the Principle

5.3.1 Introduction

In Chap. I, R. E. Prange discussed a simple and pretty argument due to Laughlin which argues for quantization of the Hall conductance in case the states near E are localized. Laughlin's gauge argument brings our attention to several of the aspects which have been the subject of the preceding section. In particular, we mention the role of the edge states and the necessity of having both extended and localized electronic levels. These aspects are not natural ingredients of the conventional studies of electronic disorder and neither is the Hall conductance itself.

A formalism which is rich enough to deal with such questions is given by means of a *replicated generating functional* for the ensemble averaged Green's functions. This formalism can be employed as a valid starting point for the development of a microscopic theory. Moreover, the derivation of effective Lagrangians, describing the singularities in the macroscopic transport phenomena, forms a suitable framework for studying problems of electron localization by means of the renormalization group calculational technique. However, besides being a calculational tool, the field-theoretic approach also forms a conceptual framework for understanding localization phenomena.

It should not be identified with the conventional, weak localization, perturbative expansions.

A very basic notion in the sequel is the observation of a *continuous symmetry* between the advanced and retarded fields in the generating functional (Wegner 1979a,b). This symmetry is related in an intimate manner to the macroscopic transport properties of the system and despite the rather abstract formulation, its consequences are far reaching. In fact, many aspects of the localization phenomenon can be formulated as direct consequences of the continuous symmetry. Its most elegant formulation can be found in the IQHE. Here it interconnects such concepts as localization, quantization of the Hall conductance, edge currents, topology and Stokes' theorem as consequences of one and the same principle.

One starts out from the standard realization that the propagators $(E - \mathcal{H} \pm i0^+)^{-1}$ can be obtained as correlations of Gaussian field variables defined through the generating function

$$\int \mathcal{D}[\bar{\psi}^+] \int \mathcal{D}[\psi^+] \int \mathcal{D}[\bar{\psi}^-] \int \mathcal{D}[\psi^-] \, e^{\mathcal{L}'} ;$$

$$\mathcal{L}' = \int d^2r \left[\bar{\psi}^+(r)(E - \mathcal{H} + i0^+)\psi^+(r) \right.$$

$$\left. + \bar{\psi}^-(r)(E - \mathcal{H} - i0^+)\psi^-(r) \right]. \quad (5.3.1)$$

The replica method enables one to perform the ensemble average over impurity distributions [Note 4]. This amounts to introducing m replications of Eq. 3.1 in conjunction with an integral over the potential $V(r)$ with given weight $P[V]$, e.g. (5.2.4). The basic starting point of the analysis is then given by

$$\mathcal{Z} = \int \mathcal{D}[V] \, P[V] \prod_{a=1}^{m} \prod_{p=\pm} \int \mathcal{D}\bar{\psi}_a^p \int \mathcal{D}\psi_a^p \, e^{\mathcal{L}} \quad (5.3.2)$$

with

$$\mathcal{L} = \int d^2r \sum_{a=1}^{m} \sum_{p=\pm} \bar{\psi}_a^p(r) (E - \mathcal{H} + i\eta s^p) \psi_a^p(r) , \quad (5.3.3)$$

Field Theory, Scaling and Localization

where $\eta = 0^+$. The index p is introduced such that the simultaneous average over the advanced and retarded quantities is realized. The subscript a runs over m replica fields and the ensemble averaged Green's functions for the electronic system are obtained in the limit $m \to 0$. We have not specified the fields $\psi, \bar{\psi}$ yet. These are preferably taken to be complex (anticommuting) fermion fields ($\bar{\psi} = i\psi\dagger$) which are defined in terms of ordinary Grassmann variables, with the usual notion of complex conjugation. For example, the one-particle Green's function can be obtained as

$$\overline{G^p(r,r')}^{\text{Av}} = \langle \psi^p_{a=1}(r) \psi^p_{a=1}(r') \rangle$$

$$= \mathcal{Z}^{-1} \int \mathcal{D}[V] \, P[V] \, G^p(r,r') \text{Det}^m[(E - \mathcal{H} + i\eta)(E - \mathcal{H} - i\eta)] \, .$$

(5.3.4)

This leads to the desired result when $m \to 0$ ($\mathcal{Z} \to 1$). The introduction of fermion field variables is a choice born out of symmetry considerations. This symmetry refers to the invariance of $\mathcal{L}$ (Eq. 3.3) under $U(2m)$ rotations performed on the ψ, i.e. the transformations which leave the quadratic form

$$\sum_a [\bar{\psi}^+_a \psi^+_a + \bar{\psi}^-_a \psi^-_a] \qquad (5.3.5)$$

invariant. The convergence term η breaks this symmetry to the level $U(m) \times U(m)$, which denotes unitary transformations on the $p = +,-$ indices separately. This symmetry was emphasized by Wegner (1979a) as an essential aspect in the development of renormalization group ideas on electronic transport. Namely, by noting that the quantity conjugate to the symmetry breaking term η, i.e. $\partial \mathcal{Z}/\partial \eta$, is proportional to the density of states, one sees that the problem exhibits a spontaneous symmetry breaking in the regime of physical interest. The matter was further elucidated by the derivation of Ward identities, which maintain their validity also in the limit of $m=0$ field components. For a discussion of this important point, the reader is referred to the original literature.

5.3.2 Source terms

A direct connection between electronic transport and symmetry is obtained from a source-term formalism for the dc conductance (Pruisken 1985b, Levine, Libby and Pruisken 1984a,b,c). This formalism will be of conceptual value throughout. One starts out by defining the generating function in the presence of a background matrix field $\mathcal{U}(r)$

$$\mathcal{Z}[\mathcal{U}] = \int \mathcal{D}[V] \, P[V] \, e^{\mathcal{L}[\mathcal{U}]};$$

$$\mathcal{L}[\mathcal{U}] = \mathrm{Tr}\, \ln[E - \mathcal{U}^{-1}\mathcal{H}_0 \mathcal{U} + V(r) + i\eta \hat{s}], \quad (5.3.6)$$

where $\mathcal{U}$ is thought of as a slowly varying $U(2m)$ rotation and where the argument of the $\mathrm{Tr}\,\ln$ is a $2m \times 2m$ matrix. In particular,

$$[\hat{s}]_{ab}^{pp'} = s^p \, \delta_{ab} \delta_{pp'} \,.$$

The conductance formulae can conveniently be obtained from the specific form

$$\mathcal{U}(x,y) = \exp[ixj_x\hat{\tau}_x + iyj_y\hat{\tau}_y] \quad (5.3.7)$$

where j_μ are the "sources" and $\hat{\tau}_\mu$ are the usual Pauli matrices acting in one replica channel only

$$[\hat{\tau}_\mu]_{ab}^{pp'} = [\tau_\mu]^{pp'} \delta_{ab}\delta_{a1} \,. \quad (5.3.8)$$

Using the expansion formulae

$$\mathcal{U}^{-1}\mathcal{H}_0 \mathcal{U} = \tfrac{1}{2} \Pi_\mu^2 - (j_\mu \hat{\tau}_\mu + j_x j_y \epsilon_{\mu\nu} r_\nu \hat{\tau}_z) \Pi_\mu +$$

$$+ \tfrac{1}{2} j_\mu^2 + O(j^3) \quad (5.3.9)$$

and

Field Theory, Scaling and Localization

$$\text{Tr } \ln [A - B]$$
$$= \text{Tr } \ln A - \text{Tr } A^{-1}B - \frac{1}{2} \text{Tr } A^{-1}BA^{-1}B - \ldots \quad (5.3.10)$$

one concludes that

$$\left.\frac{\partial^2 \mathcal{Z}}{\partial j_x \partial j_x}\right|_{j_\mu = 0} = -2\Omega \sigma_{xx} ,$$

$$\left.\frac{\partial^2 \mathcal{Z}}{\partial j_x \partial j_y}\right|_{j_\mu = 0} = 2i\Omega \sigma_{xy} , \quad (5.3.11)$$

with $\sigma_{\mu\nu}$ precisely given by Eqs. 2.23 and 2.24.

There is one important conclusion that one can draw from this. Namely, the unitary symmetry remains significant also in case of a vanishing density of bulk states near the Fermi level. That is, $\mathcal{Z}$ remains sensitive to $\mathcal{U}$ because, as previously discussed, the Hall conductance is quantized in this case. This effect, of course, must be attributed to the extended edge states in the problem.

What has been discussed above forms a simple demonstration of several new aspects in this electronic transport problem. The very notion of edge currents can immediately be exploited in order to establish the meaning of a topological classification of the field configurations (Pruisken 1985b) and a somewhat more complete discussion will be presented in the next section. Here, we discuss a still different expression for the conductance parameters, which follows from differentiating the equivalent form for $\mathcal{Z}$

$$\mathcal{Z}[\mathcal{U}] = \int \mathcal{D}[V] \, P[V] \, e^{\mathcal{L}[\mathcal{U}]} \, ;$$

$$\mathcal{L}[\mathcal{U}] = \text{Tr } \ln [E - \mathcal{H}_0 + V(r) + i\eta \hat{\mathcal{U}}s\mathcal{U}^{-1}] . \quad (5.3.12)$$

This leads to

$$\sigma_{xx} = \frac{1}{\Omega} \lim_{\eta \to 0} \eta^2 \int d^2r \int d^2r' (x-x')^2 \, \overline{G^+(r,r')G^-(r',r)}^{Av} ,$$

$$\sigma_{xy} = \frac{i}{\Omega} \lim_{\eta \to 0} \eta^2 \int d^2r \int d^2r' (xy'-x'y)^2 \overline{G^+(r,r')G^-(r',r)}^{Av}$$

(5.3.13)

which shows that both quantities can be obtained, at least in principle, from the diffusion propagator G^+G^-. We remark that these results can also be obtained from the previous expressions by using the commutation relation, Eq. 2.25.

It is a simple exercise to show that the ensemble averaged two-particle propagator in (5.3.13) is both translationally and rotationally invariant. This immediately implies that the expression for the Hall conductance vanishes identically! The argument breaks down, of course, if one includes edges in the problem. Apparently, Eqs. 3.13 express the Hall conductance entirely as an edge phenomenon.

Practically speaking, Eqs. 3.13 are less attractive to focus on because the combined effect of the edges and the limit $\eta \to 0$ appears to be very cumbersome. One way out of the edge effects would be to operate on these expressions with $\int dE' \partial/\partial E'$. One then has to pay the usual price of integrating over all states below the Fermi level and this we do not want, at least as far as the advanced-retarded combinations are concerned.

What the argument does tell us, however, is that ordinary perturbation theory presumably does not work for the diffusion propagator. That is, a straightforward expansion in powers of the impurity potential disregarding the edges leads to a vanishing Hall conductance to all orders in perturbation theory. This by itself should be a signal to worry and it is hard to visualize how topology or flux quantization could ever be part of such a theory. It should therefore not be a surprise if the Padé summation to high orders in the impurity expansion (Hikami 1984, Singh and Chakravarti 1985, Hikami and Brézin 1986) would lead to the same conclusions as the perturbative renormalization group approach: i.e., all states localized in 2D. Future work probably will tell.

Field Theory, Scaling and Localization

5.3.3 Edge currents and topology

Having noticed the direct connection between symmetry and the conductance parameters, we may next go further and ask for the possible relation between symmetry and localization. The answer to such a question, of course, should be the result of a microscopic theory. The concept of scaling will form the main vehicle for the understanding of how the system develops mobility gaps or electronic conduction at large length scales, starting from true metallic behavior at small length scales. Renormalization group computations are usually limited to the weakly localized regime, however.

On the other hand, there is the situation where the Fermi energy is located in a density of states gap which is similar in effect as, but very much simpler to deal with, than strongly localized wave functions. This motivates the search for a principle which connects the concept of localization with the notion of a continuous symmetry and which smoothly matches up weak localization theory with gap physics. In the former theory, the meaning of symmetry is predominantly contained in the renormalization group techniques. In the IQHE its role is much richer in that it unifies a variety of physical aspects of the problem.

Let us consider the gap situation in more detail. The source term formalism suggested that only the value of $\mathcal{U}$ near the boundary matters because there can only exist extended edge states in this case. So, those matrix fields $\mathcal{U}$ which reduce to an arbitrary $U(m) \times U(m)$ gauge near the edge should leave $\mathcal{Z}$ invariant. The consequences of such a statement can be worked out via a derivative expansion in the matrix field $\mathcal{U}$, which proceeds by expanding the Tr ln in Eq. 3.6 in powers of

$$\mathcal{U}^{-1} \Pi_\mu^2 \mathcal{U} - \Pi_\mu^2 = 2D_\mu \Pi_\mu - i(\nabla_\mu D_\mu) + D_\mu D_\mu;$$

$$D = -i\mathcal{U}^{-1}\nabla\mathcal{U}. \qquad (5.3.14)$$

This leads to the result (Pruisken 1985b)

$$\mathcal{Z}_{GAP}[\mathcal{U}] = \mathcal{Z}_{GAP}[\mathbb{1}]\exp\left\{\int d^2r \sum_{p,a} j^p(r)\cdot\left[D(r)\right]^{pp}_{aa} + ..\right\},$$

(5.3.15)

where only the nonvanishing terms with leading dimension have been taken into account. Furthermore

$$j^p(r) = \langle \Pi G^p(r,r) + G^p(r,r)\Pi^\dagger\rangle,\qquad (5.3.16)$$

where the expectation $\langle ...\rangle$ is with respect to $\mathcal{Z}[\mathbb{1}]$. Eq. 3.16 exactly corresponds to the expression for the ensemble averaged current density, Eq. 2.8. As discussed before, this quantity vanishes except near the sample edge. In fact, Eq. 3.15 can be rewritten as

$$\mathcal{Z}_{GAP}[\mathcal{U}] = \mathcal{Z}_{GAP}[\mathbb{1}]\exp\left\{\frac{i}{2}\sigma^{II}_{xy}\oint dr_\mu \sum_{p,a}[D_\mu]^{pp}_{aa}s^p + ..\right\},$$

(5.3.17)

with σ^{II}_{xy} given in Eq. 2.24. This quantity is integer-valued under the present circumstances.

The integral is along the perimeter of the system. With the aforementioned boundary condition on $\mathcal{U}$, this integral is easily evaluated. One concludes that only the $U(1)$ windings in each replica channel contribute, in other words, the relative phases of the advanced and retarded field variables,

$$\frac{1}{2}\oint dr_\mu \sum_{p,a}\left[\mathcal{U}^{-1}\nabla_\mu\mathcal{U}\right]^{pp}_{aa}s^p = 2\pi i\sum_{a=1}^{m} n_a,\qquad (5.3.18)$$

where n_a denotes the number of times that the relative phase in replica channel a winds around. This final result, then, is consistent with the statement that $\mathcal{Z}$ is invariant under the insertion of these background fields.

We can rewrite the contour integral (5.3.17) as a 2D integral by using Stokes' theorem,

Field Theory, Scaling and Localization

$$\frac{1}{2} \oint dr_\mu \sum_{p,a} [D_\mu]_{aa}^{pp} s^p = \frac{1}{2} \oint d^2r \; \mathrm{tr} \; \mathbf{v} \times \mathbf{D} \hat{\mathbf{s}}, \qquad (5.3.19)$$

which can also be written as

$$-\frac{1}{8} \int d^2r \; \epsilon_{\mu\nu} \; \mathrm{tr} \; \tilde{Q} \, \nabla_\mu \tilde{Q} \, \nabla_\nu \tilde{Q} \; ; \quad \tilde{Q} = \hat{\mathcal{U}} s \hat{\mathcal{U}}^{-1}. \qquad (5.3.20)$$

This result, finally, relates the existence of edge currents with a topological classification of the continuous symmetry. The matrix field $\tilde{Q}$ is a parameterization of the coset space $U(2m)/U(m) \times U(m)$ and is constant near the edge. This leads to the mathematical statement that the 2D plane is being compactified to the sphere S^2 with the edge contracted to a point. Via the homotopy theory result

$$\pi_2 \left[\frac{U(2m)}{U(m) \times U(m)} \right] = Z, \qquad (5.3.21)$$

this associates an integer Z with each field configuration. The topological invariant

$$q(\tilde{Q}) = \frac{i}{16\pi} \int d^2r \; \mathrm{tr} \; \epsilon_{\mu\nu} \, \tilde{Q} \, \nabla_\mu \tilde{Q} \, \nabla_\nu \tilde{Q} = \sum_{a=1}^{m} n_a \qquad (5.3.22)$$

is the Jacobian for the mapping of the two-sphere onto the $\tilde{Q}$ and in fact equals this integer.

5.3.4 Edge states and level crossing

We will next present an interpretation of these rather formal results in the context of the electronic transport. Let us go back to the background field $\mathcal{U}$ and ask what the $U(m) \times U(m)$ gauge implies for the physical boundary states near the Fermi surface. The contributing parts of D_μ in Eqs. 3.15,18 are nothing but a gauge transformation performed on these physical edge states. In the following we will see that this causes a specific number of edge levels to cross the Fermi energy. Together with Eqs. 3.20,22, 2.31 this leads to an identification of topology and flux.

In order to simplify things, we will continue by considering a $U(2)$ background field, which was the object of study in the source-term formalism. Because a $U(2)$ rotation is active in one replica channel only, we can rewrite the expression for $\mathcal{Z}$ (Eq. 3.6) as

$$\mathcal{Z}[\mathcal{U}] = \int \mathcal{D}[V] \, P[V] \, \mathfrak{z}^m(\mathbb{1}_2) \left[\frac{\mathfrak{z}(\mathcal{U})}{\mathfrak{z}(\mathbb{1}_2)} \right]; \quad \mathcal{U} \in U(2) \,, \quad (5.3.23)$$

where $\mathfrak{z}$ is expressed in terms of the original Grassmann fields by

$$\mathfrak{z}(\mathcal{U}) = \int \mathcal{D}\bar{\psi}^+ \int \mathcal{D}\psi^+ \int \mathcal{D}\bar{\psi}^- \int \mathcal{D}\psi^- \, e^{\mathcal{L}_2};$$

$$\mathcal{L}_2 = \int d^2r \sum_{p,p'} \bar{\psi}^p \, (E - \mathcal{U}^{-1}\mathcal{H}_0\mathcal{U} + V(r) + i\eta\tau_z)^{pp'} \psi^{p'} \,.$$

$$(5.3.24)$$

Next we introduce a complete set of orthonormal eigenstates of the Hamiltonian,

$$\mathcal{H} \, \xi_\alpha(r) = [\mathcal{H}_0 + V(r)]\xi_\alpha(r) = E_\alpha \xi_\alpha(r) \quad (5.3.25)$$

and expand the Grassmann fields in terms of the ξ_α

$$\psi^p(r) = \sum_\alpha c_\alpha^p \, \xi_\alpha(r) \,. \quad (5.3.26)$$

This all is defined, of course, for a single realization of the random potential. The $\mathfrak{z}$ can now be written as

$$\mathfrak{z}(\mathcal{U}) = \int \mathcal{D}\bar{C}^+ \int \mathcal{D}C^+ \int \mathcal{D}\bar{C}^- \int \mathcal{D}C^- \, e^{\mathcal{L}_2'};$$

$$\mathcal{L}_2' = \int d^2r \sum_{\alpha,\beta} \sum_{p,p'} \bar{c}_\alpha^p \xi_\alpha^*(r)$$

$$\times (E - \mathcal{U}^{-1}\mathcal{H}_0\mathcal{U} + V(r) + i\eta\tau_z)^{pp'} \xi_\beta(r) c_\beta^{p'} \,, \quad (5.3.27)$$

where we have used the fact that the Jacobian for the

Field Theory, Scaling and Localization

transformation $\psi \to C$ equals unity. This form allows for a discussion in terms of the physical states ξ_α. For those ξ_α which refer to edge states, we can replace the combination

$$\mathcal{U}^{pp'}(r)\xi_\alpha(r) \to \delta_{pp'} e^{i\varphi^p(r)} \xi_\alpha(r) \,, \quad (5.3.28)$$

because only the $U(1) \times U(1)$ gauge will be felt in this case. Next we limit ourselves to a fixed energy interval 2Δ around E, which we have chosen to contain only edge states. We think of Δ as being much larger than the separation between the edge levels. Under the transformation of Eq. 3.28, which is now a gauge transformation on the true spectrum of eigenstates, the set of edge states map onto itself. The manner in which this mapping proceeds can be deduced from the fact that the addition of a phase factor φ^p actually stands for the addition of magnetic flux quanta through the 2D plane for the advanced and retarded quantities separately.

This notion will next be connected up with the situation in which the system as a whole is taken out of thermodynamic equilibrium by the addition (reduction) of magnetic flux quanta. Let us go back to our earlier discussion. In Sect. 5.2 we concluded that under the present circumstances (no bulk states) a change in flux by q quanta (keeping the Fermi level fixed) leads to the removal (addition) of qi electrons from (to) the system with i the number of filled Landau levels. Each Landau level will have separate sets of edge states with which it communicates by charge transfer.

Let us limit ourselves to the situation in which E lies between the first and second Landau level. Then a change in flux by one quantum (keeping E fixed) results in a newly occupied edge state just above the Fermi level or a deletion of an edge state just underneath the Fermi level. Hence, the effect of the above discussed gauge transformation is to relabel the energies of the edge states from E_α to $E_{\alpha+q^p}$. The subscripts on E are arranged such that $E_\alpha < E_{\alpha+1}$ and the integer q^p denotes the number of flux quanta (or abelian windings) in question.

We can now perform the integration of Eq. 3.27 and we focus on the energies in the interval $[E - \Delta, E + \Delta]$,

$$\mathcal{Z}(\mathcal{U}) \Rightarrow \exp\left[\sum_{E_\alpha=E-\Delta}^{E+\Delta} \ln\left[(E - E_{\alpha+q^+} + i\eta)(E - E_{\alpha+q^-} - i\eta)\right]\right].$$
(5.3.29)

The effect of the shift can be studied by expanding $E_{\alpha+q}{}^p$ about E_α. Namely,

$$E_{\alpha+q^p} = E_\alpha + \rho_e^{-1} q^p + \ldots, \qquad (5.3.30)$$

where ρ_e denotes the density of edge states at the Fermi level, which is proportional to the size of the system L. We can now take the continuum limit

$$\sum_{E-\Delta < E_\alpha < E+\Delta} \to \int_{E-\Delta}^{E+\Delta} dE' \, \rho_e(E') \qquad (5.3.31)$$

and write for the ratio

$$\frac{\mathcal{Z}(U)}{\mathcal{Z}(1_2)} = \exp\left[\int_{E-\Delta}^{E+\Delta} dE' \rho_e(E') \left[\frac{q^+/\rho_e}{E-E'+i\eta} + \frac{q^-/\rho_e}{E-E'-i\eta}\right] + \ldots\right]$$

$$= \exp\left[\int_{E-\Delta}^{E+\Delta} dE' \, 2\pi i \delta(E-E') n\right] = \exp(2\pi i n) = 1, \quad (5.3.32)$$

where only the relative shift $(q^+, q^-) = (-n, +n)$ contributes. The higher order terms in the expansion vanish in the thermodynamic limit. We identify the integer n with the topological term found in the previous section,

$$n = \sigma_{xy} q(\tilde{Q}), \qquad (5.3.33)$$

where for the case considered $\sigma_{xy} \to \sigma_{xy}{}^{II}$ and $\sigma_{xy}{}^{II} = 1$. This equality has exactly the same meaning as Eq.

Field Theory, Scaling and Localization

2.31 and leads to the conclusion that the topological invariant should be identified with (adiabatically) applied flux quanta.

We can extend this result in several directions. First, we repeat the same argument for the Fermi energy between two higher Landau levels. The left hand side of Eq. 3.33 then becomes ni with i the number of Landau levels below the Fermi energy. In fact, we can go one step further and identify the Hall conductance with the integer i and the topological charge q with n.

Secondly, we consider the situation in which the Fermi energy is located in a region of localized bulk levels. The situation is slightly more complicated in that the exponential in Eq. 3.27 now contains three distinct pieces. One can separate the bulk localized wave functions from the edge states because by the very construction of $\mathcal{U}$ there are no matrix elements generated which mix these types of states. Our basic assumption is that the bulk part of $\mathcal{U}$ does not affect the bulk localized levels, a statement which will be given a separate justification below. Furthermore, we have to distinguish between two types of extended edge states. From our earlier analysis of the diamagnetic edge currents, we must conclude that there always exist extended edge states, i.e., whenever there are bulk levels present and also in between Landau levels.

Now, let us focus on the set of edge levels which participates in the charge transfer from the bulk to the edge and vice versa. Those states give rise to exactly the same number of level crossings as considered before and the situation remains identically the same. However, the other extended edge states also may cross the Fermi energy under a gauge transformation and hence contribute to the expansion of Eq. 3.32. These edge states can be divided into subsets, each one of which maps onto itself under a gauge transformation. Taking the effect of all of these subsets together, there will be, in general, a net number of levels crossing the Fermi energy as a result of a gauge transformation. This number is a statistical quantity as it will be different for different realizations of the random potential. So the ensemble averaged ratio of Eq. 3.32 can be translated into a statistical sum over level crossings, which schematically reads

$$\mathcal{Z}[\mathcal{U}]_{m \to 0} = \int \mathcal{D}[V]\, P[V]\, \mathcal{F}(\mathcal{U})/\mathcal{F}(\mathbb{1}_2)$$
$$= e^{2\pi i n} \sum_{\tilde{n}} \tilde{P}(\tilde{n})\, e^{2\pi i \tilde{n}}\ . \qquad (5.3.34)$$

Alternatively, however, we could introduce the flux quantum through the 2D plane as a statistical variable in order to reach the very same effect. This, then, is our interpretation of the topological invariant as a statistical object, which becomes important in the process of electronic diffusion (localization).

Finally, we mention the generalization of the $U(2)$ background field results to $U(2m)$. The only difference lies in the fact that in Eq. 3.24, one now has to provide the fluctuating fields ψ^p with a subscript a which runs from 1 to m; the same holds for the expansion coefficients $C_\alpha{}^p$. Furthermore, all of the $U(2m)$ degrees of freedom can be absorbed into a redefinition of the fields ψ^p in Eq. 3.24, except for the $U(2)$ subgroups which are active in each replica channel individually. In other words, we end up with m copies of the same theory, leading finally to a replacement of n in Eq. 3.31 by n_α, i.e., the sum of the independent n_α Fermi level crossings (including the sign) which are defined in each replica system individually. Comparing this result with Eqs. 3.18 and 3.22, we conclude that the topological charge in the $U(2m)$ theory represents the total of flux change which is adiabatically applied to m individual copies of the theory independently. From this one can gain some physical intuition as to how the replica trick manages to deal with nonperturbative phenomena.

5.3.5 Localization and continuous symmetry

The final issue is to relate the continuous symmetry with the concept of localization. Field-theoretically one interprets localization of the electronic wave functions as the occurrence of a mass gap in the theory. The absence of massless excitations (Goldstone modes) implies that the symmetry is (dynamically) restored at large length-scales (i.e. large compared to the localization length). This translates in the

Field Theory, Scaling and Localization

mathematical statement for $\mathcal{Z}$ [Eq. 3.6]

$$\mathcal{Z}(\mathcal{U}) = \mathcal{Z}(\mathbb{1}) \qquad (5.3.35)$$

for an arbitrary but slowly varying field $\mathcal{U}(r)$. The consequences of such a statement for the electronic system can be studied by means of a gradient expansion in $\mathcal{U}(r)$ as done in Sect. 5.3.3. This will serve as a further justification of the picture of level crossing of the previous section. It is a reformulation of the IQHE in field-theoretic language. The aforementioned gradient expansion proceeds in a fashion outlined in (Pruisken, 1984a) and is somewhat complex and tedious. Here we give the outcome of the analysis and make the result plausible. To second order in the derivative we can write for Eq. 3.6

$$\mathcal{Z}[\mathcal{U}] = \mathcal{Z}[\mathbb{1}]\exp[\mathcal{L}] ,$$

$$\mathcal{L} = -\frac{1}{8}\sigma_{xx} \operatorname{Tr} \nabla_\mu \tilde{Q} \nabla_\mu \tilde{Q} + \frac{1}{8}\sigma_{xy} \operatorname{Tr} \epsilon_{\mu\nu} \tilde{Q} \nabla_\mu \tilde{Q} \nabla_\nu \tilde{Q} ,$$

$$[\tilde{Q}]_{ab}^{pp'} = [\mathcal{U}\hat{s}\mathcal{U}^{-1}]_{ab}^{pp'} . \qquad (5.3.36)$$

First we remark that the result is expressed in terms of the matrix fields $\tilde{Q}$ alone. This rigorously displays the local gauge invariance of the original theory, i.e., the invariance under the transformation $\mathcal{U} \to \mathcal{U}\mathcal{V}(r)$, $\mathcal{V}(r) \in U(m) \times U(m)$. Furthermore, there is the global symmetry $\mathcal{U} \to T\mathcal{U}$ with $T \in U(2m)$ which is consistent with the original formulation.

In fact, Eq. 3.36 is the most general form in $\tilde{Q}$ that one can write down which has two derivatives and is consistent with the symmetries of the problem. For the identification of the parameters $\sigma_{\mu\nu}$ however, we have to follow a different route. Their meaning follows by inserting the special form for $\mathcal{U}$, Eq. 3.7, in which case Eq. 3.36 can be worked out to give

$$\mathcal{Z}[\mathcal{U}] = Z[\mathbb{1}]\exp[-\sigma_{xx}(j_x^2 + j_y^2)L^2 + 2i\sigma_{xy}j_xj_yL^2 + \mathcal{O}(j^3)].$$

$$(5.3.37)$$

By comparing the right hand side of Eq. 3.37 with Eq.

3.11, we conclude that the parameters $\sigma_{\mu\nu}$ in Eq. 3.36 are indeed the exact conductance formulae.

A necessary condition for Eq. 3.36 to hold is that the dissipative conductance, σ_{xx}, vanishes. This is consistent with the *ansatz* of localization. Furthermore, integer quantization of the topological charge, Eq. 3.22, implies integer quantization of the Hall conductance. Hence, we have formulated the concept of localization and the IQHE as being consequences of one and the same principle, which is that of symmetry restoration. Notice that the statements made are quite general and do not rely on the details of the random potential. Furthermore, the argument does not distinguish between the situations in which E is located in a mobility gap or in a density-of-states gap, consistent with the findings in the previous sections.

5.4 Structure of the Effective Field Theory

5.4.1 Introduction

The development of an effective field theory has been presented in (Pruisken 1984). The methodology largely follows up the exact analysis on conventional Anderson localization (Pruisken and Schaefer 1981, 1982) and we will proceed by summarizing the main steps in the σ-model approach to localization.

Going back to our original formulae, Eqs. 3.2,3, we can perform the integration over randomness [V], which for the simple white-noise distribution, Eq. 2.4, can be done explicitly. This leads to a 4-point interaction

$$\tfrac{1}{2}g \int d^2 r \sum_{pp'} \sum_{ab} \bar{\psi}_a^p(r)\,\psi_a^p(r)\,\bar{\psi}_b^{p'}(r)\,\psi_b^{p'}(r) \quad , \qquad (5.4.1)$$

which is next to be decoupled by a Gaussian transformation, introducing (bosonic) Hermitian matrix fields $Q(r)$:

Field Theory, Scaling and Localization

$$\int \mathcal{D}[Q] \exp\left[-\frac{1}{2g}\int d^2r \; \text{tr} \; Q^2 + i\int d^2r \sum \bar{\psi}_a^p Q_{ab}^{pp'} \psi_b^{p'}\right]$$

$$\propto \exp\left[\frac{g}{2}\int d^2r \sum_{pp'} \sum_{ab} \bar{\psi}_a^p(r) \psi_a^p(r) \bar{\psi}_b^{p'}(r) \psi_b^{p'}(r)\right]. \quad (5.4.2)$$

We remark that such a change of variables preserves the symmetries of the problem, i.e. $Q \to \mathcal{U}^{-1} Q \mathcal{U}$ with $\mathcal{U} \in U(2m)$ leaves the integration measure for the Q invariant. From the discussions in the previous sections, it will be clear that a formulation in terms of the Q is truly advantageous. It naturally incorporates the symmetries of the problem and allows for an explicit analysis of the Goldstone modes.

The remaining integral over fermion fields is Gaussian and leads to the result

$$\mathcal{Z} = \int \mathcal{D}[Q] \exp\left[-\frac{1}{2g} \text{Tr} \; Q^2 + \text{Tr} \; \ln[E - \mathcal{H}_0 + iQ + i\eta \hat{s}]\right].$$

$$(5.4.3)$$

The various transformations should be complemented by the introduction of source terms, from which the correlations of physical interest can be obtained in terms of the Q variables. For instance, it is a rather simple exercise to see that the expectation of the Q field involves the exact density of states in the problem [in the appropriate limit of vanishing number of field components]:

$$\frac{i}{g} \langle Q_{ab}^{pp'}(r) \rangle = \delta_{ab}^{pp'} [e + i\pi s^p \rho(E)]. \quad (5.4.4)$$

Here, the expectation is with respect to the complete theory. The quantity e is the real part of the shift, which is of secondary interest.

On the other hand, the Lagrangian exhibits a stationary point which precisely corresponds to the previously mentioned SCBA or CPA [Eq. 2.5], namely

$$i[Q_{sp}]_{ab}^{pp'} = \delta_{ab}^{pp'}(e_0 + is^p/2\tau), \quad (5.4.5a)$$

and the unknowns e_0 and τ are determined by the CPA equation

$$e_0 + is^p/2\tau = -\frac{g}{2\pi} \sum_n (E - E_n + e_0 + is^p/2\tau)^{-1} .$$

(5.4.5b)

The $1/2\pi g\tau$ will be recognized as the density of states in SCBA. This result is suggestive for the identification of the massless excitations [Goldstone modes] in the problem. One can easily check that the replacement

$$iQ_{sp} \to e_0 \mathbb{1} + (i/2\tau)[T^{-1}\hat{s}T]$$

(5.4.6)

also fulfills the saddle-point equation (for $\eta \to 0$). Hence, the notion of spontaneous symmetry breaking by a nonzero density of bulk states is naturally incorporated in the Q-field formalism. Moreover, one can next proceed in a conceptually clean fashion and separate the different pieces in the Hermitian Q fields by the following change of variables,

$$Q(r) \to T^{-1}(r)P(r)T(r),$$

(5.4.7)

where $T(r) \in U(2m)$ and

$$P(r) = P^\dagger(r) \equiv P_{ab}^p(r)$$

is block diagonal in the p-indices.

This result will serve as the basic starting point for an exact discussion on the effective field theoretic description and was introduced first in the analysis of the Anderson model (Pruisken and Schaefer 1981, 1982). In terms of the new field variables, the generating function can be written

$$\mathcal{Z} = \int \mathcal{D}[T] \int \mathcal{D}[P] \, I[P] \, e^{\mathcal{L}[P,T]} ,$$

$$\mathcal{L}[P,T] = \frac{-1}{2g} \text{Tr } P^2 + \text{Tr ln } [E - T\mathcal{H}_0 T^{-1} + iP + i\eta T\hat{s}T^{-1}] ,$$

(5.4.8)

Field Theory, Scaling and Localization

where the Jacobian of the transformation $I[P]$ takes on the form

$$I[P] = \prod_{i,j} \left[\lambda_i^1 - \lambda_j^2\right]^2 , \qquad (5.4.9)$$

with λ_i^p the eigenvalues of the matrix fields P^p. By construction, the fields $T(r)$ are recognized as the critical field components.

The longitudinal components $P(r)$ can be studied by first making the shift

$$P(r) \rightarrow P(r) + Q_{sp} \qquad (5.4.10)$$

and subsequently studying the fluctuations about the stationary point. Order by order in a loop expansion about the saddle point, one finds that the P-fluctuations give rise to short range and hence noncritical correlations (Pruisken 1984).

5.4.2 Critical fluctuations

This last result gives rise to the notion that an effective field theory for the critical fluctuations T can be constructed by formally integrating over the massive longitudinal components P. Schematically one arrives at

$$e^{\mathscr{L}_{eff}[T]} = \mathscr{Z}_0^{-1} \int \mathscr{D}[P]\, I[P]\, e^{\mathscr{L}[T,P]} . \qquad (5.4.11)$$

From Eq. 3.14 one sees that $\mathscr{L}_{eff}$ can be obtained exactly, order by order in powers of the derivative acting on the T. Such an expansion proceeds with respect to the theory involving the short range P fluctuations alone and is therefore well defined. More specifically, the zeroth order theory with respect to which the T fluctuations are weighted, can be written as

$$\mathscr{Z}_0 = \int \mathscr{D}[P]I[P]\exp\left[-\frac{1}{2g}\mathrm{Tr}P^2 + \mathrm{Tr}\,\ln(E - \mathscr{H}_0 + iP + i\eta\hat{s})\right] .$$

$$(5.4.12)$$

We remark that this theory very much resembles the one appropriate for the one-particle Green's functions. Namely, except for the measure $I[P]$, the advanced and retarded subspace are completely decoupled. In fact, in the absence of $I[P]$, one can transform the theory back to the variables $\bar{\psi},\psi$ and to the random potential $V(r)$, which now acts independently in the advanced and retarded spaces. Furthermore, we mention the fact that for those expectations, which involve only one of the p-indices, the measure $I[P]$ drops out in the limit of zero replicas. These expectations therefore reduce to the exact physical ones. We anticipate that the zeroth order theory exhibits diamagnetic edge currents and hence long range behavior along the perimeter of the system. This has direct consequences for the topologically allowed field configurations T and makes the analysis much more tricky and demanding than usual.

I will nevertheless proceed by presenting a simplified but less precise version of the derivation of (Pruisken 1984), followed by a discussion on the peculiarities involved. As mentioned before in a different context, the form of $\mathcal{L}_{eff}$ is very much restricted by symmetry considerations. Namely, if $\mathcal{L}_{eff}$ can be expressed in terms of the local variable $\tilde{Q} = T^{-1}sT$, which correctly incorporates the local $U(m) \times U(m)$ gauge invariance, then up to two gradients acting on the field $\tilde{Q}(r)$ the allowed form of $\mathcal{L}_{eff}$ can be written

$$\mathcal{L}_{eff}[T] = -\frac{1}{8}\sigma_{xx}^0 \int d^2r \; \mathrm{tr} \; \nabla_\mu \tilde{Q} \nabla_\mu \tilde{Q}$$

$$+ \frac{1}{8}\sigma_{xy}^0 \int d^2r \; \mathrm{tr} \; \epsilon_{\mu\nu} \tilde{Q} \nabla_\mu \tilde{Q} \nabla_\nu \tilde{Q} + \eta\pi\rho(E) \; \mathrm{Tr}\widehat{\tilde{Q}s} \; . \qquad (5.4.13)$$

We have added a symmetry breaking term [proportional to η] to the result, which [necessarily] involves the exact density of states in the problem. The coefficients can be obtained by making use of the source term formalism, analogous to that in Sect. 5.3.5. Namely, inserting the special expression for T, Eq. 3.7, into both sides of Eq. 4.11, it follows after differentiation that

Field Theory, Scaling and Localization

$$\sigma_{xx}^0 = -\frac{1}{\Omega} \text{Tr} \langle \Pi_x G_{aa}^+ \Pi_x G_{aa}^- \rangle_0 + \text{Tr} \langle G_{aa}^+ + G_{aa}^- \rangle_0,$$

$$\sigma_{xy}^0 = -\frac{1}{2\Omega} \text{Tr} \, \epsilon_{\mu\nu} \langle \Pi_\mu G_{aa}^+ \Pi_\nu G_{aa}^- \rangle_0$$

$$+ \frac{i}{2\Omega} \text{Tr} \langle \epsilon_{\mu\nu} r_\mu \Pi_\nu (G_{aa}^+ - G_{aa}^-) \rangle_0;$$

$$G_{ab}^p(r,r') = \langle r | [(E - \mathcal{H}_0 + iP)^{-1}]_{ab}^p | r' \rangle . \quad (5.4.14)$$

The average is with respect to the zeroth order theory, i.e. $\mathcal{X}_0$, and there is no sum on a. We note from the application of the source term formalism, that the parameters $\sigma^0_{\mu\nu}$ get the meaning of "bare" conductances as defined by the zeroth order theory.

In order to further elucidate this meaning, we need the following three advances:

1) The expression for σ^0_{xx} is not entirely written in terms of a current-current correlation function. However, from local $U(m) \times U(m)$ gauge invariance one can derive a sum rule which relates the one-particle propagator to a current-current correlation function [involving one of the p-subspaces] leading to the result (Pruisken 1984)

$$\sigma_{xx}^0 = -\frac{1}{2\Omega} \int d^2r \int d^2r'$$

$$\times \left[\langle \Pi_x [G_{aa}^+(r,r') - G_{aa}^-(r,r')] \Pi_x [G_{bb}^+(r',r) - G_{bb}^-(r',r)] \rangle_0 \right.$$

$$\left. + \sum_{p=\pm} \langle \Pi_x G_{ab}^p (r,r) \Pi_x G_{ba}^p (r',r') \rangle_0^{\text{cum}} \right] . \quad (5.4.15)$$

This form is very transparent. We mention that the additional cumulant expressions originate from a translation of statistical variables from $V(r)$ into $P(r)$.

2) The expression for the Hall conductance takes a form analogous to the exact expression [Eq. 2.24]. The term involving the magnetization operator is, as before, nonzero only by virtue of the edge currents in

the bare theory. By the same manipulations as discussed under Eq. 2.24, this can be reexpressed as a bulk quantity. The final form reads

$$\sigma^0_{xy} = [\sigma^I_{xy}]^0 + [\sigma^{II}_{xy}]^0 \;;$$

$$[\sigma^I_{xy}]^0 = -\frac{1}{2\Omega}\int d^2r \int d^2r' \; \epsilon_{\mu\nu} \langle \Pi_\mu G^+_{aa}(r,r') \Pi_\nu G^-_{aa}(r',r) \rangle_0$$

$$[\sigma^{II}_{xy}]^0 = \frac{c}{ie}\left[\int_{-\infty}^{E} dE^{(+)} \frac{\partial}{\partial B^{(+)}} \langle G^+_{aa}(r,r) \rangle_0\right.$$

$$\left. - \int_{-\infty}^{E} dE^{(-)} \frac{\partial}{\partial B^{(-)}} \langle G^-_{aa}(r,r) \rangle_0 \right] , \qquad (5.4.16)$$

where the superscripts ± on the E and B indicate that the various operations ought to be performed on the advanced and retarded subspaces separately.

3) Eqs. 4.15,16 are defined for the replicated field theory described by the zeroth order Lagrangian (5.4.12). Together with expression (5.4.13) for $\mathcal{L}_{eff}$, this constitutes the exact derivation for the effective field theory.

From this point on one needs to resort to approximations. For the parameters $\sigma^0_{\mu\nu}$ and $\rho^0(E)$ we will assume that the limit $m \to 0$ can be taken independently of the number of components of the $\tilde{Q}(r)$. In fact, one can study these parameters perturbatively, order by order in a loop expansion about the saddle point described by $\mathcal{Z}_0$. In a later section we will show that in the saddle-point approximation, one recovers the semiclassical results (see Ando et al. 1982 and Středa 1982) which are expected to give a good approximation to the exact expressions of Eqs. 4.15,16. Furthermore, one has to check whether the relation between the bulk and edge expressions for $[\sigma^{II}_{xy}]^0$ also holds in the perturbative treatment [see Eq. 2.21] such that the contact with the diamagnetic edge currents is retained order by order in the loop expansion. We will not pursue this question further here and refer to the literature (Pruisken and Wegner, unpublished).

5.4.3 Boundary conditions

One of the intriguing aspects of the problem concerns the question of the topologically allowed field configurations $\tilde{Q}(r)$ in the effective field theory. From what has been discussed before, it is clear that this question boils down to asking which boundary conditions one has to impose upon the critical fields $\tilde{Q}(r)$. We will not dwell upon this question here in purely mathematical terms, however. We have already seen that integer topological charge is special in that the corresponding field configurations are the only ones which do not perturb the [in this case diamagnetic] edge current. In fact, it is argued that these are the only ones allowed in the functional integral in order not to lose contact with the physics of the problem (Pruisken 1985a,b, Levine, Libby and Pruisken 1984a,b,c).

This statement arises from the requirement that a vanishing density of states in the bulk should be a valid limit in the field theory. In terms of the effective Lagrangian we remark that the limit under consideration consists of putting $\rho(E)$, $\sigma^0{}_{xx} = 0$ and $\sigma^0{}_{xy} = integer$. But notice that we have presented a formulation in which the specific edge effects are eliminated. Hence, the thermodynamic limit can be discussed without any reference to edges. Hence also, the presence of the massless components $T(r)$ should be immaterial in case there is no spontaneous symmetry breaking, i.e. a vanishing density of states. This then implies that only integer topological charge is allowed in $\mathcal{L}_{eff}$. This infinite space situation is modelled by a finite system with edges by imposing the previously discussed special boundary conditions upon the $\tilde{Q}$ fields. The same conclusion should of course hold for any value of the parameters $\sigma^0{}_{\mu\nu}$, a statement which physically can be understood from the fact that edge currents do exist whenever $\sigma^0{}_{xy} \neq 0$.

5.4.4 The conductance in mean field (SCBA)

To lowest order, one can insert the mean field propagators in the expressions for $\sigma^0{}_{\mu\nu}$ and neglect the cumulants. This amounts to the replacement

$$G^p_{ab}(r,r') \to \delta_{ab} \langle r|(E - \mathcal{H}_0 + e_0 + is^p/2\tau)^{-1}|r'\rangle$$

$$= \sum_{k,n} e^{ik(y-y')} \frac{\xi^*(x-k)\xi(x'-k)}{E - E_n + e_0 + is^p/2\tau} \quad . \quad (5.4.17)$$

The integrations over r and k can be worked out to lead to the following expressions

$$\sigma^0_{xx} = \frac{-1}{4\pi} \sum_{n,m} D^{n,m}_x D^{m,n}_x \sum_{p,p'} (-)^{p+p'} \frac{1}{\tilde{E}_p - E_n} \frac{1}{\tilde{E}_{p'} - E_m} \; ,$$

$$[\sigma^I_{xy}]^0 = \frac{i}{4\pi} \sum_{n,m} D^{n,m}_x D^{m,n}_y \left[\frac{1}{\tilde{E}_+ - E_n} \frac{1}{\tilde{E}_- - E_m} \right.$$

$$\left. - \frac{1}{\tilde{E}_- - E_n} \frac{1}{\tilde{E}_+ - E_m} \right] , \quad (5.4.18)$$

where $\tilde{E}_p \equiv E + e_0 + is^p/2\tau$. The two matrix elements

$$D^{n,m}_x = \int dx \, \xi^*_n(x) \frac{d}{dx} \xi_m(x)$$

and

$$D^{n,m}_y = \int dx \, \xi^*_n(x) \, x \, \xi_m(x)$$

are just those of the familiar creation and annihilation operators for the harmonic oscillator:

$$D^{n,m}_x = (\tfrac{m}{2})^{1/2} \delta_{n,m-1} - (\tfrac{m+1}{2})^{1/2} \delta_{n,m+1}$$

$$D^{n,m}_y = (\tfrac{m}{2})^{1/2} \delta_{n,m-1} + (\tfrac{m+1}{2})^{1/2} \delta_{n,m+1} \quad . \quad (5.4.19)$$

It is next a question of straightforward algebra to show that Eqs. 4.18 reduce to

$$\sigma^0_{xx} \times [\tfrac{e^2}{h}] = \frac{(E+e_0)\rho_0 e^2 \tau}{m} \frac{1}{1+(\omega_c \tau)^2}$$

$$[\sigma_{xy}^I]^0 \times [\tfrac{e^2}{h}] = -\omega_c \tau\, \sigma_{xx}^0$$

$$[\sigma_{xy}^{II}]^0 \times [\tfrac{e^2}{h}] = -ec\frac{\partial n_0}{\partial B} = -ec\int_{-\infty}^{E} dE\, \frac{\partial}{\partial B} \rho_0(E) \quad (5.4.20)$$

with ρ_0 the density of states in the saddle-point approximation. These results have been given first by Středa (1982). They reproduce the classical (kinetic theory) results in the limit $\omega_c \tau \to 0$, i.e., weak fields or strong disorder, where the density of states becomes B-independent:

$$\sigma_{xx}^0 \times [\tfrac{e^2}{h}] \simeq \frac{ne^2 \tau}{m}\, \frac{1}{1+(\omega_c \tau)^2}$$

$$\sigma_{xy}^0 \simeq -\omega_c \tau\, \sigma_{xx}^0\,; \qquad (\omega_c \tau \ll 1). \quad (5.4.21)$$

On the other hand, Eqs. 4.20 are not very transparent in the case of strong magnetic fields ($\omega_c \tau \gg 1$). However, from the saddle-point equation (5.4.5) one can derive the following result

$$\frac{\partial \rho_0}{\partial B} - \frac{\rho_0}{B} + \frac{1}{B}\frac{\partial}{\partial E}\left[(E+e_0)\rho_0\right] = 0\,. \quad (5.4.22)$$

Together with the expressions (5.4.20), this leads to the following form for the bare Hall conductance

$$\sigma_{xy}^0 \times [\tfrac{e^2}{h}] = -\frac{nec}{B} + \frac{1}{\omega_c \tau}\, \sigma_{xx}^0 \times [\tfrac{e^2}{h}]\,. \quad (5.4.23)$$

This reproduces (1.3.4) and constitutes the semiclassical theory (see Ando et al. 1982) obtained from Kubo's migration center formalism. A simple interpretation of the strong-field results in terms of the kinetic theory has been given by Ando and Uemura (1974). The first term in Eq. 4.23 can be understood from the drift velocity of the free electron when made subject to a magnetic and electric field: $v_x = cE_y/B$. The second term in (5.4.23) is interpreted as the result of a frictional force $\delta F_x = -m v_x/\tau$ due to impurity scattering which causes a drift in the Hall current by an amount $\delta j_x = \sigma^0_{xx} \delta F_x/(-e)$. The degeneracy per Landau level, $eB/\hbar c$ is such that in the strong field limit, σ^0_{xy} reduces to the number of fully occupied Landau levels (i) plus the filling fraction (v) of the partly occupied one. Furthermoe, in the same limit of strong fields, the maximum in the dissipative conductance for Landau level index n occurs at the center of the band $E = \hbar \omega_c (n + \frac{1}{2})$ and is given by

$$[\sigma^0_{xx}]^{max} = \frac{2}{\pi}(n + \frac{1}{2}) ,$$

$$\sigma^0_{xy} \simeq -i-v \quad ; \qquad (\omega_c \tau \gg 1) , \qquad (5.4.24)$$

a result which is independent of magnetic field strength and impurity potential variance g. A simple picture can be given in terms of diffusing cyclotron orbits. The diffusion is caused by jumps in the orbits due to scattering. For short-range correlated impurity potentials, the characteristic jump-distance is on the order of the cyclotron radius $\ell_n = \ell(2n+1)^{1/2}$ (with ℓ the corresponding quantity in the lowest Landau level). The characteristic time equals τ. This gives rise to a diffusion constant $D \simeq \ell_n^2/\tau$. On the other hand, the maximum in the density of states can roughly be understood from the uncertainty principle

$$\rho_0^{max} \simeq n_B/\Delta E = (1/2\pi\ell^2)(\tau/\hbar) , \qquad (5.4.25)$$

where the degeneracy per Landau level per unit area is $n_B = 1/2\pi\ell^2$. The Einstein relation between diffusion and conduction then gives

$$[\sigma^0_{xx}]^{max} \times [\frac{e^2}{\hbar}] = e^2 \rho_0^{max} D \simeq \frac{e^2}{\hbar}(n + \frac{1}{2}) , \qquad (5.4.26)$$

which agrees with Eq. 4.24 up to a numerical factor.

In summary, both limits of strong and weak magnetic fields give physically simple answers for the bare conductance parameters. In the intermediate regime one has to resort to numerical means. In Figs. 5a,b,c

Field Theory, Scaling and Localization 157

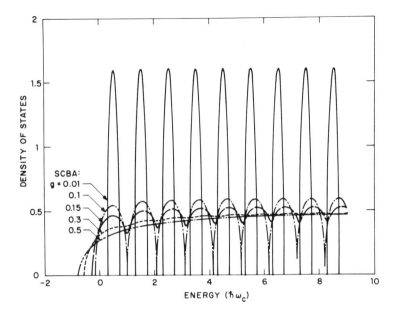

Figure 5.5a. SCBA density of states.

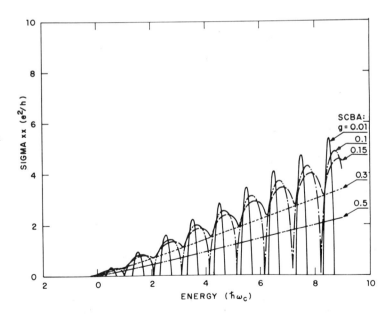

Figure 5.5b. SCBA dissipative conductance.

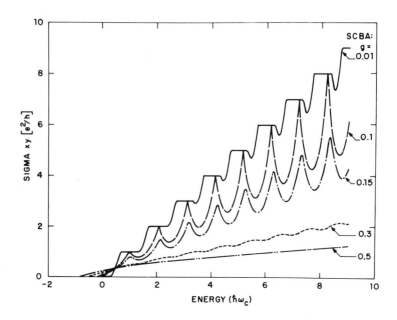

Figure 5.5c. SCBA Hall conductance for different values of the disorder parameter g.

the results are plotted for various values of the dimensionless variance (or "dirt") parameter g [$= g/2\pi\ell^2(\hbar\omega_c)^2$] as a function of Fermi energy. One observes that the bare Hall conductance shows a somewhat peculiar behavior, interpolating between the extremes of clean (or strong field) and dirty (or weak fields). We will come back to these results at a later stage.

5.5 Instantons and Scaling

5.5.1 Conventional localization

We have now completed the discussion on the derivation of the effective field theory. We next turn to the scaling implications of this theory.

The first term in $\mathscr{L}_{eff}$ is a generalization of a familiar model, the CP^{N-1} model, which can be obtained by putting $\tilde{Q}$ an element of the coset space $U(N)/U(N-1) \times U(1)$. This field theory is well known to apply to localization problems in which the time reversal symmetry is broken. The results are similar to

Field Theory, Scaling and Localization

the σ-model with orthogonal symmetries, describing potential scattering (Anderson model).

Generalizations of σ-models to coset spaces have been solved by Brézin et al. (1980) and Hikami (1981) in a $2 + \epsilon$ expansion. The basic result is contained in the renormalization group equation, known to fourth order for our unitary model

$$\beta = \frac{\partial t}{\partial \ln L} = -\epsilon t + 2mt^2 + 2(m^2+1)t^3 + (3m^2+7)m^2 t^4 + \ldots \tag{5.5.1}$$

where $\epsilon = D-2$ and $t = 1/2\pi\sigma^0_{xx}$. At this stage the analytic continuation to zero field-components $(m=0)$ can be performed. The theory is "asymptotically free" in $D=2$ and all $m \geq 0$. This gives rise to the belief in "weak" localization at short length scales (determined by the underlying short-range theory) and "strong" localization at large length scales. In $2+\epsilon$ dimensions there is a fixed point t_c of order $\epsilon^{\frac{1}{2}}$ as $m \to 0$.

The starting point for scaling in t [or σ^0_{xx}] is a smooth function of energy E such that one can associate a mobility edge E_c with this fixed point which separates metallic from insulating behavior. For t below t_c the coupling $[t]$ scales toward the infrared stable fixed point at $t=0$ after successive renormalization group transformations. This fixed point controls the Goldstone phase of broken symmetry and describes a vanishing mass or infinite localization length. On the other hand, if $t > t_c$, the coupling scales toward the strong coupling regime which is characterized by a dynamically generated mass [finite localization length] and a restoration of the continuous symmetry. Near E_c one has the scaling behavior of the conductance

$$\sigma_{xx} \sim L^{D-2} |E-E_c|^s f_0(\eta/|E-E_c|^{Dv}) . \tag{5.5.2}$$

The correlation length exponent $v = \beta'(t_c)$ and conductivity exponent $s = (D-2)v$ are given in $2+\epsilon$ dimensions by

$$v^{-1} = 2\epsilon + O(\epsilon^2); \quad s = \frac{1}{2} + O(\epsilon) , \tag{5.5.3}$$

and the situation is completely analogous to the physics of the Heisenberg ferromagnetic (Wegner 1979a,b).

In two dimensions, the fixed point reduces to $t=0$ and the Goldstone phase disappears all together. For the disordered electronic system, this leads to the well known statement that all states are localized. We mention that there is, however, an important difference with the physics of the Heisenberg ferromagnet. This difference is contained in the unconventional fluctuations in the order parameter $\rho(E)$ which have no analogue in ferromagnetic language and which are, in fact, forbidden as a consequence of the Mermin-Wagner (1966) or Coleman (1973) theorem.

We have used the notion that the density of states is noncritical. This is reproduced in the field theory in that the conjugate critical operator $\tilde{Q}$s in $\mathcal{L}_{eff}$ has a vanishing critical exponent in the limit $m=0$. Furthermore, fluctuations in the local density of states are related to the inverse of the participation ratio of the states near the Fermi level. These fluctuations should diverge as one approaches E_c from the metallic side. This has led Wegner (1979b) to an analysis of the possible singularities in the participation ratios as predicted by the field theory relevant for the original Anderson model. This would then serve as an important check to see whether the field-theoretic approach to localization phenomena gives a complete and consistent account of the physics of the problem.

Specifically, one expects the scaling behavior

$$\overline{\rho^2(E,r)}^{Av} \simeq |E-E_c|^{-\mu_2} \tilde{f}(\eta/|E-E_c|^{Dv}) \qquad (5.5.4)$$

which in σ-model language is given as the expectation of a local operator bilinear in the $\tilde{Q}$. This operator should be characterized by a negative critical exponent, a peculiarity which should be entirely a result of the replica limit. Indeed, for conventional Anderson localization, Wegner found the following result for the exponent:

$$\mu_2 = 2 + O(\epsilon)$$

making use of the field-theoretic work by Brézin et al. (1976). Recently, the critical exponent for the unitary σ-model has been calculated (Pruisken 1985a) and the answer is

$$\mu_2 = (2\epsilon)^{-\frac{1}{2}} + O(\sqrt{\epsilon}) \ . \tag{5.5.5}$$

In a sense, this result can be regarded as an argument by which conventional localization theory consistently proves its own shortcomings. That is, the field theory in the absence of the topological term was originally suggested also to apply to the localization problem in the presence of a constant magnetic field. This would again suggest complete localization in two dimensions, a result which was shared by the various schools of thought in dealing with renormalization group improved diagrammatic analysis of impurity scattering. However, from what has been discussed in the previous sections, it will be clear that the physics of the QHE is entirely contained in the topological piece of the effective Lagrangian which is not amenable to conventional analyses.

5.5.2 Localization in a magnetic field

The study of the topological invariant has been the subject of the papers by Levine, Libby and Pruisken (1983, 1984a,b,c) and Pruisken (1985b,c) and such aspects as scaling of the conductance parameters are far less obvious than in the above mentioned conventional theory. The topological term forms a 2D realization of the term which originally arose in 4D gauge theories [for a review see Rajaraman 1982]. In fact, the two-dimensional field theory shares many properties with nonabelian gauge theories (e.g. conformal invariance of the action, asymptotic freedom, instantons, a parameter N allowing for the possibility of a $1/N$ expansion).

In its explicit form, $\mathcal{L}_{eff}$ has been introduced before in a field-theoretic context, in a search for simpler field theories which show as much resemblance as possible to the world of 4D gauge theories. As one of the major advances for the condensed matter problem, it has been pointed out that the field theory supports nonperturbative topological excitations [instantons] which have to be included in the "weak" localization perturbative analyses in order to detect the presence of the Hall conductance in the problem. According to the inequality

$$\text{tr } [\nabla_x \tilde{Q} + i\tilde{Q} \nabla_y \tilde{Q}]^2 \geq 0 , \qquad (5.5.6)$$

there is a minimum for the action in each topological sector of the theory

$$\int d^2 r \text{ tr } \nabla_\mu \tilde{Q} \nabla_\mu \tilde{Q} \geq 16\pi |q(\tilde{Q})| \qquad (5.5.7)$$

where $q(\tilde{Q})$ is the previously introduced integer-valued Pontryagin index.

Field configurations for which the inequality is fulfilled as an equality are called "instantons" (or "anti-instantons"). These are *local minima* for each topological sector with finite action and furthermore are characterized by a size and an orientation within the $U(2m)$ degrees of freedom. They form a subset of solutions to the (second-order) Euler-Lagrange equations of the theory. In the weak-coupling or localization regime, we may expect that the functional integral can be treated with saddle-point techniques such that these pseudoparticles can be included as nonperturbative vacuum states. It has first been pointed out by Polyakov (1975a) that, when they dominate the functional integral, instantons may be relevant in influencing the infrared behavior of the theory [among other things]. A complete classification of saddle-point solutions, as well as the semiclassical method (saddle point plus fluctuations) turns out to be a difficult exercise in mathematics which goes well beyond the scope of this lecture. For instance, most of the conclusions from instantons in Levine, Libby and Pruisken (1983, 1984a,b,c) are wrong, not by mistake unfortunately, but by a mishandling of the problem.

In what follows, I will discuss some more recent results on instantons, as well as on scaling of the conductance parameters. Finally, experimental work will be mentioned which was specially designed to test the theory of scaling.

1) In an extended analysis of the semiclassical method by the present author (1985d), a rather transparent representation has been found for the single ($q = \pm 1$) instanton. The general solution can be written

Field Theory, Scaling and Localization

$$[Q_{INST}]_{ab}^{pp'} = [T^{-1}\Lambda T]_{ab}^{pp'} \quad ; \quad T \in U(2m) \qquad (5.5.8a)$$

$$\Lambda = \Lambda_{ab}^{pp'} = \left[\begin{array}{c|c} \Lambda^{11} & \Lambda^{12} \\ \hline \Lambda^{21} & \Lambda^{22} \end{array}\right]$$

$$= \left[\begin{array}{cc|cc} A_{11} & \underline{0} & B_{00} & \underline{0} \\ \underline{0} & \cdot_1 & \underline{0} & \cdot_0 \\ \hline B^*_{00} & \underline{0} & -A_{-1-1} & \underline{0} \\ \underline{0} & \cdot_0 & \underline{0} & \cdot_{-1} \end{array}\right], \qquad (5.5.8b)$$

where $\underline{0}$ indicates that all elements off the subdiagonals vanish and

$$A = \frac{\lambda^2 - |z-z_0|^2}{\lambda^2 + |z-z_0|^2}, \quad B = \frac{\lambda(z-z_0)}{\lambda^2 + |z-z_0|^2}, \qquad (5.5.8c)$$

where $z = x + iy$. The insertion A,B in one of the replica channels of Λ is nothing but an $O(3)$ instanton [the $O(3)$ model is obtained by putting $m=1$ and is the most popularly studied model in this context]. Hence the single-instanton solution is an $O(3)$ instanton embedded in the coset space $U(2m)/U(m) \times U(m)$ by means of the global rotation $T \in U(2m)$. The parameters λ and z_0 are recognized as the scale-size and position respectively. Together with the $U(2m)$ degrees of freedom, they form the manifold of instanton parameters which do not change the value of the action [Note 5]. In total, one finds $2m(m+2)$ instanton parameters.

2) Next, as one of the most essential aspects of the theory, one needs to compute the contribution of the single instanton to the free energy and see how and whether it survives in the replica limit $m=0$. In order to perform this task, we need several results.

First, we remark that the partition function can be expanded according to topological sectors

$$\mathcal{Z} = \int \mathcal{D}[Q] e^{\mathcal{L}_{eff}} = \mathcal{Z}^{(0)} + \mathcal{Z}^{(1)} + \mathcal{Z}^{(-1)} + \ldots \qquad (5.5.9)$$

where the superscripts refer to the topological charge of the field $\tilde{Q}$. At a classical level one has

$$\mathcal{Z}^{(n)} \propto e^{-2\pi |n| \sigma_{xx}^0}, \qquad (5.5.10)$$

so that a natural expansion scheme for the free energy can be obtained as

$$\mathcal{Z} = \mathcal{Z}^{(0)} \left(1 + \frac{\mathcal{Z}^{(1)} + \mathcal{Z}^{(-1)}}{\mathcal{Z}^{(0)}} + \ldots \right)$$

and

$$\mathcal{F} = \ln \mathcal{Z} = \ln \mathcal{Z}^{(0)} + \frac{\mathcal{Z}^{(1)} + \mathcal{Z}^{(-1)}}{\mathcal{Z}^{(0)}} + \ldots$$

$$\equiv \mathcal{F}_0 + \mathcal{F}_1 + \ldots \qquad (5.5.11)$$

In order to compute the instanton contribution to the free energy, one needs to analyze the quantum fluctuations about the single instanton. We will skip the details and just mention that a formalism has been developed, largely following conventional techniques, for the treatment of the zero modes and the regularization of the theory. The instanton contribution to the free energy has been obtained as

$$\mathcal{F}_1 = \Omega [\sigma_{xx}^0]^{2m} S_{2m}^2 C_m$$

$$\times \int \frac{d\lambda}{\lambda^3} \exp\left\{ -2\pi \sigma_{xx}^0 + 2m \ln(\lambda\mu) \right\} \cos(2\pi \sigma_{xy}^0). \qquad (5.5.12)$$

The integrand has been discussed before (Pruisken 1985c) and follows from standard arguments (e.g. Rajaraman 1982). The integral is over the scale size λ of the single instanton and shows the well known infrared divergence. In the limit m=0 this changes into an ul-

Field Theory, Scaling and Localization

traviolet divergence which now makes the instanton approach particularly meaningful. A factor Ω has been singled out as a result of the integration over the position of the instanton. C_m contains numerical factors from the $U(1)$ integration (i.e. rotation symmetry about the z-axis) and from the regularization scheme used. The phase coherence length, [conventionally called μ] is the scale at which $\sigma_{xx} = \sigma^0_{xx}$.

The factors S_{2m} are the most important ones and arise from the aforementioned generators of the broken replica symmetry. Effectively these factors represent the surface of a unit sphere in $2m$ dimensions

$$S_{2m} = 2\pi^m \Gamma^{-1}(m) \tag{5.5.13}$$

[i.e. the volume of the coset space $U(m)/U(m-1) \times U(1)$]. In the limit of small m the factor S_{2m} becomes of order m. Hence, $\mathcal{F}_1$ is proportional to m^2 and this is a necessary feature. Namely, the perturbative contribution to the free energy $\mathcal{F}_0$ has the same multiplicity, $2m^2$, as the number of generators in $\tilde{Q}$. This, then, constitutes the proof that the instanton survives the limit of zero field-components.

3) As far as the conductance parameters are concerned, much of the structure of the theory can be understood by simpler means, following up the arguments presented in Pruisken (1985b,c). As a first remark, notice that by using the source term formalism [once more!], one can obtain an expression for the renormalized conductance parameters in terms of correlations of the critical fluctuations $\tilde{Q}(r)$. In order to do so, we evaluate $\mathcal{Z}$ in the presence of a background field, $Q \rightarrow \mathcal{U}^{-1}Q\mathcal{U}$, where $\mathcal{U}$ is given by Eq. 3.7. Differentiation with respect to the sources j_μ leads to the expressions

$$\sigma_{xx} = \sigma^0_{xx} - \tfrac{1}{2} \sigma^0_{xx} \Omega^{-1} \operatorname{Tr}\left[\langle \hat{\tau}_x \tilde{Q} \hat{\tau}_x \tilde{Q}\rangle + \tfrac{1}{m} \mathbb{1}\right]$$

$$- \tfrac{1}{4} (\sigma^0_{xx})^2 \Omega^{-1} \langle \operatorname{Tr}\left[\hat{\tau}_x \tilde{Q} \, \nabla_x \tilde{Q}\right] \operatorname{Tr}\left[\hat{\tau}_x \tilde{Q} \, \nabla_x \tilde{Q}\right]\rangle \ ,$$

$$\sigma_{xy} = \sigma^0_{xy} - \frac{1}{2}\sigma^0_{xx}\Omega^{-1}\text{Tr}\langle\epsilon_{\mu\nu}\tau_\mu\tilde{Q}\nabla_\nu\tilde{Q}\hat{\tau}_z\rangle$$

$$-\frac{1}{8}(\sigma^0_{xx})^2\Omega^{-1}\langle\text{Tr }\epsilon_{\mu\nu}\left[\hat{\tau}_x\tilde{Q}\nabla_\mu\tilde{Q}\right]\text{Tr}\left[\hat{\tau}_y\tilde{Q}\nabla_\nu\tilde{Q}\right]\rangle, \quad (5.5.14)$$

where the expectation is now with respect to $\mathcal{L}_{eff}$.

Considering the limit of a gap in the density of bulk states, i.e., $\sigma^0_{xx} = 0$ and $\sigma^0_{xy} = $ integer, we conclude that this situation is correctly reproduced by the equations, as it should be. Furthermore, considering the effect of the transformation $Q(x,y) \to Q(y,x)$, we can write down the following relation between the bare and renormalized conductances

$$\sigma_{xx} = \sigma^0_{xx} + c_0 + \sum_{n=1}^\infty c_n \cos(2\pi n\sigma^0_{xy})$$

$$\sigma_{xy} = \sigma^0_{xy} + \sum_{n=1}^\infty \theta_n \sin(2\pi n\sigma^0_{xy}). \quad (5.5.15)$$

The coefficients c_n and θ_n are solely functions of the bare dissipative conductance σ^0_{xx}. This dependence, at a classical level, is through exponential factors $\exp(-4\pi|n|\sigma^0_{xx})$. These are entirely topological in origin and unobservable in the asymptotic perturbative series in $1/\sigma^0_{xx}$. A lot of the structure of the theory can be read off from the expression (5.5.15). Clearly on the lines $\sigma^0_{xy} = $ integer the Hall conductance is unrenormalized. The theory is invariant under the shift $\sigma^0_{xy} \to \sigma^0_{xy} + $ integer. Hence, on all of these lines the conclusions of conventional localization theory should hold, in particular that of Eq. 5.1.

One may next consider further the relation between localization and restoration of the continuous symmetry, as it is reasonably well understood for conventional problems, in order to elucidate the structure of the theory with σ^0_{xy} slightly deviating from integer values. This aspect can conveniently be discussed in terms of the renormalization group program as used

Field Theory, Scaling and Localization

by Polyakov (1975b) for the purpose of demonstrating asymptotic freedom of the O(3) nonlinear σ-model. This program consists of eliminating the large momentum fluctuations of the $\tilde{Q}$ and therefore reflects the spirit in which Eq. 5.14 ought to be understood.

Successive renormalization group transformations of this kind will drive the coupling into the strong coupling regime where perturbation theory ceases to be valid. In this regime, the theory develops dynamically a coupling-constant-related mass gap. For the disordered system, this means a finite bare-conductance-related localization length and a vanishing renormalized conductance.

As a direct consequence, one can state that the functional integral cannot be dependent on the very long wavelength components in $\tilde{Q}(r)$, namely those which vary slowly with respect to the localization length. In different words, the theory should be invariant under the insertion of sufficiently slowly varying background fields. This, then, yields the justification of the relation between localization and continuous symmetry, the consequences of which have been studied in Sect. 5.3.5 by means of a gradient expansion.

Let us next come back to the original discussion. If it is reasonable to assume that the theory does not describe a phase transition of σ^0_{xy} near integer values, then a finite correlation or localization length should also persist in this case. Repeating the arguments presented in Sect 5.3.5 but now for the effective field theory, one concludes the bare parameters have to scale toward $\sigma_{xx} = 0$ and $\sigma_{xy} = integer$. Furthermore, Eqs. 5.15 show that the Hall conductance remains unrenormalized along the lines $\sigma_{xy} = half\text{-}integer$. By the arguments used before, we conclude that this corresponds to delocalized wave functions near the Fermi energy.

4) In order to translate these statements into renormalization group equations, an analysis has been performed in the so-called dilute instanton gas approximation (Pruisken 1985b,c). This approximation essentially boils down to an expansion in topological charge of the type discussed under **2)** above. Here we simply give the result

$$\beta_{xx} = \frac{\partial \sigma_{xx}}{\partial \ln L} = \frac{-1}{2\pi^2 \sigma_{xx}} - \sigma_{xx} \underline{D} e^{-2\pi\sigma_{xx}} \cos(2\pi\sigma_{xy})$$

(5.5.16)

$$\beta_{xy} = \frac{\partial \sigma_{xy}}{\partial \ln L} = -\sigma_{xx} \underline{D} e^{-2\pi\sigma_{xx}} \sin(2\pi\sigma_{xy})$$

where $\underline{D}$ is a positive constant. From the discussion above, it is understood that the nonperturbative contributions to (5.5.16) essentially arise from eliminating the short distance fluctuations about the single-instanton solutions. Comparing with Eq. 5.15 we see that only the first few terms in an infinite series are accounted for. However, the very existence of the series (i.e. the diluteness of the instanton gas) means that Eq. 5.16 can be trusted qualitatively over a large range of values of σ_{xx} [the next two leading terms have exponential factors $\exp(-4\pi\sigma_{xx})$].

Extrapolating the results down to the strong-coupling regime, one obtains the renormalization group flow diagram in Fig. 5.6. The attractive feature of this diagram is the occurrence of an intermediate fixed point at half-integer values of the Hall conductance. Physically, these fixed points (together with their domain of attraction) can be associated with a delocalization of the electronic wave functions. One expects a singular part in the free energy per unit area which scales as

$$\mathcal{F}_s(\tilde{\theta},\tilde{\sigma}) = b^{-2}\mathcal{F}_s(b^y\tilde{\theta},b^{-\tilde{y}}\tilde{\sigma}); \quad y,\tilde{y} > 0. \qquad (5.5.17)$$

Here, $\tilde{\theta}$ and $\tilde{\sigma}$ denote the linear deviations from the fixed point values σ^*_{xy} and σ^*_{xx} respectively. Furthermore, one can write down the scaling behavior of the localization length as

$$\xi(\tilde{\theta},\tilde{\sigma}) = b\xi(b^y\tilde{\theta},b^{-\tilde{y}}\tilde{\sigma}) \qquad (5.5.18)$$

such that the localization length exponent ν is given as $\nu = y^{-1}$ [Note 6].

Field Theory, Scaling and Localization

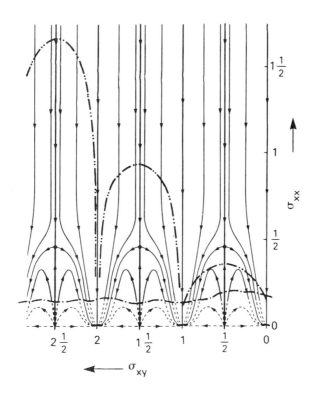

Figure 5.6 Theoretical renormalization group flow diagram showing two-parameter scaling for the IQHE (Wei et al. 1985b).

5) The flow diagram displays the more general expectation in field theory of a phase transition at $\sigma_{xy} = \frac{1}{2}$ $[\theta = \pi]$. For instance, a first order phase transition has been found in the exactly soluble large N limit of the CP^{N-1} model (Witten 1979), but with no diverging correlation length. A first order singularity in the free energy is understood in the present context by taking for the critical exponent $y = 2$ [the scaling relation for the correlation length involves a vanishing amplitude]. A mean-field type of argument has conjectured the same value in the limit of zero replicas (Levine, Libby, Pruisken 1984a). This value, together with a divergent localization length, has been reproduced in a quite different approach involving Bethe-ansatz techniques (Affleck 1985).

Finally, one can ask where the extended levels are located in the space of physical system parameters.

All one has to do is to look back at the Hall conductance in SCBA (Fig. 5c) and see where it passes through half-integer values for the various cases. For the strong field case or weak disorder, this occurs, as is known, at the Landau band center. In the opposite limit of weak fields or strong disorder, the extended levels disappear to infinite energies.

In the intermediate regime there apparently does *not* occur a continuous "floating" of extended levels (Laughlin 1984b) but rather a kind of pair production and destruction. The full meaning of this structure is not clear, but the behavior is not universal and depends on details of the imperfection system at short distances. Although the disappearance of extended levels for weak magnetic fields shows the correct trend, there remain some theoretical questions about the detailed crossover to Anderson localization (Pruisken 1984).

6) The aspect of two-parameter scaling of the conductance is amenable to experimental verification. The parabolic —··— lines in Fig. 5.6 form the loci of conductance parameters in SCBA for strong magnetic fields. These form the starting point for scaling and a qualitative overall picture of the experimental situation can be read off from this figure.

For instance, as one of the strongest predictions of 2D localization theory, there is a very sharp distinction (as a function of the bare dissipative conductance) between the quasimetallic, weak-localization regime on the one hand, and the nonperturbative regime on the other; it is only in latter that one can observe any significant scaling behavior in the Hall conductance and hence the phase transition near half integer values of the Hall conductance. A detailed discussion has been presented by Wei, Tsui and Pruisken (1986).

These authors also presented a careful experimental study of the temperature and magnetic field dependence of the transport parameters in the lowest Landau levels of InGaAs-InP heterostructures. We will not reproduce the results for the resistivities as these are qualitatively similar to those shown, e.g., in Fig. 2.2. For a sequence of temperatures between 10 K and 0.5 K, however, these measurements have been converted into conductances and plotted in the σ_{xx}, σ_{xy} plane. Fig. 6.4.10 contains the data as temperature-driven

Field Theory, Scaling and Localization

flow lines along which the magnetic field is kept constant. A distinction is made between the measurements in the regimes 10 K to 4.2 K (dashed line) and 4.2 K and 0.5 K (full lines). In the former regime, the data essentially follow the temperature dependence of the Fermi-Dirac distribution, [Eq. 2.22 with the conductance in SCBA inserted for $\sigma_{\mu\nu}$]. This notion was furthermore consistent with some numerical estimates saying that kT is larger than the Landau level impurity broadening for temperatures down to a few Kelvin. The low temperature data (full lines), on the other hand, cannot be explained by semiclassical means and is attributed to genuine scaling of the conductances.

In order to make the comparison with the theoretical flow diagram of Fig. 5.6 [Note 7], one postulates the existence of an inelastic scattering length [Thouless length]. This sets an effective temperature dependent lengthscale L_T in the problem beyond which scaling ceases to be valid [so that the system size is effectively $\Omega = L_T^2$]. The measured conductance, $\sigma_{\mu\nu}(L_T)$, should therefore directly reproduce the renormalization group flow diagram as the temperature is varied. The various aspects of the data in Fig. 6.4.10 can be successfully described this way (see Wei et al. 1986). But clearly, more experimental and theoretical work is needed, not only in order to get numerical agreement, but also to provide the necessary conceptual structure. This, I would say, is one of the major problems left for the future.

5.6 Notes

Note 1. A systematic expansion about the SCBA is discussed in Sec. 5.4. This expansion scheme is well-defined within the domain of the SCBA Landau-band, but it fails to give the Gaussian band tails (the usual draw-back of the CPA). The band tails can be obtained by the semiclassical method employed by Affleck (1983).

Note 2. In a real experimental situation, the considerations regarding edge currents changes somewhat. We have assumed throughout that the electronic disorder is perfectly homogeneous beyond a finite correlation length of the random potential. In a real system, macroscopic inhomogeneities in the electron density do occur. If one can model a realistic sample by a set of domains, each of which is truly homogeneous but has a different electronic density, then the domain walls form loci of additional equilibrium current. The average over the domains now involve a weight, equal to the volume fraction of each domain. Of course, if these macroscopic inhomogeneities form a truly random system themselves such that each random configuration is realized within the sample, then nothing new has been added and this situation is included in our previous analysis.

Note 3. It is of course of interest to check Eq. 2.17 by direct computation of both the bulk density of states and the edge current in specific examples. This has been done numerically for the lattice system and the results converge relatively rapidly toward the desired answer (Pruisken and Schweitzer, unpublished; see Fig. 2). For a strip with a width of ten cyclotron radii the results are off by only a few percent.

Note 4. The replica trick was introduced by Edwards and Anderson (1975) as a device for facilitating quenched averages over random fields. In many cases, such as the present one, the imperfections or disorder are not in thermal equilibrium, and it is necessary to average Green's functions or free energies rather than the partition functions. On the other hand, it is formally simple to average powers of the partition or generating function $\mathcal{Z}$. The trick consists of evaluating $\langle \mathcal{Z}^m \rangle_{av}$ for arbitrary [integer] m and then using

$$\mathcal{F} = \langle \ln \mathcal{Z} \rangle_{av} = \lim_{m \to 0} \langle [\mathcal{Z}^m - 1]/m] \rangle_{av} .$$

Needless to say, there is some interesting mathematics associated with this "analytic continuation".

Note 5. The insertion of A,B in one of the other replica channels does not lead to a new solution. The transformation from one replica channel into another can always be accomplished by an appropriate choice for $T \in U(m) \times U(m)$. In other words, the subgroup $U(m) \times U(m)$ of T effectively restores the replica symmetry which was initially being broken by the Λ. Hence also, the problem is characterized by a broken continuous symmetry, rather than by a broken discrete one as initially thought (Levine, Libby, Pruisken 1984a,b,c). In fact, not the entire subgroup $U(m) \times U(m)$ is being

Field Theory, Scaling and Localization

broken by Λ. One deduces the following decomposition of the T which constitutes a hierarchy of symmetry breaking

$$T \to [U(2m)/U(m) \times U(m)] \times [U(m)/U(1) \times U(m-1)]$$

$$\times [U(m)/U(1) \times U(m-1)] \times [U(1)] \times [U(1) \times U(m-1) \times U(m-1)]$$

The first factor group constitutes the usual Goldstone modes whereas the second and third constitute the generators of the broken replica symmetry. The fourth $[U(1)]$ generates a rotation of the $O(3)$ instanton about the z-axis and finally, the last, is a pure gauge which is not a part of the instanton manifold.

Note 6. The dilute gas results, of course, should not be taken seriously as far as the actual value of the exponent y or v is concerned. In fact, the quantities $\underline{D}$ in $\beta_{\mu\nu}$ are not universal and depend on the specific definition of the bare coupling (except at the critical point $t=0$) and hence on the specific regularization scheme. All this says is that the fixed point value σ^*_{xx} need not be the same for all the lines $\sigma_{xy} = half\text{-}integer$. Almost certainly it is not. However, the critical exponent y can only change due to the occurrence of marginal operators. So far there is no sign that such an effect might be present.

Note 7. The theoretical curves shown in Fig. 5.6 do not take into account the spin structure of the Landau levels. The experimental data shown in Fig. 6.4.10 were obtained on a sample in which the spin-splitting was resolved in the lowest Landau level, but not in the next higher level. This accounts for the discrepancy in the widths of the two parabolas shown in Fig. 6.4.10.

Part B

The Fractional Effect

CHAPTER 6

Experimental Aspects

ALBERT M. CHANG

6.1 Introduction

The integral quantum Hall effect and fractional quantum Hall effect are two of the most remarkable physical phenomena to be discovered in solid state physics in recent years. In many respects, the two phenomena share very similar underlying physical characteristics and concepts, for instance, the two-dimensionality of the system, the quantization of the Hall resistance in units of h/e^2 with a simultaneous vanishing of the longitudinal resistance, and the interplay between disorder and the magnetic field giving rise to the existence of extended states. In other respects, they encompass entirely different physical principles and ideas. In particular, the IQHE is believed to be a manifestation of the transport properties of a noninteracting, charged-particle system, in the presence of a strong perpendicular magnetic field, whereas the FQHE results from a repulsive interaction between particles, giving rise to a novel form of many-body ground state. Moreover, the mobile (non-localized) elementary excitations must carry fractional charge and lie at a finite energy above the ground state. These fractionally charged excitations are thought to obey unusual statistics which are neither Fermi nor Bose-Einstein.

To the author, there are two basic aspects to the fractional quantum Hall effect. One pertains to the underlying ground and excited states, and their intrinsic nature, properties, and energy structure. This aspect includes such topics as the structure of the FQHE ground state: Is it crystalline or liquid-like? Are the excited states separated from the ground state by an energy gap? If so, how does the gap vary with magnetic field? The question of a phase transition: Does the FQHE result from a condensation of a high temperature gas-like phase? There are a host of other questions.

The other aspect is in a sense more directly connected to the observability of the quantum Hall effect and deals with the nature of magneto-transport in two-dimensional (2D) systems, in the presence of disorder. This aspect is in many ways common to both the IQHE and FQHE, particularly in view of the current understanding that the QHE occurs in a 2D charged system when a mobility gap separates the ground and excited states. The topics in this category include the nature of localized and extended states, e.g., their relative proportion in number and their density of states. [Strictly speaking, this topic is appropriate for the IQHE only.] Other examples in this category are the effect of different types and strengths of impurities on the observability of the QHE, and the interrelation between the longitudinal and Hall conductances, σ_{xx} and σ_{xy}. Of course, the two aspects are intimately related: the division is helpful to the author in thinking about the QHE.

It is the aim of this lecture to review the important experimental developments in the field of the FQHE. In light of the above discussion, this lecture will be organized as follows: Section 6.2 contains a brief introduction to the FQHE, and is mainly devoted to discussions on the conditions for observing the FQHE and the semiconductor systems in which the experiments are carried out. In Section 6.3 we review experimental results in the FQHE which have a direct bearing on the nature of the ground and excited states. It contains results on the hierarchy of FQHE sequence, the demonstration of exact fractional quantization, current-voltage characteristics, and temperature dependences of the transport coefficients. Section 6.4 describes various experiments to investigate

Experimental Aspects

the nature of 2D magneto-transport. The topics addressed include measurements of the electronic density of states, current distribution within a specimen in the quantum Hall regime, and tests of the recently developed scaling theories of 2D magneto-conduction. And finally, Section 6.5 will briefly mention future work that needs to be done to bring us to an even more complete understanding of the FQHE. In relating the various experimental results, an attempt is made to give the motivation and thought behind the choice of and approach to the experiments.

6.2 Conditions for the Observation of the FQHE and the Choice of Semiconductor Systems

6.2.1 The basic phenomenon of the fractional quantum Hall effect

The FQHE was discovered in 1982 by Tsui, Störmer, and Gossard in high quality, GaAs-Al$_x$Ga$_{1-x}$As heterostructures [See Chap. I]. The basic phenomenon is remarkably, if not deceptively, similar to the IQHE. It is found that in a high quality sample (exemplified by large carrier mobility at zero B field) as very high magnetic fields are applied at very low temperatures, Hall plateaus and deep minima structures in ρ_{xx} develop at fractional fillings, $v = 1/3$ (Tsui et al. 1982) and 2/3 (Störmer et al. 1983) of the lowest Landau level, in a manner analogous to the development of the IQHE at integral fillings. See Figs. 6.2.1 and 6.2.2.

The existence of the QHE at fractional fillings came as a great surprise since it could not be understood within the framework of the then developing theory of the IQHE. It immediately suggested the possibility of fractionally charged excitations, perhaps reminiscent of the predicted fractional charges in quasi-one-dimensional systems (Su et al. 1979, Prange 1982).

6.2.2 Conditions for observing the FQHE

It is of value to obtain a rough idea of the conditions necessary for the observation of the FQHE, including requirements on sample quality, magnetic field, and temperature. The conditions are useful in

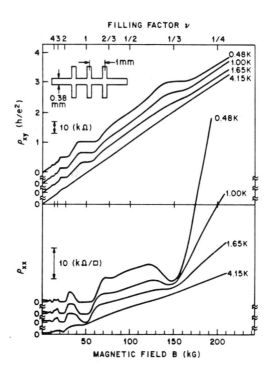

Figure 6.2.1 The fractional quantum Hall effect at 1/3 filling of the lowest Landau level as discovered by Tsui, Störmer, and Gossard (1982b).

bringing attention to various qualitative features of the phenomenon, as well as in choosing the appropriate semiconductor material systems in which to study the effect.

The QHE is a quantum-mechanical effect. Therefore, one important condition is the criterion for the delineation of the quantum regime, separating it from the classical regime. Within a semiclassical picture, where the effect of disorder is to give rise to a scattering time, τ_0, the quantum condition is:

$$\omega_c \tau_0 \gg 1 \;, \qquad (6.2.1)$$

where ω_c is the cyclotron frequency. The second condition is the temperature for the onset of the FQHE, given by:

Experimental Aspects

$$\Delta \geq k_B T \quad , \qquad (6.2.2a)$$

$$\Delta \simeq (Ce^2/2\kappa)\sqrt{n} \quad , \qquad (6.2.2b)$$

where Δ represents an energy gap, assumed to exist, separating the ground state and its current carrying excitations, κ is the dielectric constant of the medium, $n^{-1/2}$ is the average inter-particle spacing and C is a numerical constant. Since Δ arises from the Coulomb repulsion between particles, it is expected to take the form given by Eq. 6.2.2b, assuming for the time being, a strictly two-dimensional interaction, and an absence of disorder. A third condition is needed. It is crucial that the interaction between particles dominates over the effect of disorder. A reasonable estimate is given by the requirement that

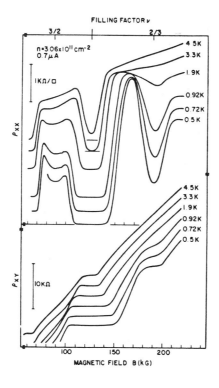

Figure 6.2.2 The FQHE at 2/3 filling of the lowest Landau level. Störmer et al. (1983c).

the energy gap Δ be larger than the disorder potential fluctuation $\langle(\Delta V)^2\rangle^{1/2}$:

$$\Delta \geq \langle(\Delta V)^2\rangle^{1/2} \approx \hbar/\tau_0 , \qquad (6.2.3)$$

where the term $\hbar/\tau_0$ is a rough estimate of the mean potential fluctuation from the uncertainty principle. In available semiconductor systems and magnetic fields, Δ is much smaller than $\hbar\omega_c$, by a factor of order 30.

6.2.3 Two-dimensional conducting systems in semiconductor materials

In Table 6.2.1, we list material parameters of relevance to the FQHE, for Si MOSFET's and III-V heterostructures. Please see Chapter II for a more complete list of materials, of relevance to the IQHE. Let us examine the suitability for studying the FQHE, based on conditions of Section 6.2.2. First, we examine the temperature condition. For the FQHE, the carrier mass plays no role, except when Landau level mixing is included [Chap. X]. The dielectric constant does not vary more than about 30% between materials. The typical Coulomb energy scale is of the order of 4K (0.3 meV) at accessible magnetic fields of around 20 Tesla, corresponding to typical carrier densities of around 2×10^{11} cm^{-2}.

Next, we examine the requirement on the sample quality, based on low-temperature carrier mobility in zero magnetic field. The conditions are given by Eqs. 6.2.1 and 6.2.3. For the FQHE, the sample quality requirement is exceedingly stringent. Even though, by now, the FQHE has been observed not only in GaAs-AlGaAs heterostructures and superlattices, but also in Si MOSFET's (Pudalov et al. 1984d, Gavrilov et al. 1985, Furneaux et al. 1985, Kukushkin and Timofeev 1985) and perhaps in InGaAs-InP superlattices as well, (Portal, et al. 1983) it is undoubtedly most accessible in GaAs-AlGaAs heterostructures, where electron mobilities in excess of 2×10^6 cm^2/Vs have been reported for a low density of 2.1×10^{11} cm^{-2} (Foxon et al. 1985). It appears from experiments that an electron mobility of around 10^5 cm^2/Vs is the minimum quality necessary in n type GaAs-AlGaAs samples of

Table 6.2.1 Material parameters for semiconductor systems with high-mobility, two-dimensional carriers. m is the carrier effective mass, μ the 4.2 K mobility, τ_0 the scattering time at zero B field, and κ the dielectric constant of the carrier host.

Semiconductor System	2D gas	m (m_e)	μ (cm^2/Vs)	τ_0 (10^{-12}s)	κ
Si-SiO$_2$	e	0.19	3,000-40,000[1]	0.33-4.4	11.9
GaAs-AlGaAs	e	0.068	20,000-2,000,000[2]	0.8-78	12.8
	h	0.38, 0.6	10,000-100,000[3]	2.8-28	
InGaAs-InP	e	0.047	6,000-35,000[4]	0.15-0.9	10.0

[1] Kukushkin and Timofeev (1985); [2] Foxon, et al. (1985); [3] Mendez et al. (1984b); [4] Wei et al. (1985a).

$\approx 1.5 \times 10^{11}$ cm^{-2} density. Because the scattering time, τ_0 is the relevant parameter for the FQHE, not the mobility, μ, which involves an extra factor of $1/m$, for p type materials, the mobility required is around five times smaller than in the n-type case.

The reason that it is possible to achieve such high quality in GaAs-AlGaAs heterostructures is as follows: In Si MOSFET's, the carriers reside near defects, and there is a large number of trap states near the interface which are able to scatter the carriers. On the other hand, in a GaAs-AlGaAs heterostructure grown by the modulation doping molecular beam epitaxy (MBE) technique (Störmer et al. 1979, Cho and Arthur 1975), the interface is sharp, to a monolayer, and the materials nearly perfectly crystalline. As a result, there are few defects to scatter the carriers. Moreover, the modulation doping technique spatially separates the dopants from the carriers confined at the interface, minimizing the scattering due to impurities. One additional factor is helpful. The band gap is larger in AlGaAs than GaAs. Therefore, the interfacial potential well resides on the GaAs side, which is a binary compound crystal and not an alloy, so scattering is reduced. In Fig. 6.2.3 we present experi-

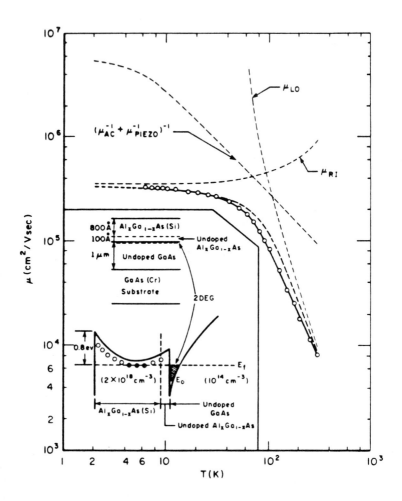

Figure 6.2.3 Temperature dependence of the 2D electron gas mobility, in a sample with a density of 5.35×10^{11} cm^{-2} and 4.2K mobility of 3.7×10^5 cm^2/Vs. Data are open circles. The light dashed lines labeled by μ_{RI}, μ_{LO}, $(\mu_{ac}^{-1} + \mu_{piezo}^{-1})^{-1}$ are the calculated mobilities for remote ionized impurity scattering, LO phonon scattering, and acoustic phonon scattering via deformation potential and piezoelectric couplings, respectively. The heavy dashed line is the combined theoretical mobility (Lin et al. 1984).

mental and theoretical results on the temperature dependence of high quality GaAs-AlGaAs heterostructures and the contributions of various scattering mechanisms (Lin et al. 1984). Nearly all the results reviewed in this article come from samples made from GaAs-AlGaAs heterostructures.

6.3 The Fractional Quantum Hall Effect

The FQHE is a manifestation of the peculiar physics associated with a 2D charged system in a perpendicular magnetic field, and the formation of a new type of correlated ground state in such a system. The phenomenon occurs as follows: A 2D gas of charged particles (electrons or holes) of density n, is embedded in a nearly impurity-free, static, neutralizing background. At low temperatures, as the magnetic field is increased, the Hall resistance develops plateaus at quantized values of h/fe^2, in the vicinity of the field $B = nhc/ef$ or filling factor $v \approx f$, where f is a rational fraction with odd denominator. At the same time, the longitudinal resistivity approaches zero. A fractional effect corresponding to a given value of f, is observed or not observed, depending on the amount of disorder and the temperature.

The original discovery (Tsui et al. 1982b) has stimulated a great deal more work. To date, the following results are known: 1) The FQHE occurs in multiple series of $f = p/q \approx v$, where p and q are integers, q odd. At present, these fractions have been observed: 1/3, 2/3, 4/3, 5/3, 7/3, 8/3, 1/5, 2/5, 3/5, 4/5, 7/5, 8/5, 2/7, 3/7, 4/7, 10/7, 11/7, 4/9, 5/9, and 13/9, some in ρ_{xx} only (Störmer et al. 1983a, Chang et al. 1984a, Mendez et al. 1983, 1984a, Ebert et al. 1984, Clark et al. 1985). 2) The fractional quantization is highly accurate (Chang et al. 1984a, Mendez et al. 1983, 1984a, Ebert et al. 1984, Clark et al. 1985, Tsui et al. 1983b). For $f = 1/3$ and 2/3, the accuracy is 3 parts in 10^5, in the vicinity of the center of the Hall plateau; for 2/5, 2.3 parts in 10^4. 3) The current-voltage (I-V) characteristics at 1/3 and 2/3 are linear down to low voltages corresponding to an average electric field of about 10^{-5}V/cm in the sample (Chang et al. 1983). This indicates that the ground state of 1/3 and 2/3 Landau level fillings are likely not a charge density wave (Yoshioka and Fukuyama 1979) since no pinning is observed. 4) The temperature evolution of the 1/3, 2/3, 4/3, and 5/3 FQHE show an activated behavior within certain temperature ranges (Chang et al. 1983, Kawaji et al. 1984, Ebert et al. 1984, Boebinger et al. 1985b). This result strongly suggests the existence of a gap in the excitation spectrum separating the ground state from the mobile, charge carrying excitation. The situation for the 2/5 and 3/5 effects is more complicated (Chang et al. 1984a, Boebinger et al. 1985d) but is consistent with the existence of a gap.

The above results are quite well established. In addition, there are important but more preliminary and tentative results: 5) A recent systematic investigation (Boebinger 1985b) of the excitation gaps for the 1/3 and 2/3 effects as functions of magnetic field and sample mobility, seems to be consistent with the notion that the actual electron-electron interaction in heterostructures is not a strictly 2D Coulomb interaction, but is modified by the finite extent of the electron wave function in the third direction (Zhang and das Sarma 1986, Girvin et al. 1986b, Yoshioka 1985b). 6) A detailed measurement of the effect of increased disorder on the 2/3 energy gap indicates the gap can decrease rapidly with increasing disorder, and suggests that a threshold exists above which the effect disappears (Chang et al. 1983). 7) The temperature dependence of ρ_{xx} and ρ_{xy} at $v = 2/3$ and vicinity suggest that, within a phenomenological picture of disorder broadened "quasi-particle" (QP) and "quasi-hole" (QH) bands (Chang et al. 1983) of the charged excitations, the QP and QH do not behave as fermions. 8) Preliminary results in a careful measurement of ρ_{xx} versus temperature, at $v = 2/3$, show that it is a smooth function of temperature. Therefore, ρ_{xx} shows no sign of a phase transition in the 2/3 FQHE. In this section, we will review these results in detail.

6.3.1 The hierarchy of fractional quantum Hall effects

In Figure 6.3.1, we show a typical magnetic field trace of the FQHE. The sample is a Hall bar of carrier density 1.45×10^{11} cm^{-2} and mobility 4×10^5 cm^2/Vs. At a temperature of 150 mK, well developed

Experimental Aspects

plateaus in ρ_{xy} are observed at Landau level fillings of 1/3, 2/3, 4/3, and 2/5, accompanied by deep minima structures in ρ_{xx}. In addition, weaker structures are

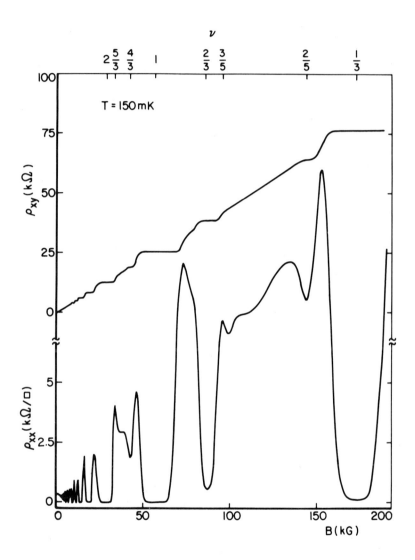

Figure 6.3.1 A typical magnetic field trace of the FQHE. The sample has a carrier density of 1.45×10^{11} cm^{-2} and a mobility of 4×10^5 cm^2/Vs. The temperature is 150mK. FQHE is observed in ρ_{xy} at $f = 1/3$, 2/3, 4/3, and 2/5, and in ρ_{xx} at $\nu = 1/3$, 2/3, 4/3, 5/3, 2/5, and 3/5 (Boebinger, Chang, Störmer, and Tsui, unpublished results).

visible in ρ_{xx} near $v = 3/5$ and $5/3$. At the lower magnetic fields, well developed IQHE at $v = 1$, 2, 3, etc. are observed. In samples of higher quality and at lower temperatures, additional fractional effects can be uncovered. In Figs. 6.3.2, 6.3.3, 6.3.4, and 6.3.5, we present experimental evidence for the presence of fractional effects near $v = 5/3$, 1/5, 7/5, 8/5, 2/7, 3/7, 4/7, 4/9, and 5/9. Structures in ρ_{xx} near $v = 7/3$, 8/3, and 4/5 have also been reported (Störmer et al. 1983a, Chang et al. 1984a, Ebert et al. 1984, Clark et al. 1985, Boebinger et al. 1985b). The data in these figures are all obtained from samples of GaAs-AlGaAs heterostructures. To obtain the data in Figures 6.3.2, 6.3.3, and 6.3.5, it has been necessary to gate the sample density using a backgate (Störmer et al. 1981). Table 6.3.1 summarizes the carrier densities and mobilities of the samples.

Figure 6.3.3 is in essence a continuation of Figure 6.3.2 to higher magnetic fields. The 70 kilogauss background magnetic field comes from the superconducting solenoid in a hybrid magnet at the National Magnet Laboratory in Cambridge, Massachusetts. The evidence for the 1/3, 2/3, 4/3, 5/3, 2/5, and 3/5 effects are convincing: the well developed Hall plateaus and deep ρ_{xx} minima firmly establish the existence of these effects. The other effects near 7/5, 8/5, 3/7, 4/7, 4/9, and 5/9 are more tentative, although their appearance is consistent with the development of the established effects at higher temperatures. Ideally, an effect should be identified by the resistance value of the quantized Hall plateau. In the absence of a well defined plateau, it is only possible to make a tentative identification, based on the position of the ρ_{xx} minimum. Data of a similar quality for $v = 2/7$ and 1/5 are presented in Figs. 6.3.4 and 6.3.5. The 1/5 effect was first observed by Mendez, Heiblum, Chang and Esaki (1983) and the 2/7 effect by Störmer, Chang, Tsui, Hwang, Gossard, and Wiegmann (1983a). A significant number of the results presented here have been reproduced by other authors in the field: Mendez et al. (1984a) in p-type systems, Ebert et al. (1984), and Kukushkin et al. (1985) in Si MOSFET's for the 1/3, 2/3, and 4/3 effects.

It is apparent that the FQHE occurs in multiple series, involving fractions of odd denominator only. Störmer et al. (1983a) were the first workers to note

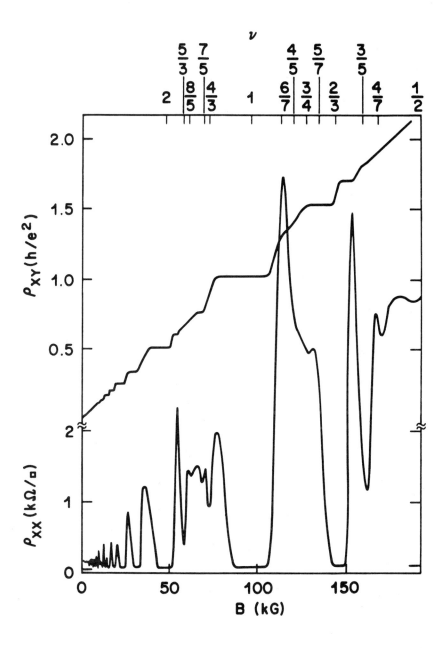

Figure 6.3.2 Magneto-transport in a high quality sample at a temperature of 100 mK. FQHE is observed at 2/3, 4/3, 5/3, 3/5, 7/5, 8/5, and 4/7. Intriguing structures are also present in ρ_{xx} near 4/5 and 3/4 (Boebinger et al. 1985a).

the emergence of multiple series of p/q. Our current understanding of the p/q series is based on a hierarchical model [Chap. VIII]. In this model, a sequence of successive ground states is derived from a parent state. Different hierarchies based on differing parent states have been proposed (Laughlin 1983a,

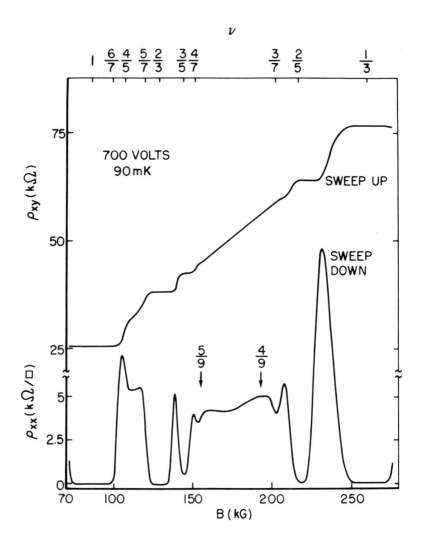

Figure 6.3.3 ρ_{xx} and ρ_{xy} at 90 mK, for a sample which shows the FQHE at v = 1/3, 2/3, 2/5, 3/5, 3/7, 4/7, 4/9, and 5/9 (Chang et al. 1984a). This trace is in essence a continuation of Figure 6.3.2 to higher magnetic fields.

Experimental Aspects

Table 6.3.1 Carrier mobility, density, and backgate voltage for the samples of Figures 6.3.1-5.

Figure	μ (cm^2/Vs)	n (10^{11} cm^{-2})	V_g (volts)
6.3.1	400,000	1.45	0
6.3.2	890,000	2.2	700
6.3.3	800,000	2.1	700
6.3.4	590,000	1.87	265
6.3.5	300,000	1.2	0

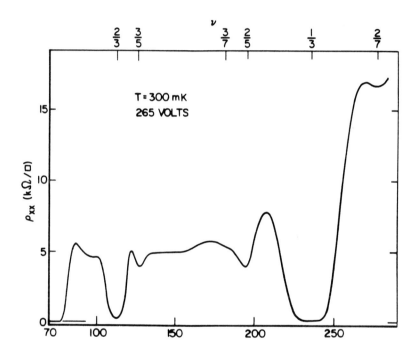

Figure 6.3.4 The 2/7 FQHE observed in ρ_{xx} at T = 300 mK (Chang, Störmer, and Tsui, unpublished).

1984a, Haldane 1983, Halperin 1984, Tao and Thouless 1983, Tao 1984, Anderson 1983). The prevailing theory, [Chap. VIII] due independently to Haldane (1983), Halperin (1984) and Laughlin (1984a) is based on a parent ground state proposed by Laughlin (1983a) [Chap. VII] for $v = 1/q$, originally intended to explain the 1/3 effect. The Laughlin state is best described as an incompressible, correlated fluid, which contains fractionally charged, quasiparticle/quasihole excitations. In this hierarchy, the quasiparticles/quasiholes are envisioned as recondensing into Laughlin-type correlated states, as a result of a mutual repulsion. Because each daughter state derived in this manner also contains a fractionally charged hierarchy, the sequence is given by the continued fraction:

$$f = \cfrac{1}{m + \cfrac{\alpha_1}{p_1 + \cfrac{\alpha_2}{p_2 + \cfrac{\cdots}{\cdots + \cfrac{\alpha_n}{p_n}}}}} \qquad (6.3.1)$$

where m is an odd positive integer, α_i equals ± 1 and p_i is an even integer. In Fig. 6.3.6, we show the hierarchies derived from the 1/3 and 2/3 parent states. The asterisks indicate effects that are experimentally observed.

Very recently, Clark, Nicholas, Usher, Harrison, and Foxon (1985) have reported very beautiful results on the FQHE from an n-type GaAs-AlGaAs heterostructure of exceedingly high quality. The sample density of 0.7×10^{11} cm^{-2} and mobility of 6×10^5 cm^2/Vs are significantly increased by illuminating with light, to 2.1×10^{11} cm^{-2} and 2.1×10^6 cm^2/Vs. At a substrate temperature of 20 mK, clear minimum structures develop in ρ_{xx} at v = 9/7, 10/7, 11/7, 5/9, and 13/9, in addition to those observed previously by other authors. New plateaus are observed in ρ_{xy} at f = 7/5, 8/5, and 4/7, therefore putting their existence on firmer ground. Intriguingly, additional features for $v > 2$ and $v > 3$ appear near even denominator filling factors, e.g., v = 9/4, 5/2, 11/4, 7/2, and 15/4. However, the claim that these represent even denominator series remains to be verified from the quantized values in ρ_{xy}. The present values associated with the weakly developed plateaus near v = 5/2 and 11/4 appear rather near 7/3 and 8/3, respectively.

It is worthwhile to point out that experiment indicates that a sequence of strong effects emanates from 1/3 (2/3) toward the center at v = 1/2. At a given temperature, the 1/3 effect is always the best devel-

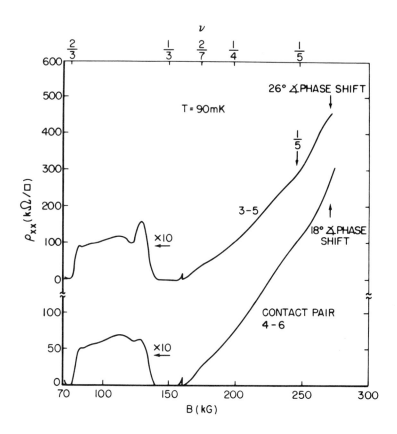

Figure 6.3.5 ρ_{xx} versus magnetic field for a sample which shows a weak structure near $v = 1/5$. The two curves represent different contact pairs on the sample. The temperature is 90 mK (Chang et al. 1984a).

oped, followed by 2/5, 3/7, and 4/9. The 2/3 sequence (3/5, 4/7, 5/9) behaves similarly. Since the magnetic field strength only varies by $\approx$ 25% within each sequence, it is reasonable to assume that the difference in field position is not responsible for the rapid reduction of the strength of the effects within a series. Moreover, from a phenomenological point of view, one expects that the closer to $v = 1/2$, where extended states should reside within a Landau band, the less affected the FQHE is by disorder, which tends to reduce the strength of an effect. Thus, the observed reduction in strength toward the center (1/2) appears to lend support to the hierarchical model.

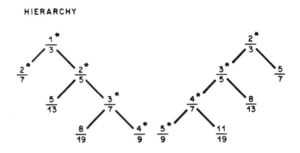

Figure 6.3.6 The hierarchical sequences of FQHE, according to Haldane (1983), Halperin (1984), and Laughlin (1984a). The asterisks indicate states which are observed experimentally.

6.3.2 Exactness of a fractional quantum number

In order to identify a FQHE, with absolute certainty, to correspond to a particular fraction f, it is necessary to establish a high degree of accuracy of Hall resistance quantization, to the value h/fe^2. An accurate fractional Hall resistance is important for two other reasons: 1) In the IQHE, the integer index, i, labels the $(i-1)$'th Landau level and therefore corresponds to an ordinary, integral quantum number. Inclusion of spin introduces a slight complication, but we may think of i in this case as a label of successive Landau bands. In the FQHE, the effects occur at fractional fillings of spin-polarized Landau bands. An accurate value of Hall quantization to h/fe^2, therefore, corresponds to the occurrence of a "fractional" quantum number. 2) In the powerful, "gauge invariance" argument [Chap. I] to explain the exactness of the IQHE quantization, even in the presence of disorder, Laughlin invokes the charge, e^* of the current-carrying elementary excitations (mobile electrons in this case, $e^* = e$), as the basic unit of quantization. In other words, ρ_{xy} is given by

$$\rho_{xy} = \frac{\Phi_0}{ie^*} = \frac{h/e}{ie^*} = \frac{h}{ie^2} \qquad (6.3.2)$$

Experimental Aspects

where i is an integer corresponding to the number of charges transferred across the Hall current flow. In the FQHE, an extension of this argument (Tao and Wu 1984) would imply the existence of charges of

$$e^* = \frac{h/i'\,e}{\rho_{xy}} = \frac{fe}{i'} \qquad (6.3.3)$$

where i' is some unknown integer corresponding to the number of quasi-particles transferred. For a minimal value of i' of 1, the fractional charge would be fe. An accurate determination of f would tell us the exact charge of the quasi-particle excitations.

The first accurate measurement of a fractionally quantized Hall resistance was made by Tsui, Störmer, Hwang, Brooks, and Naughton (1983a). They obtained a value within one part in 10^4 of exact quantization for the 1/3 Hall plateau, at a B of 185 kG, and a temperature of 140 mK. Subsequently, Chang, Berglund, Tsui, Störmer, and Hwang (1984a) have improved this value, as well as obtained accurate values for the 2/3 and 2/5 effects. Less accurate values (parts in 10^3) are also obtained for 3/5, 4/3, and 5/3 by Chang et al. (1984a), Chang, Paalanen, Störmer, Hwang, and Tsui (1984b), Ebert et al. (1984), and Boebinger et al. (1985a). Table 6.3.2 summarizes the best results available.

Table 6.3.2. A summary of the values for the Hall resistance quantization of various fractional quantum Hall effects.

Fractional Effect	Accuracy of Quantization	Sample Density (cm^{-2})	B Field (kG)	Range of Flatness (kG)
1/3	3×10^{-5}	1.53×10^{11}	190	8.2
2/3	3×10^{-5}	2.42×10^{11}	150	2.3
2/5	2.3×10^{-4}	2.13×10^{11}	220	2.9
3/5	1.3×10^{-3}	2.13×10^{11}	147	*
5/3	1.1×10^{-3}	2.06×10^{11}	53	*
3/7	3.3×10^{-3}	2.13×10^{11}	206	*

* At ρ_{xx} minimum

In the cases of 1/3 and 2/3, the accuracy is ~3 parts in 10^5, limited by experimental resolution. In the case of 2/5, it is 2.3 parts in 10^4. At a temperature of 85 mK, the Hall resistance is constant over a finite magnetic field range in the neighborhood of the filling factors 1/3, 2/3 and 2/5. Only at such a level of accuracy can we confidently rule out nearby fractions. For instance, in the 2/5 effect, we can rule out 2/7, 4/11, 5/13, 6/17, and even 200/501, etc.

A word is in order about the experimental arrangements necessary to achieve such high accuracies. In Figure 6.3.7, we show the schematics of a typical setup. A constant current is passed through a specimen and a reference decade resistance standard connected in series, thereby eliminating any discrepancy in the current. The Hall voltage across the sample and the output across the variable resistance standard are fed into two well-matched differential amplifiers with the same gain and good common-mode-rejection capability. The output of the amplifiers are in turn fed into a null detector. The decade resistance is tuned to obtain a null. The resistance value yields the Hall resistance.

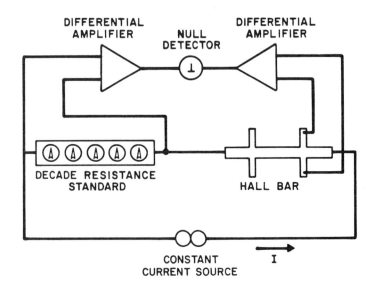

Figure 6.3.7 Schematic for high accuracy measurements ($\approx$ one part in 10^5) of the quantized value of the Hall resistance.

Experimental Aspects

Since it is difficult to obtain two differential amplifiers with exactly the same gain, it is necessary to interchange amplifiers and average the resistance readings. For a small relative difference in amplification of x and true Hall resistance value R_0, the averaging process gives a value, R

$$R = [R_0/(1+x) + R_0(1+x)]/2 \simeq R_0(1 - x^2/2) \qquad (6.3.4)$$

to order x^2. For a typical value of 3×10^{-3} for good commercial preamplifiers such as the EGG-PAR 113 or Ithaco 1201, the error is of the order 10^{-5}–10^{-6}. Alternatively, in the absence of such a sophisticated setup using matched amplifiers, one may achieve reasonably good results by using a quality digital voltmeter with 5 1/2 digits resolution and repeatedly comparing the voltage output of the Hall bar and decade resistor. Commercially available, temperature and temporally stable, high resolution decade resistance standards are accurate to $\approx 10^{-5}$.

6.3.3 Current-voltage characteristics

It is of value to obtain experimental evidence which provides some idea of what type of ground state would give rise to the FQHE. One such piece of information is obtained in the current-voltage (I-V) characteristics. In Fig. 6.3.8 we show data for the Hall voltage in the 2/3 effect at a magnetic field of 88 kG and a temperature of 60 mK. The data indicates that I and V_H are linearly related, down to a voltage of $\approx 1~\mu V$. Taking into account the sample width, this translates to an average Hall electric field of 2×10^{-5} V/cm. To this level, there is no evidence for nonlinearities, or signs indicating a pinning of the current carrying state, below some threshold electric field. Similar results have been obtained for the 1/3 effect (Chang et al. 1985).

This result indicates that the ground state associated with the 1/3 and 2/3 FQHE is most likely not a charge density wave (CDW) or a Wigner crystal. A CDW in quasi-one-dimensional systems is known to be easily pinned by impurities, even at a typical electric field of ≈ 1 V/cm (Wang et al. 1983). It is believed that a 2D CDW is also readily pinned. The ruling out of a CDW ground state is in agreement with theoretical cal-

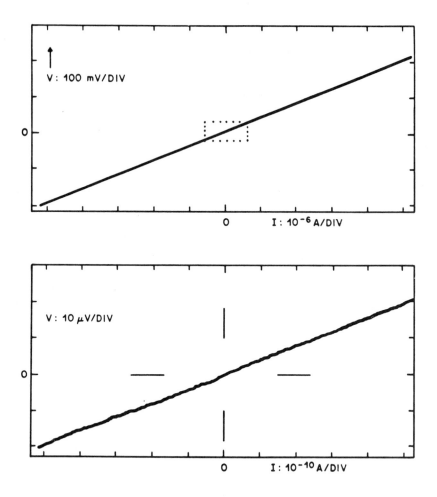

Figure 6.3.8 Current-voltage (I-V) relation for ρ_{xy} in the 2/3 FQHE. The relation is linear down to at least 1 μV, corresponding to an average Hall electric field of $\approx 2 \times 10^{-5}$ V/cm. The magnetic field is 88 kG and the temperature 60 mK (Chang, Paalanen, Störmer, and Tsui, unpublished).

culations (Yoshioka and Fukuyama 1979, Yoshioka and Lee 1983, Levesque et al. 1984, Lam and Girvin 1984, 1985) which show that a CDW state at 1/3 or 2/3 Landau level filling is not particularly energetically favored and therefore is not able to account for the experimental observations.

Experimental Aspects 197

6.3.4 Temperature evolution of the FQHE: The excitation spectrum

The temperature dependence of a physical phenomenon gives valuable information concerning the nature of the underlying physical system and/or processes. The low temperature behavior in the response of a many-body system to perturbations sensitively reflects the energy structure and nature of the low-energy elementary excitations. In systems where an energy gap separates the ground and excited states, the response functions show activated low temperature behavior, e.g., electrical conductivities of semiconductors and insulator, heat capacity of Ising spin models. If the elementary excitations are not separated by an energy gap from the ground state, a power law type of low temperature behavior is often observed; e.g., thermal conductivity and specific heat of lattice phonons, electrical and thermal conductivities of electrons in metal, and specific heat of continuous spin systems (Heisenberg, x-y, etc.). Moreover, the statistics (Bose-Einstein, Fermi, and otherwise) of the excitations is often reflected in the particular functional form of the temperature dependence.

The first studies of the temperature development of the FQHE, for the 1/3 and 2/3 effects, performed by Störmer, Tsui, Gossard, and Hwang (1983c) show the overall trend. The value of ρ_{xx} at the minimum and the magnetic field slope of ρ_{xy} decrease as the temperature is lowered, and tend to zero at very low temperatures. At higher temperatures, ρ_{xx} and the ρ_{xy} slope approach a saturation value. A subsequent study (Tsui et al. 1983a) gives an indication of an activated behavior for ρ_{xx} in the 1/3 effect. The first detailed temperature study, carried out by Chang, Paalanen, Tsui, Störmer, and Hwang (1983) gives more substantial evidence of an activated behavior. In Figure 6.3.9, we present data at two magnetic fields corresponding to 2/3 Landau level filling for two different densities. Over a dynamic range of one and a half orders of magnitude in ρ_{xx}, and a factor of five in temperature, an exponential dependence on $1/T$ is observed.

In Fig. 6.3.10, we show recent results for the 1/3, 2/3, 2/5, and 3/5 effects. First, we discuss the 1/3

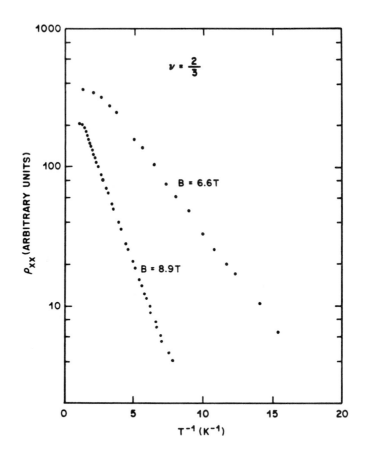

Figure 6.3.9 Activated behavior in ρ_{xx} for the 2/3 FQHE, in two different samples at two magnetic fields. The activated behavior covers one order of magnitude in ρ_{xx} and a factor of five in temperature (Chang, Paalanen, Störmer and Tsui unpublished, Boebinger et al. 1985d).

and 2/3 data. In the higher temperature portion of this low temperature regime, an activated behavior is observed, typically covering a range of one and a half orders of magnitude in ρ_{xx}. In the lower temperature portion, ρ_{xx} deviates from a simple activated behavior. The deviation is toward larger ρ_{xx}, or conductivity ($\sigma_{xx} \propto \rho_{xx}$). This trend is reminiscent of the hopping contribution to conduction in conventional disordered systems (Mott and Davis 1979) and in the IQHE (Tsui et al. 1982c, Briggs et al. 1983, Ebert et

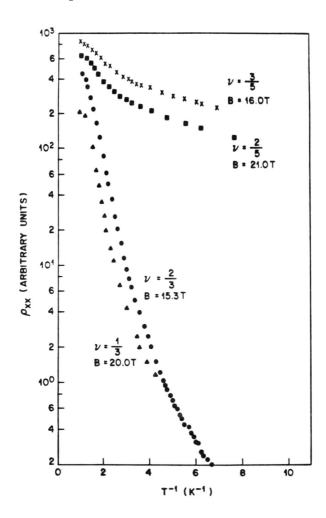

Figure 6.3.10 Semi-log plot of ρ_{xx} vs. $1/T$ for the FQHE at $v = 1/3$, 2/3, 2/5, and 3/5, at the magnetic fields 200 kG, 153 kG, 210 kG, and 160 kG, respectively (Boebinger, Chang, Störmer, and Tsui, unpublished).

al. 1983b). There is some debate as to whether the hopping conduction should behave as $\exp[-(T_c/T)^{1/2}]$ (Briggs et al. 1983, Ono 1982, Ebert et al. 1983b) or $\exp[-(T_c/T)^{1/3}]$. Although our data covers over three orders of magnitude for the 1/3 minimum, we are not able to conclusively address this question. Part of the difficulty arises from the lack of a reliable measure of the effective sample width due to density gradients (Zheng et al. 1985a). Many other groups,

Kawaji et al. (1984), Clark et al. (1985), Kukushkin et al. (1985), have also observed an activated behavior in ρ_{xx} for the 1/3 and 2/3 effects.

The 2/5 and 3/5 effects behave quite differently, in spite of the general trend of a decreasing ρ_{xx} with decreasing temperature. First of all, even though 2/5 occurs at a magnetic field very close to 1/3 and 3/5 close to 2/3 in a sample of a constant density, over a similar temperature range, ρ_{xx} varies by less than an order of magnitude ($\approx$ a factor of five), whereas it would invariably vary over two to three or more orders of magnitudes for the 1/3 and 2/3 effects. The transition between the higher temperature and lower temperature portions also appears more abrupt. See Figure 6.3.11.

The presence of two temperature ranges each of very small dynamic ranges in ρ_{xx} makes the interpretation difficult. It is possible to fit many combinations of functional forms to account for the two regions. Sensible possibilities include: exponential/exponential, exponential/$T^{-1/2}$ hopping, and exponential/$T^{-1/3}$ hopping, each with four free parameters. Therefore any statement is necessarily speculative. The author's prejudice is for the first possibility. First of all, the upper temperature portion is consistent with an activated behavior, and in accord with theoretical considerations which indicate that the existence of an energy gap is crucial to the QHE. The interpretation

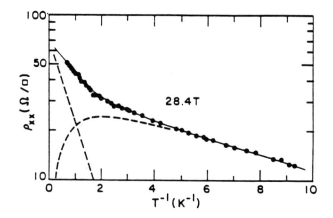

Figure 6.3.11 ρ_{xx} in the 2/5 FQHE at 284 kG magnetic field (Boebinger et al. 1985d).

Experimental Aspects

of the lower portion is more controversial. A phonon assisted, or other form of, hopping conduction is typically an enhancement mechanism to exponentially small activated conduction, in systems with a finite energy gap. It would be surprising if hopping becomes so dominant when the conductivity is still large. σ_{xx} (α ρ_{xx}) is only a factor of two to three smaller than the high-temperature value in this case. The ease of hopping between localized excitations is determined by the localization length, ξ_0, of the localized states. Since 2/5 occurs at a very similar magnetic field as 1/3 (17% difference), it is difficult to understand how hopping conduction can be so drastically modified, especially in view of the similar physical size of the quasi-particle excitations in the effects, which is of the order of the magnetic length $\ell \approx 60$ Å (Laughlin 1983a, 1984a). For particles of such a small size, the localization length is usually determined by intrinsic sample disorder. The choice of two exponentials is also in accordance with hierarchical theories, which predict the existence of more than one type of fractionally charged excitations, of different energies, for the daughter effects (Haldane 1983, Halperin 1984, Laughlin 1984a, Tao and Thouless 1983). It has also been suggested that the 1/3 and 2/3 effects show two activation energies (Kawaji et al. 1984, Tsui et al. 1982c) However, we have recently concluded that this is a non-equilibrium result (Boebinger et al. 1985b).

The observation of a mobility gap in the excitation spectrum is of fundamental importance. Its existence is requisite to the various endeavors to understand the exact Hall resistance quantization in the presence of disorder, including the gauge invariance argument (Laughlin 1981, Tao and Wu 1984) and topological considerations (Niu and Thouless 1984a, Avron et al. 1983, Tao and Haldane 1986) [see Chap. IV].

6.3.5 The energy gap

A successful theory of the FQHE should ultimately be able to predict the size of the energy gap. Better yet, it should also be able to account for the functional dependence of the gap on various parameters, such as the magnetic field and sample mobility. Experimental studies on such dependences are thus helpful. The first systematic study has been carried out by Chang, Paalanen, Tsui, Störmer, and Hwang (1983) on the 2/3 effect. Fig. 6.3.12 shows the result [Note 1]. We remark that the data does not agree with a magnetic field dependence of $B^{1/2}$ predicted by theory and suggests the existence of a mobility threshold below which the 2/3 effect does not occur. The data is taken on a single sample on which the density and mobility can be varied with a back gate. Such a threshold for the 4/3 effect has recently been observed in Si MOSFET's by Kukushkin and Timofeev (1985).

A recent comprehensive study of the energy gap of the 1/3 and 2/3 effects by Boebinger et al. (1985b,d) has yielded the results presented in Figure 6.3.13. New data are obtained from three different samples of very high mobility. Sample C is made from the same crystal as the sample of Fig. 6.3.12. Its density of

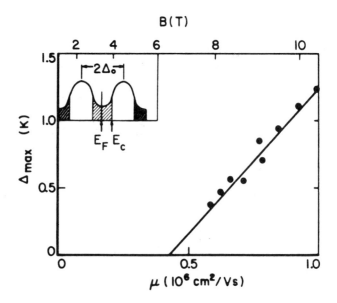

Figure 6.3.12 The $v = 2/3$ activation energy [Note 1], as a function of mobility, μ, and magnetic field B. The solid line is a guide to the eye. The data is taken from a sample in which the density and mobility are continuously varied with a back gate. The data suggests a mobility threshold below which the 2/3 FQHE does not exist (Chang et al. 1983).

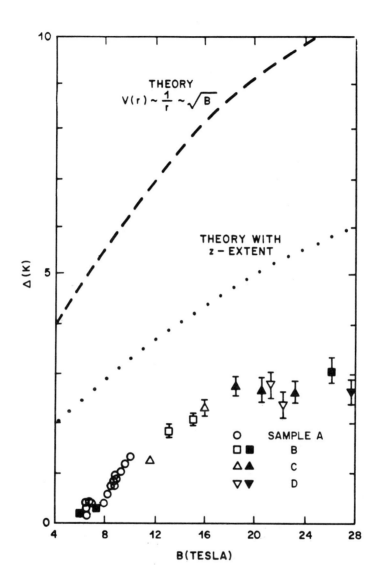

Figure 6.3.13 A collection of data for the 1/3 (4/3), solid symbols, and 2/3 (5/3), open symbols, activation energies (Boebinger et al. 1985b). Current theoretical results (C = 0.1) are included for comparison (Levesque et al. 1984, Morf and Halperin 1985, Halperin 1985, Yoshioka et al. 1983, Su 1985, Haldane and Rezayi 1985, Girvin et al. 1985, 1986b, Zhang and das Sarma 1985).

1.5×10^{11} cm^{-2} and mobility of 4.5×10^5 cm^2/Vs can be gated to 2.1×10^{11} cm^{-2} and 9×10^5 cm^2/Vs. Samples B and D have densities of $\approx 2.0 \times 10^{11}$ cm^{-2} and mobilities of $\approx 10^6$ cm^2/Vs. In addition, the data in Fig. 6.3.12 (sample A is Fig. 6.3.13) and data from Boebinger et al. (1985a) on 4/3 and 5/3, are all included for comparison and for completeness. In spite of a significant amount of scatter in the data, especially near the threshold region, all the data appear to fall on a single curve [Note 2]. The larger scatter near the threshold is understandable, since the magnetic field threshold must depend on the quality of the various samples. Away from the threshold, in the high magnetic field and sample mobility region, the values show less scatter, with a tendency to saturate. One expects intuitively that in this high mobility region, the values should approach those predicted by theory in which disorder is not taken into account. Theoretical calculations based on Laughlin's wave function (Laughlin 1983a, Girvin et al. 1985a, 1986a,b, Levesque et al. 1984, Morf and Halperin 1986, Halperin 1985) and on exact diagonalization of systems with small numbers of particles (Yoshioka et al. 1983, Su 1985, Haldane and Rezayi 1985a) for a 2D, repulsive $1/r$ interaction between electrons gives a functional form for the energy gap, Δ, of

$$\Delta = C\ e^2/2\kappa\ell \qquad (6.3.5)$$

with a proportionality constant C which ranges from 0.06 to 0.11 for the 1/3 and 2/3 effects. Here, the dielectric constant κ is about 13 in GaAs. Using a value of 0.1 for C, the gaps obtained are at best a factor of 3 larger than experiments. In Fig. 6.3.13, we have included a curve using a current value of C = 0.10 (Halperin 1985) for comparison. Three factors have been invoked to account for the discrepancies: 1) The finite extent of the electrons in the third direction in heterostructures. 2) The mixing of higher Landau levels, neglected in Laughlin's wave function, which tends to reduce Δ. 3) The effect of disorder, which can drastically reduce Δ.

The extent, b, of the electronic wave function in the third (z) direction in GaAs-AlGaAs heterostructure is about 100 Å. Since the separation between electrons in the experimental situation is typically

Experimental Aspects

around 250 Å, one expects some effect from the finite z extent. A more realistic potential than a $1/r$ potential is perhaps $1/(r^2+b^2)^{1/2}$ for $r \gtrsim b$, and $(-1/b)\ln(r/\ell)$ for $r \ll b$, $b \approx 100$ Å. This type of modification is found to reduce Δ by $\approx 40\%$ in the B-field range of interest (Zhang and das Sarma 1986, Girvin et al. 1986b, Yoshioka 1985b) The latter author (1984, 1985b) has performed numerical calculations taking into account the mixing of higher Landau levels. He finds a reduction of about 20% for Δ. It appears that the above two factors are not sufficient to account for the discrepancy between theory and experiment. The remaining difference is likely an effect of disorder. MacDonald, Liu, Girvin and Platzman (1985) recently considered the effects of remote ionized impurities on the 1/3 and 2/3 energy gap. In a self-consistent calculation, they succeeded in obtaining the qualitatively correct behavior of a disorder threshold for the occurrence of the effects. Even though the calculation is performed for an exciton (a quasiparticle/quasihole bound pair) - a neutral object which in principle cannot carry DC current -- a similar result may be expected for an exciton whose constituents are 'bound' at infinite separation, i.e., a quasiparticle and quasihole pair separated at infinity which may be considered as two free objects.

Figure 6.3.14 shows available data on the activation energies of 2/5 and 3/5 effects [Note 1]. The 2/3 and 3/5 effects represent particle-hole conjugates of the same state and therefore should have the same excitation energy under identical conditions. The data is less complete than for the 1/3 and 2/3 effects. Nevertheless, they indicate a typical value of around 0.5-2 K for the larger of the two activation energies discussed above, at magnetic fields of 16-25 T. These values are also lower than, but consistent with, theoretical values.

6.3.6 Quasiparticle statistics--are they Fermi-Dirac?

In both the IQHE and FQHE, important information concerning the filling behavior of available states can be obtained by studying the temperature dependence of the transport coefficients ρ_{xx} and ρ_{xy}, as the filling factor, v, is varied in the vicinity of a particular QHE step. In the IQHE, experimental (Paalanen et al.

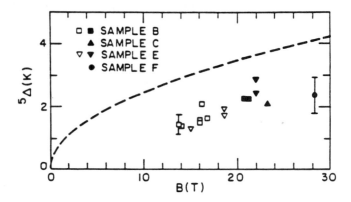

Figure 6.3.14 The activation energies [Note 1] of the 2/5 (solid symbols), and 3/5 (open symbols) FQHE, as functions of B (Boebinger et al. 1985b,d).

1982) and theoretical (Trugman 1983, Joynt and Prange 1984, Levine et al. 1983, Giuliani et al. 1983, Aoki and Ando 1981) observations demonstrate that within a Landau band, only a minuscule fraction of the electronic states, near the band center are extended, with all others localized. This result suggests that the transport coefficients should show an activated behavior not only at a complete filling of a Landau band, but also in the immediate neighborhood about a complete filling. By measuring the variation of the activation energy with filling factor it is possible to deduce the single-particle density of localized states (Wei et al. 1985a, Weiss et al. 1985). This topic is discussed in detail in Section 6.4.1. In the FQHE, it is possible to carry out a similar experiment, in which the temperature dependence of ρ_{xx} and the deviation of ρ_{xy} from its quantized value ($\Delta\rho_{xy} = \rho_{xy} - h/fe^2$) are studied, as the filling factor is varied about $v = f$, either by varying the magnetic field or sample density. The first such study has been carried out by Chang et al. (1983). Figure 6.3.15, shows the results for a sample with $v = 2/3$ at 92.5 kG. Within our experimental resolution ρ_{xx} and $\Delta\rho_{xy}$ are proportional to each other, at each v about 2/3, for $0.6 < v < 0.71$. v is varied by changing the density. At each v, ρ_{xx} and $\Delta\rho_{xy}$ appear activated. One may plot the activation energy, Δ, as a function of v, as in Fig.

Experimental Aspects

6.3.15b. The resulting curve is reminiscent of a bell shaped curve. It is almost symmetric about the maximum at $v = 2/3$. Similar curves for 1/3, 2/3, 2/5, and 3/5 have been obtained in other works (Kawaji et al. 1984, Boebinger et al. 1985c). The observed dependence of Δ on v may be understood in terms of a phenomenological picture of disorder-broadened quasiparticle/quasihole bands (Chang et al. 1983) in direct analogy with the situation in the IQHE (Paalanen et al. 1984). See Section 6.4.1. The picture goes as follows: According to theory, at 2/3, a gap 2Δ develops between the centers of 1/3-charged quasiparticle and quasihole bands. Each band is highly degenerate. If an electron is added (removed), three quasiparticles (quasiholes) are created within the band. Disorder broadens the bands and gives rise to a sharp mobility edge separating the extended and localized states. In an experiment, the observed activation energy measures the energy difference between the mobility edge, E_c, and the quasiparticle "Fermi-energy". See the inset in Figure 6.3.12. As the filling factor is increased from 2/3, the "Fermi-energy" moves toward E_c and a reduced activation energy is obtained, eventually reaching zero as it crosses E_c.

It is of importance to point out that if we assume the energy gap is correctly predicted by the prevailing theories, for an ideal system without disorder, the results of Fig. 6.3.15 would indicate that the quasiparticles do not behave as fermions, so that it is possible to place more than one quasiparticle in the same state. The argument goes as follows: First, we choose the lowest value of C consistent with the various calculations, and take into account the gap reduction factor [points (1) and (2) in Section 6.3.5] to arrive at a low value of $C = .04$. Next, we observe that disorder should not alter the separation of the quasiparticle/quasihole band center separation, which is given by the ideal theoretical gap value. Such an observation is consistent with experimental results on the IQHE. The total number of available states in each band is $(1/3)(eB/hc)$ (Laughlin 1983a). Our Δ vs. v has a width, Δv, of 0.1 between the mobility edges in the particle and hole band, or 0.05 between the mobility gap center ($v = 2/3$), and each mobility edge. This implies that at the quasiparticle (quasihole) mobility edge, the quasiparticle (quasihole) band holds three times 0.05 or 0.15 eB/hc quasiparticles (quasiholes), i.e., 45% of the available states within the band. If the quasiparticles (quasiholes) were

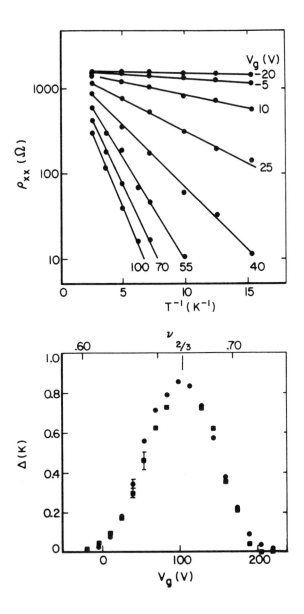

Figure 6.3.15 (a) ρ_{xx} vs. $1/T$ at $B = 92.5$ kG, at various V_g, for ν around 2/3. (b) Activation energy for ρ_{xx} and $\Delta\rho_{xy} = \rho_{xy} - (3/2)(h/e^2)$ vs. V_g and ν. Solid circles: ρ_{xx}, squares: $\Delta\rho_{xy}$ (Chang et al. 1984b).

Experimental Aspects

fermions, this would imply that 90% of the band is localized. However, at $B = 92.5$ kG, using $C = 0.04$, a separation of 2.8 K is predicted between the center of the mobility gap and the band center, whereas the measured maximum Δ, at $v = 2/3$, is 850 mK. Even if we assume the extreme scenario wherein the bands have a flat density of available states, we would necessarily conclude that 90% of the fermions are residing in, at most, the (850 mK)/(2.8 K) = 30% band-tail region. We are forced to conclude that the quasiparticles (quasi-holes) are not fermions.

Theoretical considerations have also pointed to the possibility that the quasiparticles are not fermions, but obey a new type of statistics, fractional statistics (Halperin 1984), in which exchange of identical quasiparticles give rise to a multi-valued phase factor. This type of exotic statistics was discovered by field theorists (Wilczek and Zee 1983, Jackiw and Rebbi 1976, Wu 1984, Goldin et al. 1985) and is believed to be obeyed by magnetic monopoles (Ringwood and Woodward 1984) and vortices in 2D superfluids (see however, Haldane and Wu 1985).

6.3.7 Phase transition

Laughlin's theory indicates that the ground state of the FQHE is an incompressible electron fluid. The question naturally arises as to whether the formation of the fluid constitutes a phase transition phenomenon in which an electron gas condenses into the correlated fluid, as the temperature is lowered. This question is first raised in a paper by Anderson (1983) in connection with the Tao-Thouless (1983) theory of the FQHE, which is based on the idea of a discrete broken symmetry. The paper points out the possibility that Laughlin's wave function may be obtained from the Tao-Thouless wave function by adiabatically turning on a residual electron-electron interaction, and would also contain a discrete broken symmetry. It is further pointed out that the fractionally-charged quasiparticle excitations are domain walls between discrete broken symmetry ground states, much like domain walls in a 1D ferromagnetic Ising model with $1/r^2$ interactions, topological defects in 2D crystals, or vortices in 2D superfluids. In the latter systems, the domain wall (vortex) and anti-domain wall (anti-vortex) attractive interaction is logarithmic in their separation, and a Kosterlitz-Thouless (1973) marginal phase transition is known to occur. This phase transition corresponds to the unbinding of domain wall/antidomain wall pairs above a critical temperature.

In the present system, regardless of the controversy surrounding the existence of a discrete broken symmetry, it is clear that the fractionally charged quasiparticles (quasiholes) are topological entities, since the wave function differs from the ground state wave function in such a way that a quasiparticle localized at the origin can be detected at infinity, at the boundaries of the many body system [see Chap. X]. It is well known that the excitation energy for a free, unbound, quasiparticle/quasihole pair is finite, unlike the situation in the Kosterlitz-Thouless system where it is logarithmically divergent [see Chaps. IX, X]. Therefore one can argue, using a free energy consideration on the stability of the ground state, that a phase transition does not occur. The argument is as follows (Chang 1985, Girvin et al. 1986a,b): If the ground state condensate is thermodynamically favored, it must be stable against the formation of free quasiparticles/quasihole pairs. At $v = 2/3$, it costs an energy of 2Δ to create a free pair. However, the number of available pairs to be created is $N = (1/3)(eB/hc)A$, where A is the total sample area. As a result, the free energy difference for creating free pairs compared to the ground state free energy is

$$\Delta F = 2\Delta - k_B T \ln N \qquad (6.3.6)$$

Since Δ is finite, in the thermodynamic limit of $N \to \infty$, the ground state is favored only at $T = 0$, and a phase transition does not occur at a finite temperature.

The preliminary but careful experimental study is in accord with the above discussion. Since ρ_{xx} is a response function, one may expect that a phase transition would manifest itself in the form of a structure in ρ_{xx}, as the temperature sweeps through the critical temperature. Figure 6.3.16 shows our result for the 2/3 effect. The sample density is 1.5×10^{11} cm^{-2} and the magnetic field 92 kG. The two disjoint curves, crosses and dots, represent data in liquid He^3 and He^4, respectively, in order to cover the temperature regions of interest. The experimental resolution

Experimental Aspects

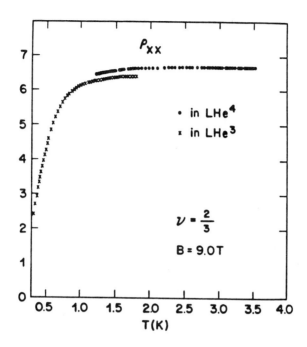

Figure 6.3.16 ρ_{xx} vs. T at $v = 2/3$ and $B = 90$ kG. ρ_{xx} appears smooth. The activation energy measured for ρ_{xx} is 850 mK (Chang, Störmer, and Tsui, unpublished).

within a continuous temperature sweep is represented by the scatter in the data. Within our resolution, there is no evidence of a phase transition. As a reference, the activation energy in ρ_{xx} has been measured to be about 850 mK.

6.4 Magneto-Transport in Two Dimensions

In this section, we address experiments which seek to gain a more complete understanding of strong (magnetic) field magneto-transport in two dimensions. As pointed out in the introduction, to a large extent, the IQHE and FQHE find a common ground in this subject. The topics covered include: 1) The 2D electronic density of states in a high magnetic field, deduced from magnetization and transport measurements. Strictly speaking, the concept of a (single-particle) density of states is meaningful only in the IQHE, in which the electrons can be considered as noninteracting. However, the relevance to the FQHE becomes clear within the phenomenological picture of impurity-broadened excited-state bands, discussed in Section 6.3.6. 2) Hall potential distribution within a sample in the quantized Hall regime. The results give an idea of the nature of the current-carrying, extended states and their spatial distribution within a sample. 3) The scaling theory of magneto-conduction. We discuss what we believe to be evidence consistent with or in support of the scaling theories, for both the IQHE and FQHE.

6.4.1 Electronic density of states

A microscopic theory of strong-field magneto-transport poses considerable technical difficulties, associated with the large degeneracy N_B within an impurity-free Landau level. As a result, most microscopic theories (Joynt and Prange 1984, Giuliani et al. 1983, Halperin 1982, Thouless 1981) and calculations of the density of states (Ando and Uemura 1974, Brézin et al. 1984, Wegner 1983, Affleck 1983) neglect inter-Landau level mixing, assuming from the start that the strength of the random potential fluctuation is small. The density of states obtained tends to have exponentially small or vanishing density between broadened Landau bands, especially near the middle of the gap. Experimentally, it has always seemed that a non-negligible density exists between Landau bands. This conclusion results from the observation that at an integer filling factor of even i, e.g., $i = 2$ IQHE, an activation energy very nearly equal to $\hbar\omega_c/2$ is observed in ρ_{xx} (Tausenfreund and von Klitzing 1984). If the density of states is negligible between Landau bands, it would be difficult to tune the filling factor so precisely so as to pin the Fermi level at the middle to give an activation energy of $\hbar\omega_c/2$, and a small deviation of the magnetic field or carrier density would drastically reduce the observed activation energy.

Experimentally, there has been a variety of work to deduce the density of states, from specific heat (Gornik et al. 1985a,b), magneto-capacitance (Smith et al. 1985b), magnetization measurements (Störmer et al. 1983d, Eisenstein et al. 1985b) and the temperature

Experimental Aspects

dependence of ρ_{xx} (Weiss et al. 1985) and ρ_{xy} (Wei et al. 1985a), all with rather similar results. It is gratifying to find that results deduced from equilibrium thermodynamic measurements are consistent with those deduced from nonequilibrium, transport measurements. Here, we review the magnetization and transport experiments.

The magnetization, M, of a (free) Fermi system is given by (Störmer et al. 1983d):

$$M(B) = \frac{dE}{dB} = \frac{d}{dB} \int_0^\infty \frac{\rho(E,B) \, E \, dE}{1 + e^{(E-E_F)/k_B T}} \qquad (6.4.1a)$$

with the constraint,

$$n = \int_0^\infty \frac{\rho(E,B) \, DE}{1 + e^{(E-E_F)/k_B T}} . \qquad (6.4.1b)$$

Therefore, if one is able to measure $M(B)$, one can attempt to choose the best functional form of the density of states $\rho(E,B)$ to fit $M(B)$. In contrast to a transport measurement, a magnetization measurement is not sensitive to a distinction between localized and extended states. The first attempt to measure the magnetization was carried out by Störmer, Haavasoja, Narayanamurti, Gossard, and Wiegmann (1983d), using a commercial SQUID magnetometer of $\approx 10^{-10}$ J/T sensitivity. Oscillatory behavior is observed in M as B is varied, in synchronization with oscillations in ρ_{xx} and plateaus in ρ_{xy}. The samples are typically GaAs-AlGaAs superlattices of 172 layers, with a mobility of 1.9×10^4 cm^2/Vs, a density of 8.2×10^{11} cm^{-2} per layer, and a total effective area of 33 cm^2. At $T = 1.5$ K, the maximum signal in the range of 20 kG < B < 60 kG is about 5×10^{-10} J/T. From this work, the authors are able to deduce a Landau band width of 2 meV around 5 T, which is significant compared to an $\hbar\omega_c$ of 9 meV.

To improve the signal to noise ratio and approach the sensitivity needed to detect the magnetization of a single layer, an ingenious, DC torsional magnetometer was invented by Eisenstein (1985). The device is depicted in Fig. 6.4.1a. A wire of Pt-W alloy, 37 microns in diameter and 2 cm long, is used as the tor-

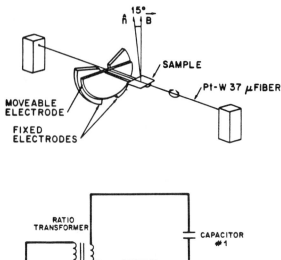

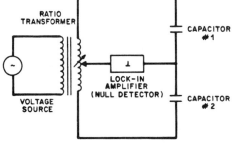

Figure 6.4.1 (a) Schematic diagram of a high sensitivity torsional magnetometer (Eisenstein 1985a,b). (b) The ratio-transformer and null detection circuitry to measure capacitance variations due to a rotation in the torsional magnetometer.

sional element. It is held under tension between plastic supports in a direction perpendicular to the magnetic field. A 1 cm radius plastic disk, one half of it (π radians) coated with gold, is mounted onto the wire with its plane perpendicular to the wire. A second plastic disk, on which two identical pie-shaped electrodes are evaporated, is held fixed at a separation of 150 microns from the first disk. In this arrangement, two capacitors are formed, one each between a fixed pie-shaped electrode and the gold electrode on the disk attached to the wire. A null circuitry (Figure 6.4.1b) is used to detect any rotation of the torsional wire, employing a ratio transformer and an AC lock-in technique, achieving a sensitivity of 10^{-7} radian. In an actual experiment, the sample is mounted onto a plastic extension of the movable disk, at an

Experimental Aspects

angle of 15° from the B-field direction. A torque, $\tau = \mathbf{m} \times \mathbf{B}$, is exerted on the torsional wire, giving rise to a static rotation of the wire, after the transient oscillations are damped out by eddy currents in a silver tape. Because the magnetometer detects magnetic field induced torque, background magnetic moments induced in the bulk, sample holder, etc., in the B direction, are not detected [$\mathbf{B} \times \mathbf{B} = 0$], while the 2D magnetization of the heterostructure (superlattice) is constrained to lie at 15° from the $\mathbf{B}$ direction, and is readily detected. In this manner a sensitivity of about 10^{-12} J/T is obtained at 50 kG. Fig. 6.4.2 shows the results on a superlattice of 50 layers and a single heterostructure of GaAs-AlGaAs, and the density of states at $B = 5\ T$ deduced for the superlattice sample. It is apparent that there is a very significant density between Landau bands.

In a transport measurement to deduce the density of states, one relies on a phenomenological picture of impurity broadened Landau bands (Wei et al. 1985a, Weiss et al. 1985, Paalanen et al. 1984). Experiment and theory strongly suggest that extended states exist only near the center of a Landau band. Therefore, one may make the assumption that the mobility edge, E_c, which separates the extended and localized states, resides at the band center. If the Fermi level lies in the localized region, at low temperature, the transport should show an activated behavior, in which the activation energy, Δ, measures the separation between E_F and the nearest E_c; i.e., $\Delta = |E_F - E_c|$. If we vary the electron density n, E_F, and hence Δ, varies. One therefore measures $\Delta(n)$. We have:

$$\rho(E_F) = \frac{dn}{d\Delta} = \left[\frac{d\Delta(n)}{dn} \right]^{-1} . \quad (6.4.2)$$

In an actual experiment, E_F is often varied by varying B, not n. What one measures instead, is a quantity related to $\rho(E_F)$. Wei et al. have deduced the density of states for a sample of InGaAs-InP heterostructure, using the activated behavior in $\Delta\rho_{xy}$. Weiss et al. have made a similar deduction by studying ρ_{xx} in GaAs-AlGaAs heterostructures of varying mobilities and densities. [See Figures 6.4.3 and 6.4.4.]

The results are similar to Fig. 6.4.2c. For low-mobility samples, around 2-3 $\times\ 10^4$ cm^2/Vs, a density of $\approx 10^{10}$ cm^{-2}/meV has been deduced at the middle of the mobility gap at around 35-70 kG (Wei et al. 1985a, Weiss et al. 1985), while it is about 2×10^9 cm^{-2}/meV for a high mobility sample at 72 kG (Weiss et al.).

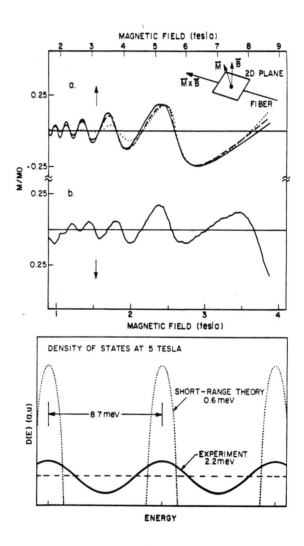

Figure 6.4.2 Normalized magnetization (Eisenstein et al. 1985b) for: (a) GaAs-AlGaAs superlattice of 50 periods. (b) GaAs-AlGaAs single heterostructure. (c) Comparison of model density of states, solid line, used to fit the data in (a) at 50 kG, with the short-range theory of Ando.

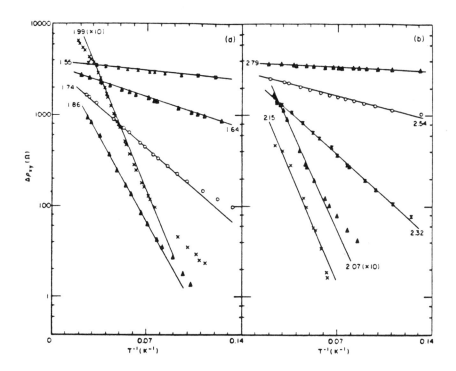

Figure 6.4.3 $\Delta\rho_{xy} = \rho_{xy} - h/2e^2$ vs. $1/T$ in a semi-log plot, for various filling factors about $v = 2$, in an InGaAs-InP heterostructure. An activated behavior is observed at each v which is varied by changing B (Wei et al. 1985a).

6.4.2 Hall potential distribution in the QHE

The distribution of the Hall potential is intimately connected with the distribution of the Hall current, as well as the local charge density. These relationships result from the requirement of local current balance and Maxwell's equation relating the electrostatic potential and the charge distribution (MacDonald and Středa 1984):

$$j(r) \times B = \rho(r) \, E(r)$$

$$\nabla \cdot E(r) = 4\pi \, \rho(r)/\kappa ,$$

(6.4.3)

where $j(r)$, $\rho(r)$, and $E(r)$, are local current density,

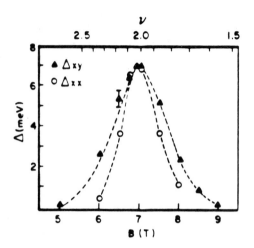

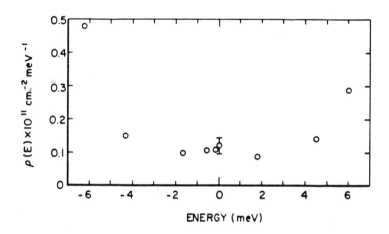

Figure 6.4.4 (a) Activation energy, Δ, vs. filling factor, ν, for: $\Delta\rho_{xy}$ of Fig. 6.4.3 (solid triangles), ρ_{xx} (open circles). (b) Electronic density of states deduced from Δ for $\Delta\rho_{xy}$ in (a) (Wei et al. 1985a).

charge density, and electric field, respectively, and κ is the dielectric constant. In an ideal 3D system of uniform charge density and spatial dimensions L_x, L_y, L_z, $(L_x, L_z) \gg L_y$, it is simple to envision the distribution of the Hall potential. See Figure 6.4.5. Let the magnetic field point in the z direction and the applied current, I, be in the x direction. Then the Hall electric field is uniform in the y direction

Experimental Aspects

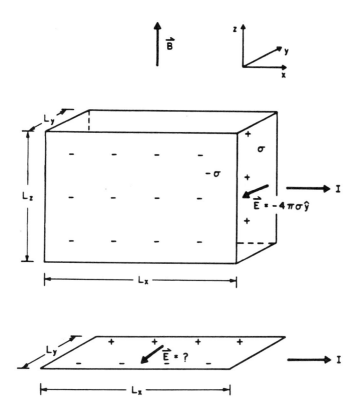

Figure 6.4.5 (a) The charge and electric field distributions in a 3D sample in the Hall effect. A Hall voltage drop, $V_H = 4\pi\sigma L_y$, occurs between the two y-faces of the specimen. (b) The situation in 2D.

with $E(r) = V_H/L_y$, where V_H is the total Hall voltage, and the current density is also uniform with $j(r) = I/L_yL_z$. The local density $\rho(r)$ is uniformly zero (neutralized by a positive background) in the bulk, but two sheets of opposite charges accumulate, one at each y-face of the 3D system, one sheet arising from the residual neutralizing charge after a depletion of the carriers. Two large, if not infinite, parallel sheets of opposite charge give rise to a constant electric field in between. The consistency requirement in Eq. 6.4.3 is naturally satisfied. The situation is more complicated in two dimensions (MacDonald and Středa 1984). Charge accumulation at the y edges only will not give rise to a constant Hall electric field in the interior. In addition, in the QHE, the presence of localization effects makes a determination of the Hall potential distribution even more difficult. In order to sort out the relative contributions of edge states and bulk states (Halperin 1982, Kazarinov and Luryi 1982) to current conduction, and the nature of the extended states, it is of importance to measure the Hall potential distribution in the interior of a sample, in the quantized Hall regime.

Several groups, Zheng et al. (1985a), Ebert et al. (1985), Simon et al. (1985), and Sichel et al. (1985) have measured the quantum Hall potential or current distributions, with similar results. First of all, it is found that the source impedance of the voltage probe contacts in the interior of a specimen becomes very large in the well-quantized regime. In fact, it should scale as $1/\sigma_{xx}$, and diverge as $T \rightarrow 0$. For this reason, in voltage measurements it is necessary to use an electrometer with large input impedance ($\geq 10^{13}$ ohms) to avoid loading down the source voltage. Secondly, it is found that macroscopic density inhomogeneities determine the Hall potential and current distributions, giving rise to a bunching effect near the sample edge in the Hall plateau region. See Figure 6.4.6a. Because the resistance in a local region of a sample is minimum when the quantization condition, $n(r)hc/eB = i$, is satisfied, and because the density inhomogeneities often arise in the form of a gradient in a particular direction, the Hall potential drop switches from one edge to the other as the magnetic field sweeps through the center of the Hall plateau. Zheng et al. 1985, Ebert et al. 1985, Sichel et al. 1985). Furthermore, Zheng et al. showed that if the density gradient is reduced, the bunching effect is reduced; see Fig 6.4.6b. In Figure 6.4.7, we plot the Hall potential distribution of Figure 6.4.6b, for the $i = 2$ IQHE, as a function of the radius from the sample center. It appears that the potential is well distributed throughout the sample interior and is not confined to the edges. The ability of the total Hall voltage across the sample to remain quantized, regardless of the macroscopic density inhomogeneities, seems to suggest that a percolation picture of the extended states, wherein the extended states follow contours of constant potential, is appropriate in the QHE.

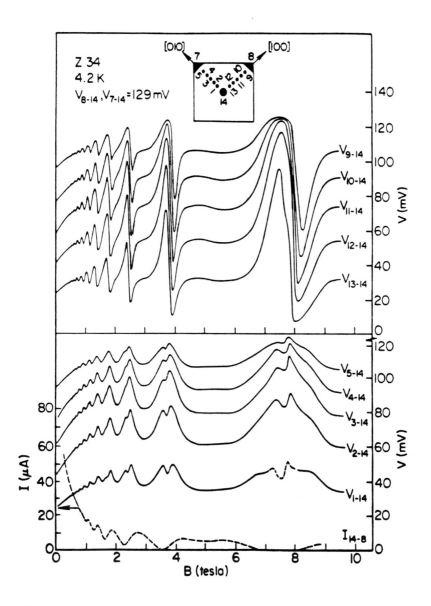

Figure 6.4.6 The potential as a function of B, measured at different interior contacts with respect to the center contact (14), of a Corbino-like sample of GaAs-AlGaAs heterostructure. In one direction, in which the sample density gradient is the most pronounced (top figure), bunching of the potential is observed in the quantized Hall region. In the perpendicular direction, (bottom figure) bunching is greatly reduced. (Zheng et al. 1985).

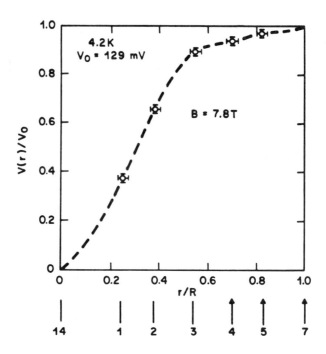

Figure 6.4.7 The distribution of the Hall potential plotted against the radius from the sample center, for Fig. 6.4.6 (bottom) at a magnetic field of 78 kG corresponding to $v = 2$. The potential is well distributed throughout the sample bulk.

6.4.3 Scaling theories of 2D magneto-conduction

The theoretical demonstration of the existence of extended states, for white-noise potentials (Pruisken 1984, Levine et al. 1983), for long-range potential fluctuations (Trugman 1983) and δ-function scatterers (Aoki and Ando 1985), is a crucial result in magneto-transport in two dimensions. Using this result and calculations on nonlinear sigma models [Chap. V], theorists (Pruisken 1985a, Khmel'nitzkii 1983) have invented a two-parameter scaling theory relating the change in σ_{xx} and σ_{xy} as the sample size is varied, with a result depicted in Figure 6.4.8. Although some of the essential features may possibly be expected from experiment, for instance the presence of the fixed points at $\sigma_{xx} = 0$ and $\sigma_{xy} = ie^2/h$, this theory (Laughlin 1984b) is able to predict a shift in the

Experimental Aspects

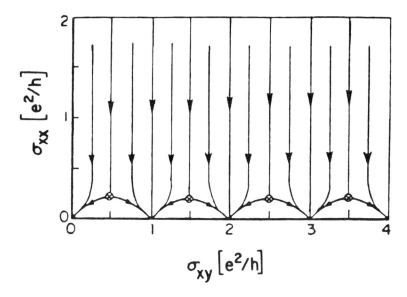

Figure 6.4.8 A renormalization group flow diagram showing two-parameter scaling of the IQHE (Wei et al. 1985b).

position of the extended states in the energy spectrum, when the disorder becomes comparable to or larger than the Landau level separation. More recently, Laughlin, Cohen, Kosterlitz, Levine, Libby, and Pruisken (1985) (the gang of six!?) have extended this theory by surmising the effects due to electron-electron interaction. In this three-parameter theory, an order-disorder transition is found which corresponds to the destruction of the FQHE. In addition, a shift in the energy position of extended quasiparticle (quasihole) states is also predicted, when the disorder becomes large and comparable to the excitation gap.

Experimentally, two aspects of the scaling theories are accessible and can be readily tested. One is the scaling curve relating σ_{xx} and σ_{xy}, and the other, the shift in the energy position of the extended states. The simplest attempt to deduce the scaling diagram is to map out the temperature dependence of σ_{xx} and σ_{xy} at various filling factors v. The connection between the values obtained and the theoretical values from changing the length scale is made by supposing the existence of an inelastic length scale, the Thouless length (Thouless 1977, Abrahams et al. 1979, Aoki and Ando 1985), at each given temperature. At sufficiently low temperatures, where the inelastic lengths are expected to be long, true scaling may be expected. The shift of the extended state is observed as a displacement in the magnetic field position of the QHE in a sample where the density is fixed, while the magnetic field is changed to vary the filling factor. The displacement may be in the central position of the quantized Hall plateau or the onset position, i.e., the edge of the plateau as $T \rightarrow 0$, or both. In principle the displacement can either be upward or downward, depending on the extent of mixing of neighboring higher or lower Landau (quasiparticle) levels. See Figure 6.4.9.

Wei, Chang, Tsui, and Rezaghi (1985a) first noted the possibility of the connection between their experimentally measured ρ_{xx} and ρ_{xy} and the scaling theory. Subsequently, Wei, Tsui, and Pruisken (1986) and colleagues (1985b) completed a more extensive study. The system which is studied is an InGaAs-InP heterostructure with a density of 3.4×10^{11} cm^{-2} and mobility of 3.5×10^4 cm^2/Vs. The temperature ranges from 0.5 K to 50 K. The resulting scaling diagram is shown in Figure 6.4.10. At low temperatures, $T < 4K$, a considerable similarity is observed between the experimental curves and theory (Figure 6.4.8). At higher temperatures, the data has been explained in terms of a temperature smearing of the Fermi function, (Wei et al. 1986) within a self-consistent Born approximation. More recently, Ando (1985) has carried out a numerical calculation to deduce the scaling diagram. It is found that for short-ranged scatterers, the values of σ_{xx} and σ_{xy}, at different Fermi energies and length scales, all fall on a single curve approximated by

$$(\sigma_{xx}/\sigma^n_{SCBA})^2 + (h\sigma_{xy}/e^2 - n - 1/2)^2 = 1,$$

where σ^n_{SCBA} is the value of σ_{xx} in a self-consistent Born approximation for the nth Landau level. Such a relationship is tantalizingly suggested by the low temperature data in Figure 6.4.10 [see Note 5.6].

A word is in order about comparing theory and experiment and whether or not the concept of a Thouless length is well-defined in strong magnetic fields.

Experimental Aspects

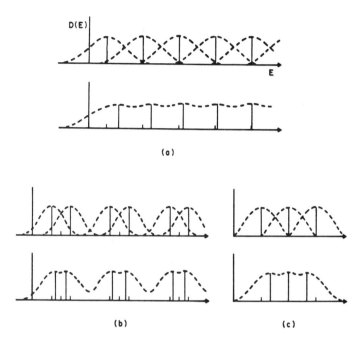

Figure 6.4.9 Shift of the energy position of the extended state in the energy spectrum due to excessive disorder: (a) Landau levels, (b) spin-split levels in each Landau level, and (c) 1/3 and 2/3 FQHE quasi-particle/quasi-hole bands.

For one thing, the kinetic energy is quenched and a Fermi velocity has no meaning. This is an important but difficult question. It would seem that an inelastic length scale could well exist for E_F near the center of a Landau band ($v = n+1/2$). Here, the electronic states relevant to conduction, i.e., those which are extended on the inelastic length scale, are bunched together energetically, and therefore may behave similarly. In the event a percolation picture correctly describes such states, they are also spatially close together, and perhaps an average drift velocity can be defined, over the total width of these "extended" states, to be used as a velocity scale. Away from $v = n+1/2$, the situation is nebulous. Normally, as in the weak localization regime of 2D transport in the absence of a magnetic field, the inelastic length is determined by inelastic processes which transfer energy, ω, of the order of kT (Thouless 1977). In the present

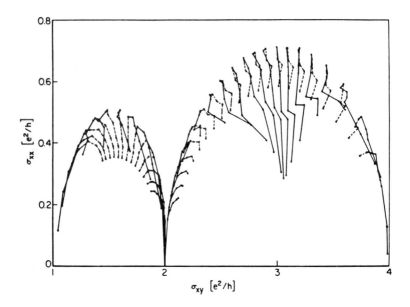

Figure 6.4.10 Experimental $\sigma_{xx}(T)$ versus $\sigma_{xy}(T)$ plotted as T-driven flow lines from 10 K to 0.5 K. The dashed portion covers 10 K to 4.2 K and the full lines from 4.2 K to 0.5 K (Wei et al. 1985b).

case, in the same low-temperature regime where scaling is expected, activated transport is observed, signaling that transport is beginning to be dominated by energy transfer processes with ω significantly larger than kT. It curious and remarkable that the results of experiment look so similar to the scaling theory result.

Interesting results have been obtained on the energy positional shift of the extended states due to large disorder in both the IQHE (Paalanen et al. 1982) and FQHE (Chang et al. 1984b). In Figures 6.4.11 and 6.4.12, we show available data. The dashed curves in the figures indicate the ideal QHE at $T = 0$, i.e. a transition at $v = i+1/2$ for the IQHE and $v = 1+5/6$ and $1+1/2$ for the 5/3 FQHE (Laughlin et al. 1985). The positional shift in the 5/3 effect is striking, and well outside of experimental uncertainties. In the IQHE, the narrowing of the $i = 5$, spin-split QHE is evident, corresponding to the situation in Figure 6.4.9b. These results seem to support the predictions of the scaling theories.

Experimental Aspects

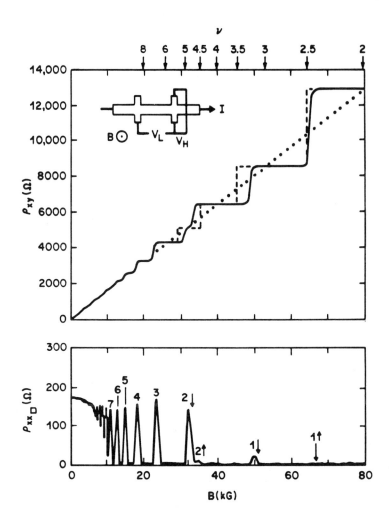

Figure 6.4.11 Experimental evidence for the energy position shift of the extended states due to disorder in the IQHE. The dashed curve indicates the ideal Hall resistance in which the transition between plateaus occurs at $v = i+1/2$. Note in particular the shrinkage of the $i = 5$ plateau, corresponding to the situation in Fig. 6.4.9b. The data is from Paalanen et al. (1982).

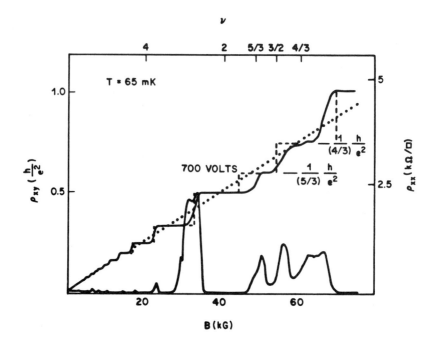

Figure 6.4.12 Energy positional shift of the extended quasi-particle/quasi-hole states in the 5/3 FQHE. In the ideal situation, the transition from the $\nu = 2$ to $\nu = 5/3$ plateau should occur at $\nu = 11/6$ (dashed line). The data is taken from Chang et al. (1984b). Note that an error in the ρ_{xx} trace in this reference has been corrected.

6.5 Future Experiments

It is apparent from the experimental results presented in Sections 6.3 and 6.4 that a great deal is now known about the phenomenon of the FQHE. Nevertheless, our current picture is not complete, and a host of important but technically difficult experiments need to be carried out. Among these are experiments designed to directly determine the structure of the ground state, the (fractional) charge of the mobile elementary excitations, and the k-dispersion of the neutral elementary excitation spectrum. In addition, it is of interest to determine at what low value of the filling factor the ground state crosses over to a Wigner crystal (and therefore no longer exhibits the FQHE) and what new effects exist near $v = 1/2$.

6.5.1 X-ray diffraction structure factor of the FQHE ground state

Even though the prevailing theory of the FQHE, due to Laughlin (1983a, 1984a) and others (Haldane 1983, Halperin 1984, Girvin et al 1985, 1986a,b), exhibits a number of desirable properties and is able to qualitatively and quantitatively (order of magnitude-wise) explain the experimental results, there exist other theories, based on apparently different ground-state wave functions, which are competitive in terms of ground-state energy. Various authors have raised the possibility of a crystal-type ground state (Tosatti and Parrinello 1983, Emery 1983, Luryi and Kazarinov 1985, Kivelson et al. 1986). The wave function discussed by Chui, Ma, and Hakim (1985) represents a 2D pseudocrystal in which the structure factor shows a power law divergence at a wave vector corresponding to the inverse of the average inter-particle spacing (Chui 1985). On the other hand, the structure factor for the Laughlin fluid wave function should show a broad peak only. If a pseudocrystal ground state is correct, because of the presence of disorder, the power law singularity is expected to be significantly broadened. Nevertheless, it is still possible that a residual sharp peak can be present.

In order to conclusively distinguish between a fluid and a pseudocrystal state, it is necessary to directly measure the structure factor by x-ray diffraction. Since the system is two-dimensional, a high intensity, synchrotron source will be necessary, as well as a grazing incidence diffraction technique (Marra et al 1982). One expects little background from the substrate at wavelengths corresponding to typical inter-particle separations of 250 Å. A major difficulty should arise from the small x-ray scattering cross section due to the small, unit charge of the electron (S. M. Girvin, private communication).

6.5.2 Quasiparticle charge

The existence of fractionally charged elementary excitations is expected in the FQHE. The observation of exact fractional quantization of the Hall resistance does not constitute a definitive proof that such excitations exist. Various workers including Störmer and Tsui have suggested the possibility that low temperature shot noise in the FQHE should give a signature for the charge of the current carrier. This is an intriguing idea. One possible geometry for attempting such an experiment is the Corbino geometry. A voltage is applied across the width of the Corbino ring, between the inner and outer edges, causing a Hall current to circulate around the ring. Since the current across the width of the ring should come from carriers excited from localized quasiparticle states to extended states, the shot noise of this current should reflect the fractional charge of the carriers. Unfortunately, in the QHE, the Johnson noise, $(k_B TR\Delta f)^{1/2}$ [R is the source impedance, Δf the measurement bandwidth], becomes very large at low temperatures, since R scales as $1/\sigma_{xx}$. The reader is encouraged to come up with alternative possibilities.

6.5.3 k-dependent collective excitations

The activation energy measurement in DC and low frequency magneto-transport involves charged excitations. Theorists have uncovered a branch of k-dependent neutral excitations [see Chap. IX] (Girvin et al. 1985, 1986a,b, Haldane and Rezayi 1985a, Kallin and Halperin 1984, Laughlin 1985), in which a roton minimum is present at a finite k vector. Following conventional wisdom, such neutral entities should be probed by phonon scattering. Perhaps, at the roton minimum, the large density of states would compensate for the small amount of material in this 2D system and give rise to an appreciable cross section.

6.5.4 Wigner crystallization at low filling factors and higher order FQHE hierarchies near $n = 1/2$

Theory (Laughlin 1983a, Levesque et al. 1984, Lam and Girvin 1984) indicates that at a low filling factor of between 1/6 and 1/11, a FQHE state no longer exists and a gapless Wigner crystal becomes stable. Experimental results are inconclusive as to where the transition actually takes place (Mendez et al. 1983, 1984b). Definitive evidence for the observation of a

Wigner crystal is also not yet available. The major obstacle is disorder. Excess disorder is known to destroy the FQHE and thought to aid the formation of a Wigner glass [see Chap. IX]. A significant improvement in sample quality is needed before a clear cut demonstration of Wigner crystallization can be made. As an intuitive guess, a 4.2 K mobility of 10^6 cm^2/Vs at a density of 0.6×10^{11} cm^{-2} would be a good starting point.

The largest denominator FQHE observed thus far are the ninth's – 4/9, 5/9, and 13/9. It is an intriguing question whether the hierarchies near $v = 1/2$ should go on indefinitely, to the eleventh's, thirteenth's, etc. These ever higher-order states should give a better idea of the hierarchical sequence, and serve as a more stringent test of the theories. Again, high quality materials are necessary. A reasonable target would be a 4.2 K mobility of 3×10^6 cm^2/Vs, at a density of 2×10^{11} cm^{-1}.

6.6 Acknowledgements

It is a pleasure to acknowledge the contributions of my colleagues: D. C. Tsui, H. L. Störmer, G. S. Boebinger, M. A. Paalanen, H. P. Wei, H. Z. Zheng, J. P. Eisenstein, P. Berglund, J. C. M. Hwang, A. C. Gossard, W. Wiegmann, M. Razeghi and G. Weimann.

6.7 Notes

Note 1. The activation energy parameter plotted in Fig. 6.3.12 is defined by means of the convention $\rho_{xx} \sim \exp(-\Delta/k_B T)$. The data shown in Fig. 6.3.14 is from Boebinger et al. (1985b,d) who use the expression $\rho_{xx} \sim \exp(-\Delta/2k_B T)$. While the former is more natural from an experimental point of view, the latter (which is the form used in Chaps. II and IX) is motivated from a theoretical point of view by the law of mass action. See Chap. IX.

Note 2. Data from recent works on the activation energies of 1/3 and 2/3 effects in lower-mobility samples show substantial deviation from the curve (G.

Boebinger, H. Störmer, D. Tsui unpublished, J. Wakabayashi, S. Kawaji, J. Yoshino, and H. Sakaki 1986).

CHAPTER 7

Elementary Theory: the Incompressible Quantum Fluid

ROBERT B. LAUGHLIN

7.1 Introduction

In this lecture, I shall outline what I believe to be the correct fundamental picture of the fractional quantum Hall effect. The principal features of this picture are that the 1/3 state and its daughters are a new type of many-body condensate, that there are analogs of electrons and holes in the integral quantum Hall effect which in this case carry fractional charge, that the Hall plateau observed in experiments is due to localization of these fractionally charged quasiparticles, and that the rules for combining quasiparticles to make daughter states and excitons are simple. The key facts the theory has to explain are these:

1. The effect occurs when electrons are at a particular *density*, determined by the magnetic field strength. The separation between neighboring electrons locks in at particular values.

2. The fractional quantum Hall effect looks to the eye like the integral quantum Hall effect, except that the Hall conductance, in the case of the 1/3 step, is $1/3\ e^2/h$. There is a *plateau*. Changing the electron density a small amount does not alter the Hall conductance, but changing it a large amount does.

3. The 1/3 effect turns on in GaAs at about 1 K when the magnetic field is 15 Tesla.

4. The effect occurs only in the cleanest samples. Excessive dirt destroys it.

5. Other fractions (2/5, 2/7, ...) also occur but they are more easily destroyed by dirt and require lower temperatures to be observed. The most 'stable' states have small denominators.

The most radical feature of this picture, the existence of fractionally charged quasiparticles, is also the least controversial. It is now understood that the integral quantum Hall effect may be viewed as a spectroscopy of the electron charge. (See Chap. I, Laughlin 1981, and Halperin 1982.) If the fractional and integral effects are to be analogous, the former must be interpreted as a spectroscopy of a quasiparticle charge that is fractional. If they are not analogous, one needs to invent a totally new and plausible mechanism for plateau formation. To the best of my knowledge, no such mechanism has yet been proposed. The occurrence of fractional plateaus does not prove the existence of fractionally charged quasiparticles, but is rather accounted for by their existence. If one can make a convincing argument that such quasiparticles exist and behave more or less like electrons in the lowest Landau level, then the effect is explained.

7.2 Quantized Motion of Small Numbers of Electrons

It is customary to discuss the fractional quantum Hall effect within the context of an idealized Hamiltonian of the form

$$\mathcal{H} = \sum_j \left[\frac{1}{2m_e} \left[\frac{\hbar}{i} \nabla_j + \frac{e}{c} \mathbf{A}_j \right]^2 + V(r_j) \right]$$

$$+ \frac{1}{2} \sum_{j \neq k} \frac{e^2}{|r_j - r_k|} , \qquad (7.2.1)$$

where ∇_j is the gradient with respect to the position of the jth electron, $\mathbf{A}_j$ is the vector potential evaluated at the position of the jth electron, and V(r) is the potential generated by a neutralizing background of density ρ_0:

$$V(r) = -\rho_0 \int_{\text{sample}} \frac{e^2}{|r - r'|} d^2r' . \qquad (7.2.2)$$

The electrons are confined to the x-y plane. There is a magnetic field of strength B normal to this plane, so that the vector potential for the symmetric gauge is

$$A = \frac{B}{2} (y\hat{x} - x\hat{y}) . \qquad (7.2.3)$$

The spin degrees of freedom are assumed to be frozen out by the magnetic field. Note that in comparing the predictions of this Hamiltonian with experiment, one needs to substitute the band effective mass (0.07 electron masses, for the conduction band of GaAs) for m_e and to screen the Coulomb forces with the bulk dielectric function ($\kappa = 13$ for GaAs) in the manner $e^2 \to e^2/\kappa$.

The fractional quantum Hall effect is caused, in part, by the massive degeneracy associated with two-dimensional electron motion in a magnetic field. The eigenstates of the single-body Hamiltonian

$$\mathcal{H}_{SB} = \frac{1}{2m_e} \left[\frac{\hbar}{i} \nabla + \frac{e}{c} A \right]^2 , \qquad (7.2.4)$$

which may be written

$$\psi_{m,n}(x,y) = e^{\frac{1}{4}(x^2+y^2)}$$
$$\times \left(\frac{\partial}{\partial x} + i\frac{\partial}{\partial y} \right)^m \left(\frac{\partial}{\partial x} - i\frac{\partial}{\partial y} \right)^n e^{-\frac{1}{2}(x^2+y^2)} , \qquad (7.2.5)$$

with the magnetic length $\ell = (\hbar c/eB)^{\frac{1}{2}}$ set to unity, have energy eigenvalues that depend on the quantum number n only, in the manner

$$\mathcal{H}_{SB} \, \psi_{m,n} = \left[(n + \tfrac{1}{2}) \hbar \omega_c \right] \psi_{m,n} \, , \qquad (7.2.6)$$

where $\omega_c = eB/m_e c$. The manifold of states with the same n, and thus with the same energy, is called the nth Landau level. Occupying this system with non-interacting electrons therefore always results in a degenerate ground state, except when the fermi energy lies between Landau levels, a rare occurrence.

It must be emphasized that the random potential present in all real samples removes this degeneracy in principle. Whether it removes it sufficiently to destroy the FQHE depends on its size. If the random potential is weak, it is properly thought of as a perturbation to the fractional quantum Hall state, stabilized by Coulomb interactions, which localizes the fractionally charged quasiparticles and facilitates observation of the effect. If it is strong, the potential is properly thought of as localizing the electrons, subsequently perturbed by the 'weak' Coulomb interaction, and destroying the effect. The border between the two regimes is delicate and controversial (see Laughlin et al. 1985, Girvin et al. 1986b, MacDonald, Liu et al. 1986, Chap. IX).

Electrons in the fractional quantum Hall effect like to have only certain distances between them. The easiest way to understand physically why this occurs is to consider the two-electron problem. Suppose, for simplicity, that the Coulomb potential is 'small' compared to the cyclotron energy $\hbar \omega_c$. It is then a good approximation to require the two-body wave function to be comprised solely of single-body wave functions lying in the lowest Landau level. The state must also be an eigenstate of internal angular momentum, since the Hamiltonian is azimuthally symmetric. However, up to unimportant center-of-mass motion, the only states satisfying these two conditions are of the form

$$\psi_k = (z_1 - z_2)^{2k+1} \, e^{-\tfrac{1}{4}(|z_1|^2 + |z_2|^2)} \, , \qquad (7.2.7)$$

where $z_j = x_j + i y_j$ is the location of the jth electron expressed as a complex number. These are therefore the sought energy eigenstates. Note that the exponent $2k+1$ must be *odd* to satisfy the Pauli principle. Inter-electronic spacing is quantized because the eigenstates of the two-body Hamiltonian do not depend on the potential repelling the electrons, provided this potential is sufficiently small.

That these considerations are pertinent to conditions in actual fractional quantum Hall experiments may be seen in Fig. 1, where the exact and approximate (Eq. 2.7) wave functions are compared for the case $k=0$ and $\hbar\omega_c = e^2/\ell$. This is a legitimate worst-case limit, for the condition $\hbar\omega_c = e^2/\ell$ is achieved in GaAs when the magnetic field is 6T, whereas the experiments are typically done near 15T. Also, this disparity decreases with larger values of k because the particles then orbit farther apart and thus interact less. When Eq. 2.7 is a good approximation, one has

$$\langle |z_1 - z_2|^2 \rangle \cong 8k + 8 . \quad (7.2.8)$$

This is an equality only in the limit that the pair potential is zero. Keep this in mind when considering the fractional quantum Hall ground state, for its inter-particle spacing is exact even when the potential is not zero.

Electrons in the fractional quantum Hall effect also like to be far apart. In most cases, particles accomplish this optimally by crystallizing. It comes as no surprise, therefore, that the Hartree-Fock Wigner-crystal ground state for this system is extremely favorable energetically (Yoshioka and Fukuyama 1979, Yoshioka and Lee 1983), and indeed, is thought to be a correct variational description of the electrons when the Landau level is nearly empty. That the fractional quantum Hall ground state is not a crystal, at least one in the usual sense, is indicated by its ability to carry currents with no resistive loss (Tsui et al. 1983a, see also Chap. VI). Ordinary crystals, by definition, have 'lattice positions' near which the electrons have a larger-than-average probability to be found, and thus a large ground-state degeneracy corresponding to the different ways the lattice can be centered. Transport of electric current usually cannot occur without motion of the lattice as a whole, or, more precisely, via a collective excitation associated with the multiplicity of ground states. To convince

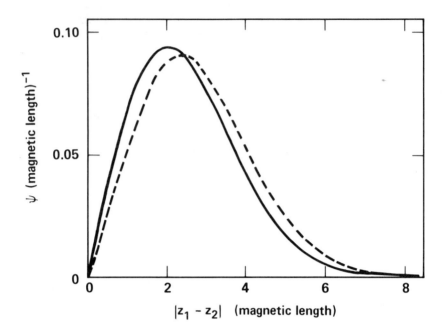

Figure 7.1 Comparison of exact two-electron wave function (dashed) with approximate one (solid) given by Eq. 2.7 for the case $\hbar\omega_c = e^2/\ell$ and $k=0$.

oneself that this occurs in the present case, at least when impurities are absent, imagine creating the crystal at rest and then Lorentz transforming it. The transformation induces both a current and an electric field perpendicular to the current, and it clearly results in motion of the lattice. Thus, if the crystal is also capable of carrying current without moving, then there must be two current-carrying mechanisms with identical Hall conductances. This is not inconceivable, but it is unusual and would require explanation. Given that the transport of current in this crystal is no different from that in any other, one would expect any impurities present to break the ground-state degeneracy, pin the crystal, and prevent the electrons from carrying current. All known charge density wave systems exhibit this behavior. Since the fractional quantum Hall ground state does not do this, it is either not a crystal or one with extremely peculiar properties.

Elementary Theory: Incompressible Quantum Fluid

Even though the ground state is not crystalline, the behavior of small numbers of electrons, for which numerical calculations are feasible (See Haldane Chap. VIII, Laughlin 1983b, Yoshioka et al. 1983, Yoshioka 1984a,b, Haldane and Rezayi 1985a) indicates that it is 'almost' crystalline. I shall illustrate this with the case of three electrons, as it influenced my own thinking and because it is simple. Note, however, that more thorough calculations now exist.

Let the Hamiltonian be

$$\mathcal{H} = \frac{1}{2m_e}\left[\frac{\hbar}{i}\nabla_1 + \frac{e}{c}\mathbf{A}_1\right]^2 + \frac{1}{2m_e}\left[\frac{\hbar}{i}\nabla_2 + \frac{e}{c}\mathbf{A}_2\right]^2$$
$$+ \frac{1}{2m_e}\left[\frac{\hbar}{i}\nabla_3 + \frac{e}{c}\mathbf{A}_3\right]^2 + \frac{e^2}{r_{12}} + \frac{e^2}{r_{23}} + \frac{e^2}{r_{31}}, \quad (7.2.9)$$

and let the vector potential be expressed in the symmetric gauge. Define center-of-mass and internal coordinates in the manner,

$$\bar{z} = \frac{z_1 + z_2 + z_3}{3}$$

$$z_a = \sqrt{\frac{2}{3}}\left[\left(\frac{z_1 + z_2}{2}\right) - z_3\right]$$

$$z_b = \frac{1}{\sqrt{2}}(z_1 - z_2), \quad (7.2.10)$$

so that the unimportant center-of-mass motion may be removed, giving an internal Hamiltonian of the form

$$\mathcal{H}_{internal} = \frac{1}{2m_e}\left[\frac{\hbar}{i}\nabla_a + \frac{e}{c}\mathbf{A}_a\right]^2 + \frac{1}{2m_e}\left[\frac{\hbar}{i}\nabla_b + \frac{e}{c}\mathbf{A}_b\right]^2$$
$$+ \frac{e^2}{\sqrt{2}}\left[\frac{1}{|z_b|} + \frac{1}{\left|\frac{1}{2}z_b + \frac{\sqrt{3}}{2}z_a\right|} + \frac{1}{\left|\frac{1}{2}z_b - \frac{\sqrt{3}}{2}z_a\right|}\right].$$

$$(7.2.11)$$

The Pauli principle is satisfied if the wave function is *odd* in z_b and *symmetric* under rotation by $\pm 120°$ in the a-b plane. This is accomplished automatically if one adopts the orthonormal basis functions

$$|m,n\rangle = \frac{1}{\sqrt{2^{6m+4n+1}(3m+n)!\,n!\,\pi^2}}$$

$$\times \left[\frac{(z_a+iz_b)^{3m}-(z_a-iz_b)^{3m}}{2i}\right]\left[z_a^2+z_b^2\right]^n e^{-\frac{1}{4}(|z_a|^2+|z_b|^2)},$$

(7.2.12)

where again the magnetic length $\ell = (\hbar c/eB)^{1/2}$ has been set to unity. These span the lowest Landau level, satisfy the symmetry criteria, and are eigenstates of angular momentum with eigenvalue $M = 2n + 3m$. They are also very close to the correct energy eigenstates, and it is correct to think of them as such. For $M = 9$, for example, the first instance of degeneracy, the expectation values of $|3,0\rangle$ and $|1,3\rangle$ across $\mathcal{H}_{internal}$ are $0.722\ e^2/\ell$ and $0.867\ e^2/\ell$, respectively, while the correct energy eigenvalues are $0.717\ e^2/\ell$ and $0.872\ e^2/\ell$. The correct values are obtained trivially once the off-diagonal matrix element

$$|\langle 3,0|\mathcal{H}_{internal}|1,3\rangle| = 0.0277\ e^2/\ell$$

is computed, since $\mathcal{H}_{internal}$ conserves M.

The energy eigenstates of three electrons approximated by $|m,0\rangle$ have unusual stability and display behavior expected of fractional quantum Hall ground states. This is shown in Fig. 2, where the angular momentum and 'area' of the electron ground state are plotted as a function of the strength of a background potential given by the expression

Figure 7.2 Total angular momentum M and root-mean-square area of the ground state of a three-electron cluster as a function of $1/\alpha$. The 'pressure' α and area operator A are defined in Eqs. 2.13 and 2.15.

Elementary Theory: Incompressible Quantum Fluid 241

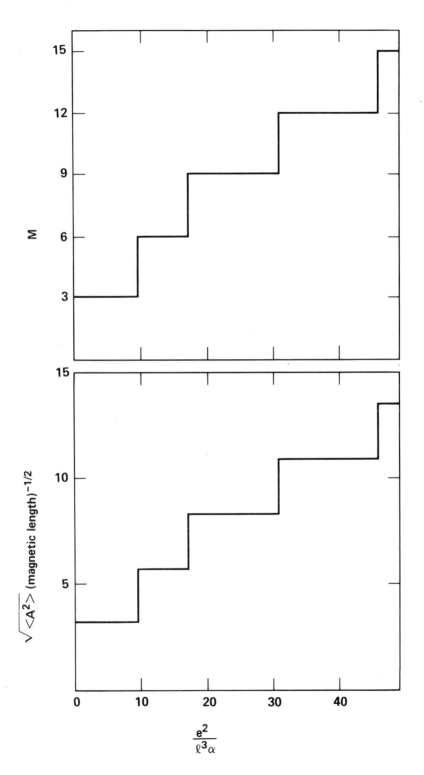

$$V = \frac{\alpha}{2}(|z_1|^2 + |z_2|^2 + |z_3|^2)$$
$$= \frac{3}{2}\alpha|\bar{z}|^2 + \frac{\alpha}{2}(|z_a|^2 + |z_b|^2) \quad (7.2.13)$$

which is added to the Hamiltonian to hold the electrons together. The functional form of V is chosen for convenience and is not critical. It satisfies

$$\langle m,n|\frac{\alpha}{2}(|z_a|^2 + |z_b|^2)|m',n'\rangle = \delta_{mm'}\cdot\delta_{nn'}\cdot(M+2)\alpha ,$$

$$(7.2.14)$$

and thus has no effect on the form of wave functions, but only lowers the small-M states energetically relative to those with large M. The area of the triangle whose vertices are the positions z_1, z_2, and z_3 is an operator given by

$$A = \frac{1}{2}\,\mathrm{Im}\left[\left[\left(\frac{z_1+z_2}{2}\right) - z_3\right](z_1-z_2)^*\right]$$

$$= \frac{\sqrt{3}}{4i}(z_a z_b^* - z_b z_a^*) . \quad (7.2.15)$$

Its matrix elements are

$$\langle m,n|A|m',n'\rangle = 0 \quad (7.2.16)$$

and

$$\langle m,n|A^2|m',n'\rangle = \delta_{mm'}\cdot\delta_{nn'}\cdot\frac{3}{4}[(3m)^2 + (M+2)] .$$

$$(7.2.17)$$

One sees in Fig. 2 discontinuous jumps in the angular momentum in successive integral multiples of three as α is decreased. Multiples of three occur because at any given value of α, the lowest energy state has $n = 0$. Why this is so may been seen in Fig. 3, where the probability to find either electron 1 or 2 at z, given that electron 3 is fixed at the position marked X and the center of mass is fixed at the origin, is

Elementary Theory: Incompressible Quantum Fluid 243

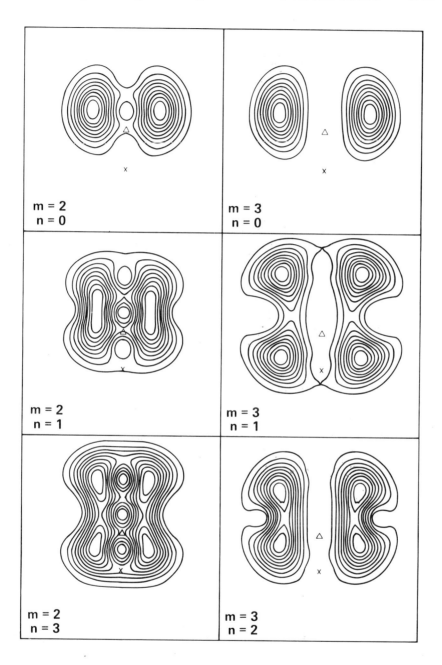

Figure 7.3 Charge densities of states $|m,n\rangle$ defined in Eq. 2.12 given that the center of mass lies at the origin (▲) and that one of the electrons lies at $y=-3\ell$ (**X**). Contours lie at integral multiples of 0.1 times the maximum charge density, given the two constraints.

plotted. The states with $n = 0$ alone cause all three electrons to avoid one another in a way that looks remarkably 'crystalline.'

7.3 Variational Ground State

The central feature of my picture of the fractional quantum Hall effect is a variational ground-state wave function (Laughlin 1983a), the form of which is motivated by the observation that $|m,0\rangle$ approximates the ground state of three particles well. This state's only unique feature is its tendency to exclude probability from the region near $z_b = 0$. This suggests that the principal characteristic of a good trial wave function should be *deep nodes*, as opposed to crystallinity. This property does not preclude crystallinity, but neither does it guarantee it, except at low density.

The variational ground-state wave function is constructed by appealing to precedents from the study of liquid helium, another condensed-matter system in which the particles want badly to avoid one another. It is well known (Feenberg 1969) that a good variational energy for liquid ^{3}He may be obtained using a single Slater determinant multiplied by the factor

$$\prod_{j<k}^{N} f(z_j - z_k) ,$$

where f is a function that goes to zero as its argument goes to zero. This reduces the amplitude for *pairs* of particles to approach one another, which is usually adequate for computational purposes because the approach of three or more particles is improbable and thus variationally unimportant. In the present case, since the magnetic field effectively turns off the kinetic energy, nothing is gained by keeping the Slater determinant, so I remove it, at the same time making the Jastrow product antisymmetric. I also factor out an exponential for each particle in the manner

$$\Psi(z_1,\ldots,z_N) = \prod_{j<k}^{N} f(z_j - z_k) \exp\left\{-\frac{1}{4}\sum_{\ell}^{N} |z_\ell|^2\right\} .$$

(7.3.1)

Elementary Theory: Incompressible Quantum Fluid

There is no loss of generality in doing this, since

$$\prod_{j<k}^{N} f(z_j - z_k) \exp\left\{-\frac{1}{4}\sum_{\ell}^{N} |z_\ell|^2\right\}$$

$$= \prod_{j<k}^{N}\left[f(z_j - z_k)\, e^{-\frac{1}{4N}|z_j - z_k|^2}\right] \exp\left\{-\frac{N}{4}\left|\frac{1}{N}\sum_{\ell}^{N} z_\ell\right|^2\right\}.$$

(7.3.2)

The last factor describes center-of-mass motion. One now wishes to vary f until the expectation value of this wave function across the Hamiltonian is minimized. However, this step is trivial in the present case because the wave function must satisfy three constraints which, in conjunction with the assumption of a Jastrow form, completely remove all variational freedom. These constraints are:

> 1. That the many-body wave function be comprised solely of single-body wave functions lying in the lowest Landau level. This is accomplished only if $f(z)$ is an *analytic* function of its argument z.
> 2. That the wave function be totally antisymmetric. Thus $f(z)$ must be *odd* $[f(z) = -f(-z)]$.
> 3. The wave function is an eigenstate of total angular momentum. Note that this condition can be misleading if improperly applied. For example, even though the degenerate ground states of a crystal can be superimposed to make an eigenstate of angular momentum, the ground state is still degenerate and thus still crystalline. Conservation of angular momentum implies that
>
> $$\prod_{j<k}^{N} f(z_j - z_k)$$
>
> must be a polynomial in the particle posi-

tions $z_1, \ldots, z_N$ of degree M, where M is the total angular momentum.

The only function satisfying all three constraints is $f(z) = z^m$, with m odd. Thus the wave function is completely determined once one has assumed it to be of the Jastrow form.

To allay fears that this assumption might be too restrictive, even though one doubts this on physical grounds, one may compare the variational wave function with the exact one for small numbers of particles. Table 1 lists the overlap of the three-particle variational wave function with the exact wave function (see Sec. 7.2), as calculated numerically using three different inter-electronic repulsive potentials. The near perfect overlap, particularly at large values of m, is nontrivial, and gives one confidence that the physics is right. The ground-state wave function is, in fact, exceedingly rigid and unresponsive to variations in the form of the inter-electronic repulsive potential, exactly as the assumption of a Jastrow-type wave function predicts.

The search for a variational minimum thus reduces to investigation of a discrete set of wave functions characterized by an odd integer m:

$$|m\rangle = \Psi_m(z_1, \ldots, z_N) = \prod_{j<k}^{N} (z_j - z_k)^m \exp\left\{-\frac{1}{4}\sum_{\ell}^{N} |z_\ell|^2\right\}.$$

(7.3.3)

Since $|m\rangle$ is an eigenstate of angular momentum with eigenvalue

$$M = \frac{N(N-1)}{2} m \;,$$

mixtures of these states are not allowed. One discovers which state $|m\rangle$ has the lowest energy by interpreting the square of the wave function as a classical probability distribution function, in the manner:

$$|\Psi_m(z_1, \ldots, z_N)|^2 = e^{-\beta\phi(z_1, \ldots, z_N)} \;. \quad (7.3.4)$$

Elementary Theory: Incompressible Quantum Fluid

Table 7.1 Projection of variational three-body wave functions Ψ_m on the state Φ_m in the manner $\langle\Psi_m|\Phi_m\rangle/(\langle\Psi_m|\Psi_m\rangle\langle\Phi_m|\Phi_m\rangle)^{1/2}$. Φ_m is the lowest-energy eigenstate of angular momentum $3m$ calculated with $V=0$ and an inter-electronic potential of either $1/r$, $-\ln(r)$, or $\exp(-r^2/2)$.

m	$1/r$	$-\ln(r)$	$\exp(-r^2/2)$
1	1	1	1
3	0.999 46	0.996 73	0.999 66
5	0.994 68	0.991 95	0.999 39
7	0.994 76	0.992 95	0.999 81
9	0.995 73	0.994 37	0.999 99
11	0.996 52	0.995 42	0.999 96
13	0.997 08	0.996 15	0.999 85

In the present case, I set the fictive temperature $1/\beta$, which is totally arbitrary, to m so that the physical significance of the effective classical potential energy ϕ is more transparent:

$$\phi(z_1,\ldots,z_N) = -2m^2 \sum_{j<k} \ln|z_j - z_k| + \frac{m}{2} \sum_{\ell}^{N} |z_\ell|^2 . \tag{7.3.5}$$

This is the potential energy of a two-dimensional one-component plasma: Particles of 'charge' m repelling one another via logarithmic interactions, the natural 'Coulomb' interaction in two dimensions, and being attracted to the origin via the same logarithmic interactions by a uniform neutralizing background of 'charge' density $\rho_1 = 1/(2\pi\ell^2)$. A plasma's most compelling urge is to be electrically neutral everywhere. Thus, we know that the electron density of the quantum state

$$\rho(z_1) = \frac{N \int \cdots \int |\Psi_m(z_1,\ldots,z_N)|^2 \, d^2z_2 \cdots d^2z_N}{\langle m|m\rangle} \tag{7.3.6}$$

must be equal to $1/m$ times the 'charge' density of the equivalent plasma and thus uniform and equal to

$$\rho_m = \frac{1}{2\pi m \ell^2}.$$

Since the states $|m\rangle$ have significantly different charge densities, the one with the lowest energy is clearly the one most closely reproducing the (true) neutralizing background density ρ_0 of the original problem (cf. Eq. 2.2). The remaining variational step is therefore trivial. At high densities, the $m = 1$, or full Landau level wave function, is most stable. As ρ_0 is decreased, in rough analogy with the decrease of α in Fig. 2, first the $m = 3$ state becomes stable, then the $m = 5$, and so on. The 1/3 effect corresponds to the first nontrivial state in this sequence, $m = 3$. Stability at $m = 5$ may have been observed experimentally (see Chap. VI, Störmer et al. 1983a, Mendez et al. 1983), but this is not clear. That $m = 1$ corresponds to a full Landau level is easily seen by expanding the Vandermonde polynomial in the manner

$$\prod_{j<k}^{N} (z_j - z_k) = (-1)^{\frac{N(N-1)}{2}} \sum_{\sigma} \text{sgn}(\sigma)$$
$$\times z_{\sigma(1)}^0 z_{\sigma(2)}^1 z_{\sigma(3)}^2 \cdots z_{\sigma(N)}^{N-1}, \qquad (7.3.7)$$

where σ is a permutation of N things and $\text{sgn}(\sigma)$ is its sign. Ψ_1 is a single Slater determinant in which the first N single-body orbitals of the form $z^k \exp(-|z|^2/4)$ are filled with electrons.

The uniformity guaranteed by 'electrical' neutrality is only on length scales of the inter-particle spacing and larger. To guarantee uniformity on small length scales as well, one needs to know that the equivalent plasma is a liquid. This is true, however, for the cases of interest because the fictive temperature is sufficiently high. The plasma is characterized by a dimensionless coupling parameter, Γ, which measures the ratio of potential energy to temperature. For this problem, $\Gamma = 2m$. Extensive Monte Carlo studies (Caillol et al. 1982) of the two-dimensional one-

Elementary Theory: Incompressible Quantum Fluid

component plasma have shown it to be a solid for $\Gamma \gtrsim 140$ and a liquid otherwise. Thus for $\Gamma = 2, 6, 10, \ldots$, the cases of interest for the fractional quantum Hall effect, the equivalent plasma is unquestionably a liquid.

An important test of the correctness of the theory is how well its energy compares with that of competing variational wave functions. Wave functions of the Jastrow form have the important property that their expected energies can be calculated to high accuracy using conceptually simple computational machinery. The quantity one needs is the radial distribution function, which is $|\Psi_m|^2$ with all but two of its coordinates integrated out:

$$g(|z_1 - z_2|) = \frac{N(N-1)}{\rho_m^2}$$

$$\times \frac{\int \cdots \int |\Psi_m(z_1,\ldots,z_N)|^2 \, d^2z_3 \cdots d^2z_N}{\langle m|m\rangle} \quad . \quad (7.3.8)$$

Disregarding the unimportant $\hbar\omega_c/2$, the total energy per electron is given in terms of g by

$$U_{total} = \frac{1}{2} \rho_m \int \frac{e^2}{|z|} \Big[g(|z|) - 1\Big] d^2z \quad . \quad (7.3.9)$$

The radial distribution functions of Ψ_m for $m = 1, 3,$ and 5 are shown in Fig. 4. Note that they are plotted against the scaled variable $x = |z|/(2m)^{1/2}$. One sees in each case characteristic liquid-like behavior: a region of size one around the origin, the exchange-correlation hole, from which electrons are excluded, and rapid healing and convergence to one beyond this region. Note also that the deviations from unity persist to larger values of x as m is increased. This is because the equivalent plasma, and, by implication, the quantum state it describes, is becoming more 'strongly coupled' and thus more difficult to distinguish from a crystal. The similarity between crystal and liquid is shown more explicitly in Fig. 5, where I compare $g(|z|)$ for Ψ_5 with the equivalent quantity for a Wigner crystal ground state of the form

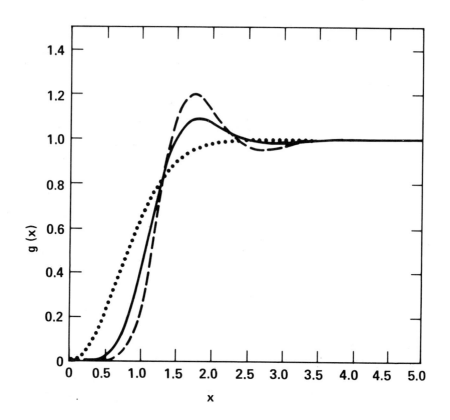

Figure 7.4 Radial distribution function for Ψ_m, as defined in Eq. 3.8, for $m=1$ (dotted), $m=3$ (solid), and $m=5$ (dashed) plotted against the reduced variable $x = |z_1 - z_2|/(2m)^{1/2}$.

$$\Psi_W(z_1, \ldots, z_n) = \sum_\sigma \text{sgn}(\sigma)$$

$$\times \varphi_{j_1 k_1}[z_{\sigma(1)}] \cdots \varphi_{j_n k_n}[z_{\sigma(n)}], \quad (7.3.10)$$

where the orbitals $\varphi_{jk}[z]$ are Gaussians

$$\varphi_{jk}[z] = e^{-\frac{1}{4}|z_{jk}^{(o)}|^2} e^{\frac{1}{2} z^* z_{jk}^{(o)}} e^{-\frac{1}{4}|z|^2} \quad (7.3.11)$$

Elementary Theory: Incompressible Quantum Fluid

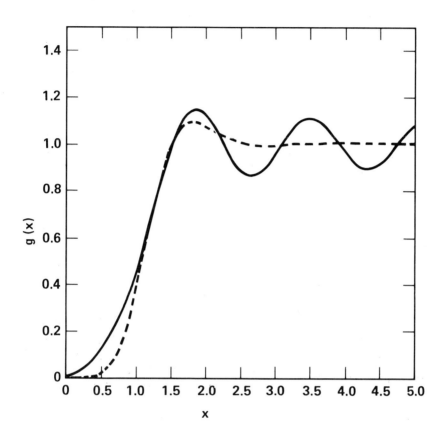

Figure 7.5 Comparisons of radial distribution function of Ψ_m (dashed) with that of Hartree-Fock Wigner crystal (solid) at the same density, for $m=3$, plotted against reduced variable $x=|z_1-z_2|/(2m)^{1/2}$.

centered at hexagonal lattice sites $z_{jk}^{(o)}$:

$$z_{jk}^{(o)} = \sqrt{\frac{4\pi m}{\sqrt{3}}} \left[j + (\frac{1}{2} + \frac{i\sqrt{3}}{2})k \right] . \qquad (7.3.12)$$

One sees that the two are almost identical for small values of the scaled parameter x, and that there are vestiges of the near-neighbor and second-neighbor structure in the liquid. The slight difference between the two behaviors at very small x -- $g(x)$ goes to zero rigorously as x^{2m} for the liquid but only as x^2 for the crystal -- is the main reason the liquid is variationally superior. The radial distribution function for Ψ_1 is known exactly (Jancovici 1981) and equals

$$g(|z|) = 1 - e^{-\frac{1}{2}|z|^2} . \qquad (7.3.13)$$

Those for Ψ_3 and Ψ_5 were calculated using the modified hypernetted-chain technique of Caillol et al. (1982) which, in turn, is known to agree with Monte Carlo results. The total energies obtained from these are

$$U_1 = -\sqrt{\frac{\pi}{8}} \, e^2/\ell , \qquad U_3 = -0.416 \pm 0.001 \, e^2/\ell ,$$

and

$$U_5 = -0.334 \pm 0.003 \, e^2/\ell .$$

[Superior calculations of these energies now exist. See Levesque et al. (1984) and Morf and Halperin (1986).] These are compared with the Hartree-Fock Wigner crystal results of Yoshioka and Lee (1983) in Fig. 6. The liquids Ψ_3 and Ψ_5 are visibly superior variationally to the Hartree-Fock state. It is primarily on the basis of this comparison that I consider Ψ_3 to be a legitimate candidate for the correct 1/3 ground state, and that I have predicted the existence of a 1/5 state when the interactions are Coulombic. The stability of states Ψ_m for $m = 5$ and beyond is currently a matter of debate. Lam and Girvin (1984, 1985) have made a strong case that Ψ_m is unstable to crystallization for $m > 7$.

The dashed line in Fig. 6 is a crude interpolation formula for the energy of Ψ_m of the form

$$U_{total}(m) \cong \frac{0.814}{\sqrt{m}} \left[\frac{0.230}{m^{(0.64)}} - 1 \right] e^2/\ell , \qquad (7.3.14)$$

which is derived in the following way. U_{total} is a smooth function of $\Gamma = 2m$ which must limit at large expansions to the crystal Madelung energy. This is very nearly the ion-disc energy, given by

Elementary Theory: Incompressible Quantum Fluid

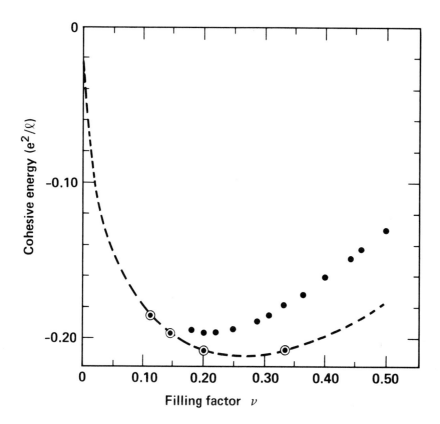

Figure 7.6 Comparison of cohesive energy per electron for charge density wave (dots) [from Yoshioka and Lee 1983] and Ψ_m (open circles) plotted vs. filling factor $v=1/m$. The cohesive energy is $U_{total} + (\pi/8)^{1/2} v e^2/\ell$. The dashed curve is a plot of Eq. 3.14.

$$U_{ion-disc} = -\rho_m \int_{|z|<R} \frac{e^2}{|z|} d^2z$$

$$+ \frac{\rho_m^2}{2} \iint_{\substack{|z_1|<R \\ |z_2|<R}} \frac{e^2}{|z_1-z_2|} d^2z_1 d^2z_2 \, , \qquad (7.3.15)$$

where $\pi R^2 \rho_m = 1$, which yields the result

$$U_{\text{ion-disc}} = \left[\frac{4}{3\pi} - 1\right]\sqrt{\frac{\pi}{8}}\, e^2/\ell .$$

At $m = 1$, U_{total} must equal

$$-\sqrt{\frac{\pi}{8}}\, e^2/\ell = 0.814 \times (0.230 - 1)\, e^2/\ell .$$

The exponent 0.64 is obtained from a fit to the value of U_{total} at $m = 3$. The ability of the formula to predict U_{total} at $m = 5$ fairly well, even though it was not fit to this value, gives one confidence that it is roughly correct. A more accurate formula has recently been developed by Levesque et al. (1984).

It should be remarked that the ground-state density distribution need not be disc-shaped. Ψ_m may be caused to have any shape desired, so long as the area is preserved, by multiplying it by

$$\exp\left\{\sum_{\ell}^{N} f(z_\ell)\right\} ,$$

where f is an arbitrary analytic function. This procedure leaves the density and $g(r)$ invariant.

7.4 Computational Methods

Because the reader may wish to reproduce some of these results for himself, I shall describe briefly how hypernetted-chain and Monte Carlo calculations are performed.

The radial distribution function $g(r)$ of a classical liquid with a short-range pair-potential $v(r)$ and constant external potential is given exactly by the solution of the equations (see Stell 1964)

$$h(r) = g(r) - 1 , \tag{7.4.1}$$

$$h(|\mathbf{r}|) = c(|\mathbf{r}|) + \rho_m \int h(|\mathbf{r} - \mathbf{r}'|)\, c(|\mathbf{r}'|)\, d^2 r' \tag{7.4.2}$$

Elementary Theory: Incompressible Quantum Fluid

$$h(r) = \exp[-\beta v(r) + h(r) - c(r) + b(r)] \quad , \quad (7.4.3)$$

where $1/\beta$ is the temperature and $b(r)$ is the 'bridge function', a quantity with a formal diagrammatic definition that is difficult to calculate. The bridge function is known to be short-ranged and small in the cases of interest, so we set it to zero as a first approximation, and obtain thereby the hypernetted-chain equations.

In order to be used with the long-range forces found in plasmas, the hypernetted chain equations must be regularized. This amounts physically to considering the 'Coulomb' potentials to be limits of short-range potentials. Since they are not, unless the 'charge' density of the plasma exactly equals that of the uniform neutralizing background, the assumption of 'electrical' neutrality is tacit in the procedure. The simplest short-range potential limiting to a 'Coulomb' interaction is

$$-\frac{1}{\pi}\int \frac{e^{i\mathbf{q}\cdot\mathbf{r}}}{q^2 + q_o^2} d^2q = -2K_o[q_o r] = 2\ell n\left[\frac{q_o r}{2}\right] + \ldots \quad (7.4.4)$$

where K_0 is a modified Bessel function of the second kind. One regularizes the equations using this potential by substituting $-2mK_0[q_0 r]$ for $\beta v(r)$ in Eq. 4.3 and then looking for a solution in the limit $q_0 \to 0$. A solution exists in this limit because the divergence of the potential at $q \to 0$ is canceled by a similar divergence in c. That this must be true may be seen by observing, in Eq. 4.3, that $h(r)$ cannot be short-ranged unless $c(r)$ equals $-\beta v(r)$ at large r. The solution is conveniently expressed in terms of $c_s(r)$, the difference between $c(r)$ and the long-range part of $-\beta v(r)$. One subtracts the long-range part of $-\beta v(r)$ as opposed to all of it, to form $c_s(r)$ only to increase numerical stability. Defining

$$c_s(r) = c(r) + \frac{m}{\pi}\int \frac{Q^2 e^{i\mathbf{q}\cdot\mathbf{r}}}{(q^2+q_o^2)(q^2+Q^2)} d^2q \quad (7.4.5)$$

where Q is a 'cutoff' momentum, one obtains

$$\beta v(r) + \hat{c}(r) = \frac{m}{\pi} \int \frac{q^2 e^{i\mathbf{q}\cdot\mathbf{r}}}{(q^2+q_c^2)(q^2+Q^2)} d^2q + c_s(r) , \quad (7.4.6)$$

and thus

$$g(r) = \exp[h(r) - c_s(r) - 2mK_0(Qr)] , \quad (7.4.7)$$

in the limit $q_0 \to 0$. Also, with Fourier transforms defined in the manner

$$\hat{c}_s(q) = \int c_s(r) e^{i\mathbf{q}\cdot\mathbf{r}} d^2r \quad (7.4.8)$$

$$= 2\pi \int c_s(r) r J_0(qr) dr ,$$

where J_0 is a Bessel function, one obtains for Eq. 4.2

$$\hat{h}(q) = \frac{\hat{c}(q)}{1 - \rho_m \hat{c}(q)} , \quad (7.4.9)$$

with

$$\hat{c}(q) = \hat{c}_s(q) + \frac{1}{\pi} \frac{mQ^2}{q^2(q^2+Q^2)} . \quad (7.4.10)$$

The regularized equations are 4.1, 4.7, 4.9 and 4.10. Note that the $g(r)$ one obtains by solving them does not depend on Q.

The regularized hypernetted chain equations are conveniently solved in the following way (Rogers 1980): One represents functions of position or momentum by their values at the grid points $r_j = (j-1)\Delta r$ and $q_k = (k-1)\Delta q$. One then initializes $c_s(r_j)$ to zero. Given $c_s(r_j)$ either from initialization or from the previous iteration, one calculates (using Eq. 4.8)

$$\hat{c}_s(q_k) = 2\pi \Delta r \sum_j c_s(r_j) r_j J_0(r_j q_k) , \quad (7.4.11)$$

and converts this to $\hat{h}(q_k)$ using Eqs. 4.9 and 4.10. One then back-transforms to get a new value for $h(r_j)$:

$$h(r_j) = \frac{\Delta q}{2\pi} \sum_k \hat{h}(q_k) J_0(r_j q_k) . \qquad (7.4.12)$$

This is combined with $c_s(r_j)$ in Eq. 4.7 to produce a new $g(r_j)$. The value of $h(r_j)$ obtained by subtracting unity from $g(r_j)$ (cf. Eq. 4.1) is different from the one produced by the back-transform (Eq. 4.12) unless the guess for $c_s(r_j)$ was correct. If it was not, one uses the difference to refine $c_s(r_j)$, in the manner

$$c_s^{new}(r_j) = c_s^{old}(r_j) - h^{new}(r_j) + g^{new}(r_j) - 1 . \qquad (7.4.13)$$

This is essentially the procedure used to generate Fig. 4. It converges rapidly, usually after three to five iterations.

The hypernetted-chain, and known analytic methods generally, must be considered unreliable until tested against Monte Carlo 'experiments'. (See Levesque et al. 1983.) The nature of these is best illustrated with an example.

Suppose one wishes to calculate the charge density of Ψ_m for a finite number of electrons as a function of position, as in Eq. 3.6. The radial distribution function is a special case of this in which the position of one electron is held fixed. The task is to perform a multi-dimensional definite integral which, up to an overall normalization, is equivalent to a 'thermal' average over an ensemble of electron configurations. The Monte Carlo method replaces this ensemble average with a 'time' average, but with 'time' evolution equations that are artificial and chosen solely for convenience. To calculate $\rho(r)$, one partitions the radial axis into bins with boundaries r_n determined by

$$\pi r_n^2 = n\Delta A \qquad (7.4.14)$$

and, at the end of each 'time' step, adds one to a bin if an electron lies in it. As 'time' progresses, the number of electrons in a bin will become proportional to the charge density. The choice of evolution method is not critical, at least in principle. The most commonly used is the following: At each 'time' step one scans through the particles and attempts to move each once. The attempt involves picking a random direction and calculating the change ΔU to the system's 'energy' that would occur if the particle were moved a distance Δr in this direction. The parameter Δr is chosen to optimize the convergence rate of the algorithm. If ΔU is negative, the move is accepted, and the particle is given the new location. If ΔU is positive, the move is accepted conditionally. One picks a random number between zero and one and accepts the move if this number is smaller than $\exp(-\beta\Delta U)$. Metropolis algorithms of this kind work because they satisfy detailed balance. Any time evolution characterized by transition rates r_{jk} to go from configuration k to configuration j will produce a correct thermal average if

$$\frac{r_{jk}}{r_{kj}} = e^{\beta[\epsilon_k - \epsilon_j]} \qquad (7.4.15)$$

where ϵ_j and ϵ_k are the 'energies' of the states. It works well because it discriminates against improbable configurations, which in this case are almost all of them. Fig. 7 compares a random configuration of 1000 particles with one 'typically' sampled when describing Ψ_m with $m = 3$ and $\Delta r = 0.5\ell$. The random configuration is 10^{6080} times less probable than the typical one, and is therefore 'never' sampled. The configuration typical of Ψ_m is actually useful for visualizing the state. It is evident in Fig. 7 that the particles avoid one another efficiently and that long-wavelength fluctuations are strongly suppressed.

Figure 7.7 Comparison of a random distribution of electrons (top) with a 'typical' one for electrons described by Ψ_m with $m=3$. The number of electrons is 1000. The lower configuration is more probable by a factor of 10^{6080}.

Elementary Theory: Incompressible Quantum Fluid 259

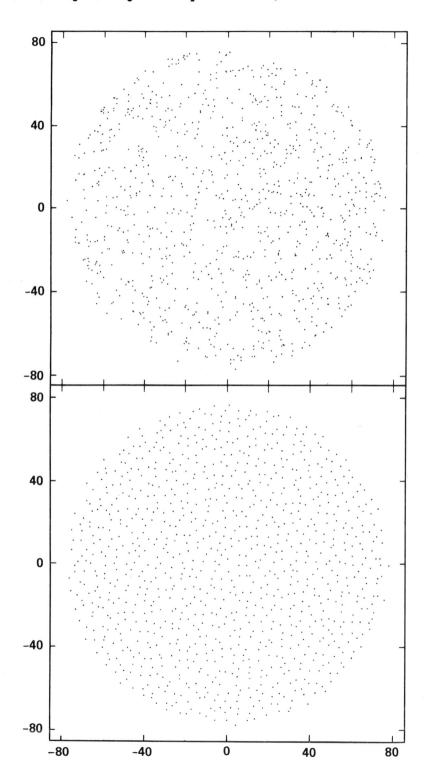

7.5 Fractionally Charged Quasiparticles

The existence of fractionally charged elementary excitations in this system, the least controversial aspect of the theory, follows automatically from properties of the ground state, *provided there is no degeneracy*. Our belief that the fractional quantum Hall ground state is not degenerate comes mainly from trying to find other ground states and failing. If other ground states are eventually discovered, these arguments will become academic.

The reasoning leading to the existence of fractionally charged quasiparticles is illustrated in Fig. 8. Imagine piercing the exact many-body ground state at location z_0 with an infinitely thin magnetic solenoid and adiabatically inserting through this solenoid a flux quantum $\phi_0 = hc/e$. As the flux is added, the wave function evolves adiabatically so as to always be an eigenstate of the changing Hamiltonian. However, once an entire flux quantum has been added, the Hamiltonian has, up to a gauge transformation, evolved back to its value at zero flux. Thus, the solenoid threaded by ϕ_0 may be gauged away, leaving behind an *exact* excited state of the original Hamiltonian. This excitation is a quasiparticle.

The charge of the quasiparticle may be determined by imagining the solenoid to be surrounded by a large Gauss's law box. Near the solenoid, the response of the ground state to the addition of flux is complicated and difficult to calculate accurately. Far away from the solenoid, however, the ground state just moves over. The single-body wave functions out of which the many-body wave function is built respond to the addition of flux ϕ through a solenoid at the origin in the manner

$$(re^{i\chi})^k e^{-\frac{1}{4}r^2} \to r^{|k+\phi/\phi_0|} e^{ik\chi} e^{-\frac{1}{4}r^2} , \quad (7.5.1)$$

where $re^{i\chi} \equiv z$. Thus, the effect on these wave functions of adding a whole flux quantum, and then gauging, is to increment or decrement their total angular momentum by one

$$z^k e^{-\frac{1}{4}|z|^2} \to z^{k\pm 1} e^{-\frac{1}{4}|z|^2} \quad (7.5.2)$$

Elementary Theory: Incompressible Quantum Fluid

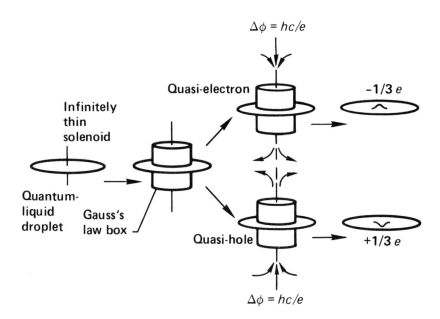

Figure 7.8 Illustration of thought experiment generating fractionally charged quasiparticles. This shows the charge to be $\pm 2\pi\ell^2$ times the ground-state charge density, regardless of the form of the interelectronic repulsions, so long as the ground state is non-degenerate.

according as the sign of the flux addition. Note that the state with $k = 0$ evolves under removal of flux to $z^*\exp(-|z|^2/4)$, a state in the next Landau level. With this one exception, each state in the lowest Landau level transforms into its neighbor, in rough analogy with a shift register. Such an operation will simply shift the ground state, transporting in or out of the Gauss's law box the average charge per state, which we know to be $1/m$. The validity of this 'Schrieffer counting argument' (Su and Schrieffer 1981) implies that the charge of the quasiparticle and the charge *density* of the ground state are essentially the same thing.

I have assigned approximate wave functions to the quasihole and quasielectron centered at z_0 of the form

$$S_{z_0}|m\rangle = e^{-\frac{1}{4}\sum_\ell^N |z_\ell|^2} \prod_i^N (z_i - z_0) \prod_{j<k}^N (z_j - z_k)^m \tag{7.5.3}$$

and

$$S^\dagger_{z_o}|m\rangle = e^{-\frac{1}{4}\sum_\ell^N |z_\ell|^2} \prod_i^N (2\frac{\partial}{\partial z_i} - z_o^*) \prod_{j<k}^N (z_j - z_k)^m \ ,$$

(7.5.4)

reasoning in the following way. Most electrons in the sample, those far from the solenoid, are affected trivially by the addition of flux. This behavior is independent of the Coulomb forces, and, indeed, would occur even if these forces were turned *off* before flux was added. It is sensible, therefore, to evolve the ground state as though the Coulomb forces actually *were* turned off and interpret the approximate quasiparticle wave function thus generated *variationally*. This wave function is inaccurate near the solenoid, but probably not severely. The state Ψ_m may be written

$$\Psi_m(z_1,\ldots,z_N) = \sum_{k_1\ldots k_N} a_{k_1\ldots k_N} z_1^{k_1} \ldots z_N^{k_N} e^{-\frac{1}{4}\sum_\ell^N |z_\ell|^2} \ .$$

(7.5.5)

Considered as an eigenstate of the Hamiltonian with the Coulomb interactions turned off, Ψ_m evolves exactly to

$$\Psi_m \to \sum_{k_1\ldots k_N} \frac{a_{k_1\ldots k_N}}{\sqrt{2^N(k_1+1)\ldots(k_N+1)}} z_1^{k_1+1} \ldots z_N^{k_N+1}$$

$$\times\ e^{-\frac{1}{4}\sum_\ell^N |z_\ell|^2}$$

(7.5.6)

under the operation that generates a quasihole at the origin. The operation that generates a quasielectron acts similarly, except that a configuration with any $k_j = 0$ acquires a projection on the next Landau level. Since the true quasielectron wave function has insig-

Elementary Theory: Incompressible Quantum Fluid

nificant Landau level mixing, I interpret the operation as annihilating these configurations. The normalization factors in Eq. 5.6 make computation difficult, so I remove them in the manner

$$\Psi_m \to \sum_{k_1\ldots k_N} a_{k_1\ldots k_N} z_1^{k_1+1} \cdots z_N^{k_N+1} e^{-\frac{1}{4}\sum_\ell^N |z_\ell|^2} ,$$
(7.5.7)

justifying this on the grounds that the normalization factors have little effect on this particular wave function and that the wave function is only variational in the first place. Removing these factors has no effect at all for $m = 1$, when Ψ_m is a single Slater determinant. I take the quasielectron creation to be the Hermitian adjoint operation:

$$\Psi_m \to \sum_{k_1\ldots k_N} \left[2^N k_1\ldots k_N\right] a_{k_1\ldots k_N} z_1^{k_1-1} \cdots z_N^{k_N-1}$$

$$\times e^{-\frac{1}{4}\sum_\ell^N |z_\ell|^2} .$$
(7.5.8)

This has the useful feature that it annihilates configurations for which any $k_j = 0$. Both approximate wave functions generalize in a straightforward way to Eqs. 5.3 and 5.4 when the solenoid is at positions z_0 other than the origin. Numerical calculations indicate these approximate wave functions are rather accurate, particularly for the quasihole (Haldane and Rezayi 1985a, Morf and Halperin 1986). MacDonald and Girvin (1986) have recently calculated properties of quasiparticle wave functions with the normalization factors left in (Eq. 5.6) and found them to be equivalent to mine.

Figure 9 shows results of hypernetted-chain calculations of the electron density as a function of distance from z_0 for the quasihole and quasielectron wave functions $S_{z_0}|m\rangle$ and $S^+_{z_0}|m\rangle$ for $m = 3$. One sees in each case a region of size ℓ centered at z_0 containing a total 'excess' charge of $\pm 1/3$ e. My original

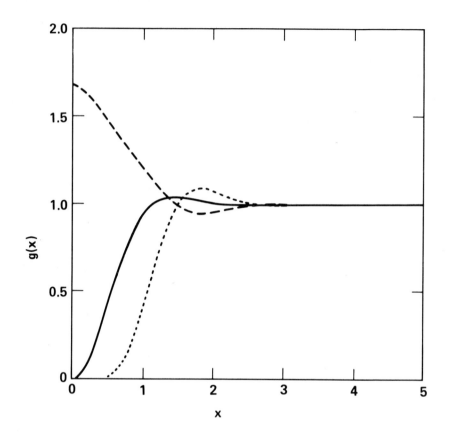

Figure 7.9 Electron density, in units of $1/(2\pi m)$, as a function of $x = |z-z_0|/(2m)^{1/2}$, the distance from the center of the quasihole (solid) and quasielectron (dashed), for the wave function given by Eqs. 5.3 and 5.4 with $m=3$. The radial distribution function of the ground state (dots) taken from Fig. 7.5 is shown for comparison. The quasiparticle wave functions accumulate an excess charge $\pm e/m$ at z_0.

estimates of the creation energies for these states (Laughlin 1983a) were 0.022 e^2/ℓ and 0.025 e^2/ℓ, respectively. More recent calculations of Haldane and Rezayi (1985a) and Morf and Halperin (1986) indicate the quasielectron charge density has a slight dip at the origin and that the correct creation energies associated with these wave functions are 0.026 e^2/ℓ and 0.073 e^2/ℓ. At a magnetic field strength of $B = 15T$

Elementary Theory: Incompressible Quantum Fluid

one has $0.03\, e^2/\kappa\ell = 0.50$ meV, which corresponds to a temperature of 5.8 K. This compares very favorably with the experimental observation that the fractional quantum Hall effect turns on at a temperature of roughly 1 K in such fields. The quasiparticle creation energy is appropriately considered an *energy gap*, such as that occurring in a superconductor or in an insulator.

I calculate the quasihole properties in the following way (Laughlin 1984d). As was the case with the ground state, the square of the wave function $S_{z_0}|m\rangle$ may be interpreted as a statistical-mechanical probability distribution function, in the manner

$$\left|S_{z_0}\Psi_m(z_1,\ldots,z_N)\right|^2 = e^{-\beta\phi'} \qquad (7.5.9)$$

with $\beta = 1/m$ and

$$\phi' = -2m^2\sum_{j>k}^{N} \ell n |z_j - z_k| + \frac{m}{2}\sum_{\ell}^{N} |z_\ell|^2$$

$$- 2m\sum_{i}^{N} \ell n |z_i - z_0| \, . \qquad (7.5.10)$$

But for the absence of a term of the form $|z_0|^2/2$, which corresponds to an unimportant normalization factor for $S_{z_0}|m\rangle$, this is the potential energy of N electrons of 'charge' m and one phantom particle of 'charge' 1 moving in a background of 'charge' density $1/2\pi$. The normalization integral $\langle m|S^+_{z_0}S_{z_0}|m\rangle$ is therefore the partition function of a plasma with one of its particles held *fixed*. This is proportional to the probability to find the particle at z_0 if it is *not* fixed, a quantity we know to be constant. Thus, we have

$$\langle m|S^\dagger_{z_0}S_{z_0}|m\rangle = \langle m|S^\dagger_0 S_0|m\rangle\, e^{\frac{1}{2m}|z_0|^2} , \qquad (7.5.11)$$

and

$$Z = \int \cdots \int e^{-\beta[\phi' + \frac{1}{2}|z_o|^2]} d^2z_1 \cdots d^2z_N \, d^2z_o$$

$$= 2\pi m N \langle m | S_o^\dagger S_o | m \rangle \, , \quad (7.5.12)$$

for the plasma partition function. I calculate the two correlation functions

$$g_{11}(|z_1 - z_2|) = \frac{(2\pi m N)^2}{Z}$$

$$\times \int \cdots \int e^{-\beta[\phi' + \frac{1}{2}|z_o|^2]} d^2z_3 \cdots d^2z_N \, d^2z_o \quad (7.5.13)$$

and

$$g_{12}(|z_1 - z_0|) = \frac{(2\pi m N)^2}{Z}$$

$$\times \int \cdots \int e^{-\beta[\phi' + \frac{1}{2}|z_o|^2]} d^2z_2 \cdots d^2z_N \quad (7.5.14)$$

which represent the electron density, divided by the asymptotic value $1/2\pi m$, with an electron or the phantom held fixed, using the two-component hypernetted-chain equations (cf. Eqs. 4.1 – 4.3):

$$h_{\alpha\beta}(r) = g_{\alpha\beta}(r) - 1 \quad (7.5.15)$$

$$h_{\alpha\beta}(|r|) = c_{\alpha\beta}(|r|) + \sum_\gamma \rho_\gamma \int h_{\alpha\gamma}(|r-r'|) \, c_{\gamma\beta}(|r'|) d^2r'$$

$$(7.5.16)$$

$$h_{\alpha\beta}(r) = \exp\left[-\beta v_{\alpha\beta}(r) + h_{\alpha\beta}(r) - c_{\alpha\beta}(r)\right] \, .$$

$$(7.5.17)$$

Elementary Theory: Incompressible Quantum Fluid

The greek indices in these equations run over the two types of particles, electrons and phantoms. The quantities $\rho_1 = 1/2\pi m$ and $\rho_2 = \rho_1/N$ are the particle densities. I solve these equations perturbatively using ρ_2 as the small parameter. To zeroth order in ρ_2, $g_{11}(r)$ is the $g(r)$ satisfying Eqs. 4.1 - 4.3, a property of the ground state one already knows. In addition, $g_{12}(r)$ is given by the solution of the equations

$$h_{12}(r) = g_{12}(r) - 1 \qquad (7.5.18)$$

$$h_{12}(r) \cong c_{12}(r) + \rho_1 \int h_{11}(|r-r'|) \, c_{12}(|r'|) \, d^2r' \qquad (7.5.19)$$

$$g_{12}(r) = \exp[-\beta v_{12}(r) + h_{12}(r) - c_{12}(r)] \,. \qquad (7.5.20)$$

which is readily obtained using methods described in Sec. 7.4. I then calculate the small change to $g_{11}(r)$ resulting from the presence of the phantom by solving the linearized equations

$$\delta h_{11}(r) = \delta g_{11}(r) \,. \qquad (7.5.21)$$

$$\delta h_{11}(r) = \delta c_{11}(r) + \int \Big\{ \rho_1 [\delta h_{11}(|r-r'|) \, c_{11}(|r'|)$$
$$+ h_{11}(|r-r'|) \, \delta c_{11}(|r'|)]$$
$$+ \rho_2 \, h_{12}(|r-r'|) \, c_{21}(|r'|) \Big\} d^2r' \,, \qquad (7.5.22)$$

and

$$\delta g_{11}(r) = g_{11}(r)[\delta h_{11}(r) - \delta c_{11}(r)] \,. \qquad (7.5.23)$$

This is used to calculate the quasihole creation energy in the manner

$$\Delta_{\text{quasihole}} = \frac{N\rho_1}{2} \int_0^\infty \frac{e^2}{r} \delta g_{11}(r) \, 2\pi r \, dr \,. \qquad (7.5.24)$$

Note that since Eqs. 5.21 - 5.23 are linear, the factor of N in Eq. 5.24 is exactly canceled by the factor of $1/N$ implicitly present in Eq. 5.22 through ρ_2. Thus $\Delta_{quasihole}$ does not depend on N, so long as it is large.

Calculations for the quasielectron are more difficult but similar in spirit. One observes that the expression for the electron density around z_0, the quantity described by g_{12} in the quasihole calculation, may be integrated by parts to yield

$$\rho(z_1) = \frac{N}{\langle m|S_{z_0} S_{z_0}^\dagger|m\rangle}$$

$$\times \int \ldots \int |S_{z_0}^\dagger \Psi_m(z_1,\ldots,z_N)|^2 \, d^2z_2 \ldots d^2z_N$$

$$= \frac{1}{2\pi m} e^{-\frac{1}{2}|z_1|^2} \left[2\frac{\partial}{\partial z_1} - z_0^*\right] \left[2\frac{\partial}{\partial z_1^*} - z_0\right]$$

$$\times \left[\frac{e^{\frac{1}{2}|z_1|^2}}{\left[|z_1-z_0|^2 - 2\right]} \tilde{g}_{12}(|z_1 - z_0|)\right], \qquad (7.5.25)$$

where

$$\tilde{g}_{12}(|z_1-z_0|) =$$

$$\left[\frac{2\pi m N}{\langle m|S_{z_0} S_{z_0}^\dagger|m\rangle}\right] \int \ldots \int e^{-\frac{1}{4}\sum_\ell^N |z_\ell|^2} \prod_i^N \left[|z_i-z_0|^2 - 2\right]$$

$$\times \prod_{j<k}^N |z_j-z_k|^{2m} \, d^2z_2 \ldots d^2z_N \bigg]. \qquad (7.5.26)$$

Elementary Theory: Incompressible Quantum Fluid 269

But for the substitution $|z_i - z_0|^2 \to |z_i - z_0|^2 - 2$, Eq. 5.26 is exactly the expression for the quasihole g_{12}. I therefore use the quasihole machinery to calculate this quantity, and then numerically differentiate it to obtain the quasielectron charge profile shown in Fig. 9. This is formally allowed because the diagrammatic sums [see Stell (1964)] leading to the hypernetted chain equations are defined even when $\exp(-\beta v_{12}(r))$ is occasionally negative.

The quasielectron creation energy may also in principle be calculated using this approach. Unfortunately, the computation is complicated by the need for a three-point correlation function of the integrated-by-parts problem, for which reliable analytic expressions do not exist. Rather than perform this calculation, I originally estimated the quasielectron energy by using the quasielectron g_{12} in Eqs. 5.21 - 5.24, arguing that the energy should be the Coulomb repulsion of the quasielectron charge with itself. This was correct, but the error is now known to be unacceptably large.

Insofar as the quasiparticles are correctly described by $S_{z0}|m\rangle$ and $S^+_{z0}|m\rangle$, they have many properties in common with electrons in the lowest Landau level. For example, their size, charge, and energy do not depend on z_0, exactly as is the case with the single-electron wave function

$$\psi_{z_0}(z) = e^{-\frac{1}{4}[|z|^2 + |z_0|^2]} e^{\frac{1}{2}zz_0^*}, \qquad (7.5.27)$$

(cf. Eq. 3.11). In fact, $S_{z0}|m\rangle$ for $m = 1$ is exactly the particle-hole conjugate of ψ_{z0} [particle-hole conjugation is described in detail in Girvin (1984a)]:

$$[\psi_{z_0}]^\dagger(z_1,\ldots,z_N) = \sqrt{N+1}\,\frac{1}{\langle 1|1\rangle}\int\cdots\int \prod_{j<k}^{N+1}(z_j-z_k)$$

$$\times\, e^{-\frac{1}{4}\sum_\ell^N |z_\ell|^2}\,\psi_{z_0}^*(z_{N+1})\,d^2z_{N+1}$$

$$= 2\pi\sqrt{N+1} \; \frac{e^{-\frac{1}{4}|z_0|^2}}{\langle 1|1\rangle} \prod_i^N (z_i - z_0) \prod_{j<k}^N (z_j - z_k) \; e^{-\frac{1}{4}\sum_\ell^N |z_\ell|^2}.$$

(7.5.28)

In evaluating this integral, I have used the fact that

$$\int e^{-\frac{1}{2}|z|^2} e^{\frac{1}{2}z^* z_0} f(z) d^2 z = 2\pi f(z_0) ,$$

(7.5.29)

for any analytic function f (see Girvin and Jach 1984). The identification of $S_{z0}|m\rangle$ for $m = 1$ with the adjoint of ψ_{z0} should not be surprising, for the thought experiment generating quasiparticles is correct even when the Landau level is full, for which we know the quasiparticles to be ordinary holes. In general it is useful and appropriate to think of $S_{z0}|m\rangle$ and $S^+_{z0}|m\rangle$ in terms of ψ_{z0}.

The quasiparticle size is invariant because it is the distance over which the equivalent plasma screens. If this were a weakly-coupled plasma ($\Gamma \ll 1$), which it is not, this distance would be the Debye length [see Ichimaru (1973)] $\lambda_D^2 = \ell^2/2$. Since it is strongly coupled, a better estimate is the ion-disc radius associated with 'charge' $1/m$, $(\lambda_{ion-disc})^2 = 2\ell^2$. Either estimate gives a quasiparticle size of ℓ, regardless of the value of m. Quasiparticles have, in this sense, the same 'size' as electrons in the lowest Landau level.

The quasiparticle charge is exactly $\pm 1/m$ because the equivalent plasma screens *perfectly*. This is easily understood in the case of the quasihole because it is exactly equivalent to a phantom particle of 'charge' 1 in a plasma of electrons of 'charge' m. Its screening 'charge' -1 corresponds to $-1/m$ electrons. In the case of the quasielectron, the effective electron-phantom 'interaction' $-2m\ell n[|z-z_0|^2 - 2]$, gives rise to an electron-phantom distribution $\tilde{g}_{12}(r)$ reflecting a screening 'charge' of the wrong sign:

Elementary Theory: Incompressible Quantum Fluid

$$\frac{1}{2\pi} \int_0^\infty [\tilde{g}_{12}(r) - 1] \, 2\pi r \, dr = -1 \, . \qquad (7.5.30)$$

This is because the net 'charge' accumulated around the phantom is determined solely by the long-range behavior of the 'interaction', which is visibly the same as that in the quasihole problem. However, the actual $g_{12}(r)$ is related to $\tilde{g}_{12}(r)$ by (cf. Eq. 5.22)

$$g_{12}(r) = \left[\frac{\partial^2}{\partial r^2} + \frac{1}{r}\frac{\partial}{\partial r} + 2r\frac{\partial}{\partial r} + r^2 + 2 \right] \left[\frac{\tilde{g}_{12}(r)}{r^2 - 2} \right] . \qquad (7.5.31)$$

In light of Eq. 5.30 and the fact that $\tilde{g}_{12}(r) \to 1$ at large r, one has

$$\frac{1}{2\pi} \int_0^\infty [g_{12}(r) - 1] \, 2\pi r \, dr = +1 \, . \qquad (7.5.32)$$

The quasiparticle creation energy is invariant because, up to small changes in the electron-electron correlations near the quasiparticle, it is the Coulomb repulsion of the accumulated charge with itself. One can physically understand its size by approximating the quasiparticle as a disc of area $2\pi\ell^2$ on which a charge $\pm e/m$ is uniformly spread. The electrostatic potential energy of this disc is

$$\Delta_{disc} = \left[\frac{(e/m)^2}{2\pi\ell^2} \right] \iint_{disc} \frac{d^2r \, d^2r'}{|r - r'|} = \frac{4\sqrt{2}}{3\pi} \frac{1}{m^2} \frac{e^2}{\ell} \, , \qquad (7.5.33)$$

which gives a value of $0.067 \, e^2/\ell$ for $m = 3$. An important implication of this formula is that the energy gap decreases *very* rapidly with quasiparticle charge.

Quasiparticles also act like electrons in the lowest Landau level in that they can be placed into eigenstates of angular momentum k analogous to

$$\psi_k(z) = z^k e^{-\frac{1}{4}|z|^2} . \quad (7.5.34)$$

These describe a quasiparticle moving in a circular orbit, as ψ_k does, but with an orbit radius $m^{1/2}$ larger than that appropriate for Ψ_k, much as though the quasiparticle were an eigenstate of the effective Hamiltonian

$$\mathcal{H}_{effective} = \frac{1}{2m_e} \left| \frac{\hbar}{i} \nabla + \frac{e^*}{c} \mathbf{A} \right|^2 , \quad (7.5.35)$$

with $e^* = e/m$. The quasiparticle in this sense couples to *magnetic* fields as a fractionally charged object, just as it does to electric fields. The operators which generate these states are the elementary symmetric polynomials $S_k(z_1,\ldots,z_N)$, defined by

$$\prod_i^N (z_i - z_o) = \sum_{k=0}^N z_o^k S_k(z_1,\ldots,z_N) . \quad (7.5.36)$$

and their Hermitian adjoints $S_k^\dagger$. The wave functions $S_k|m\rangle$ and $S^\dagger{}_k|m\rangle$ may be considered superpositions of the degenerate states $S_{z_0}|m\rangle$ and $S^+{}_{z_0}|m\rangle$ of the form (cf. Eq. 5.29)

$$S_k|m\rangle = \frac{1}{(2m)^{k+1} \pi k!} \int z_o^{*k} e^{-\frac{1}{2m}|z_o|^2} S_{z_o}|m\rangle d^2 z_o \quad (7.5.37)$$

and

$$S_k^\dagger|m\rangle = \frac{1}{(2m)^{k+1} \pi k!} \int z_o^k e^{-\frac{1}{2m}|z_o|^2} S_{z_o}^\dagger|m\rangle d^2 z_o , \quad (7.5.38)$$

just as $\psi_k(z)$ is a superposition of the degenerate states ψ_{z_o} (Eq. 5.27) of the form

$$\psi_k(z) = \frac{1}{2\pi} \int z_o^{*k} e^{-\frac{1}{4}|z_o|^2} \psi_{z_o}(z) d^2 z_o . \quad (7.5.39)$$

Elementary Theory: Incompressible Quantum Fluid

The charge densities of the first few states $S_k|m\rangle$ for $m = 3$ calculated using the Monte Carlo technique described in Sec. 7.4 are compared with those for $m = 1$ in Fig. 10. One sees in each case a strong dip at radius $(2mk)^{1/2}$ exactly as expected of a hole described by Eq. 5.35.

I shall now show analytically that the orbit of the quasihole state $S_k|m\rangle$ is $m^{1/2}$ larger than that of $\psi_k(z)$. The quasielectron calculation is essentially the same. It is first necessary to normalize $S_k|m\rangle$. This is tractable because $\langle m|S^+_{z_A}S_{z_B}|m\rangle$, an analytic function of z_A^* and z_B, can be determined everywhere by analytic continuation once its value at $z_A = z_B$ is known. Thus, from Eq. 5.11 one has

$$\langle m|S^\dagger_{z_A} S_{z_B}|m\rangle = \langle m|S^\dagger_o S_o|m\rangle \, e^{\frac{1}{2m} z_A^* z_B} \qquad (7.5.40)$$

and

$$\langle m|S^\dagger_{k_A} S_{k_B}|m\rangle =$$

$$\left[\frac{1}{(2m)^{k_A+1} \pi k_A!}\right]\left[\frac{1}{(2m)^{k_B+1} \pi k_B!}\right] \langle m|S^\dagger_o S_o|m\rangle$$

$$\times \iint z_A^{k_A} z_B^{*k_B} e^{\frac{1}{2m} z_A^* z_B} e^{-\frac{1}{2m}[|z_A|^2+|z_B|^2]} d^2z_A d^2z_B$$

$$= \frac{\delta_{k_A k_B}}{(2m)^{k_A} k_A!} \langle m|S^\dagger_o S_o|m\rangle . \qquad (7.5.41)$$

The matrix elements of the electron density operator $\rho(z)$ are similarly continuable from the known electron distribution around the quasihole:

$$\langle m|S^\dagger_{z_o} \rho(z) S_{z_o}|m\rangle = \frac{\langle m|S^\dagger_{z_o} S_{z_o}|m\rangle}{2\pi m} g_{12}(|z-z_o|) .$$

$$(7.5.42)$$

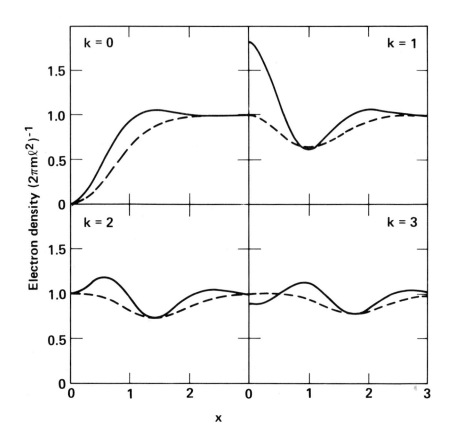

Figure 7.10 Electron density, in units of $1/(2\pi m)$, vs. distance from the origin ($x=|z|/(2m)^{1/2}$) for the states $S_k|m\rangle$ defined by Eq. 5.37 for $m=3$ (solid) and $m=1$ (dashed). Each shows a strong dip at the orbit radius $x=k^{1/2}$. Quasiparticles execute cyclotron motion like that of electrons in the lowest Landau level, except that the orbit radius is $m^{1/2}$ times larger, reflecting the particles' fractional charge.

Finally defining the 'enhanced' or 'unsmoothed' function $g_{12}^e(|z|)$ in the manner

$$g_{12}(|z|) = \frac{1}{2\pi m}\int e^{-\frac{1}{2m}|z-z_c|^2} g_{12}^e(|z_c|)\, d^2 z_c\ ,\quad (7.5.43)$$

I obtain

Elementary Theory: Incompressible Quantum Fluid

$$\frac{\langle m|S_k^\dagger \rho(z) S_k|m\rangle}{\langle m|S_k^\dagger S_k|m\rangle} = \frac{1}{(2\pi m)^3}\left[\frac{1}{(2m)^{k+1}\pi k!}\right]\iiint g_{12}^e(|z-z_c|)$$

$$\times e^{-\frac{1}{2m}[z_A^*-z_c^*][z_B-z_c]}\ z_A^k z_B^{*k}\ e^{\frac{1}{2m}z_A^* z_B}$$

$$\times e^{-\frac{1}{2m}[|z_A|^2+|z_B|^2]}\ d^2z_A d^2z_B d^2z_c$$

$$= \frac{1}{2\pi m}\int g_{12}^e(|z-z_c|)\left[\frac{1}{(2m)^{k+1}\pi k!}|z_c|^{2k}\ e^{-\frac{1}{2m}|z_c|^2}\right]d^2z_c.$$

(7.5.44)

Note that even though g_{12}^e is difficult to calculate accurately, perhaps impossible in some cases, the moment in this expression is always well defined, for it equals the finite series

$$\frac{\langle m|S_k^\dagger \rho(r) S_k|m\rangle}{\langle m|S_k^\dagger S_k|m\rangle} = \frac{k!}{2\pi m}\left\{\sum_{j=0}^k \frac{\left[\frac{m}{2}\left[\frac{\partial^2}{\partial r^2}+\frac{1}{r}\frac{\partial}{\partial r}\right]\right]^j}{(k-j)!\,(j!)^2}\right\}g_{12}(r).$$

(7.5.45)

Enhancement commutes with subtracting unity, thus one has *exactly* for the mean-square orbit radius

$$\langle r^2 \rangle = -m\int\left[\frac{\langle m|S_k^\dagger \rho(z) S_k|m\rangle}{\langle m|S_k^\dagger S_k|m\rangle}-\frac{1}{2\pi m}\right]|z|^2\,dz$$

$$= -\frac{1}{2\pi}\int h_{12}^e(|z|)\,[|z|^2+2m(k+1)]\,d^2z$$

$$= -\frac{1}{2\pi} \int h_{12}(|z|) \, [|z|^2 + 2mk] \, d^2z$$

$$= 2mk + 2 , \qquad (7.5.46)$$

where $h_{12}^e = g_{12}^e - 1$. In evaluating this expression I have used the neutrality

$$\frac{1}{2\pi} \int h_{12}(|z|) \, d^2z = -1 , \qquad (7.5.47)$$

and constant-screening

$$\frac{1}{2\pi} \int h_{12}(|z|) |z|^2 \, d^2z = -2 \qquad (7.5.48)$$

sum rules, both of which are consequences of the $1/q^2$ divergence in the 'Coulomb' potentials of the equivalent plasma.

It may be seen from this analysis that quasiparticles are also like holes in an otherwise full Landau level in having degeneracy N where N is the number of electrons. This is reflected both in the maximum number of available polynomials S_k (cf. Eq. 5.36) and in the impossibility of having a quasihole orbit outside the sample ($2mk + 2 < 2mN$). In addition, the superposition of the charge densities associated with the states $S_k|m\rangle$ is uniform.

7.6 Exactness of the Fractional Quantum Hall Effect

The integral quantum Hall effect is exact in the limit of low temperature and large sample size because it is a measurement of the electron charge, a fundamental quantum of nature. This is not transparently the case for fractionally charged quasiparticles, and thus the question of errors in the fractional quantum Hall effect is much more difficult to address. I wish to argue that the fractional effect is also inherently exact, but only in the limit of large sample sizes, much as localization is truly complete only at arbitrarily large length scales. The validity of this argument implies that the quasiparticle charge is also a fundamental quantum of nature.

Elementary Theory: Incompressible Quantum Fluid

The quasiparticle charge, or, equivalently, the ground state charge density, is manifestly *not* exactly quantized when the sample is small. There are many reasons for this, but the most worrisome is the mixing into the ground state of states from higher Landau levels. The wave function for *two* electrons, the smallest imaginable 'sample', is given by Eq. 2.7 only in the limit that the magnetic field is infinite. When the field is not infinite, the wave function is a superposition of states from all Landau levels having the same internal angular momentum in (cf. Eq. 2.5):

$$\Psi(z_1, z_2) = \left\{ \sum_{n=0}^{\infty} \alpha_n |z_1 - z_2|^n \right\} (z_1 - z_2)^m$$

$$\times\, e^{-\frac{1}{4}[|z_1|^2 + |z_2|^2]} \quad . \tag{7.5.49}$$

When the magnetic field is large but finite, so that the amplitudes α_n for $n \neq 0$ are small, Landau level mixing has an insignificant effect at small separations $|z_1 - z_2|$, where the probability is large, but an enormous effect at large separations. This is significant because the plasma sum rules responsible for quantization of the charge density of Ψ_m (cf. Eqs. 3.3 – 3.5), and thus the quasiparticle charge, follow from the *long-range* behavior of the 'Coulomb' interaction, which in turn, may be identified with the behavior of a pair wave function at large separations. Since this behavior is not exact, there is no *prima facie* reason to expect the sum rules or their implications, quantization of the charge density and of quasiparticle charge, to be exact either. The demonstrable inaccuracy of quantization at small length scales implies that perfect quantization, if it exists, must be a property of the macroscopic limit.

I begin the argument for exact quantization with the assertion that quantization is exact when Landau level mixing is turned off, in the limit of large length scales. This is based primarily on the observation by Yoshioka, Halperin and Lee (1983) that, when Landau level mixing is forbidden, the real wave function differs from the approximate one (Eq. 3.3) only in having slightly nondegenerate zeros. When Landau level mixing is forbidden, the real wave function is forced to take the form

$$\Psi(z_1,\ldots,z_N) = F(z_1,\ldots,z_N)\, e^{-\frac{1}{4}\sum_\ell |z_\ell|^2}, \qquad (7.5.50)$$

where F is an homogeneous polynomial of degree $mN(N-1)/2$. Since it is a polynomial, F is determined globally once its zeros are known. These have a particularly simple form in the approximate wave function Ψ_m: When the positions of $N-1$ electrons are fixed, the Nth electron sees a third-order zero located at each of the other electrons. In the real wave function, these zeros are located not on the $N-1$ fixed electrons but near them, a few percent of ℓ away. One assumes that they do not stray far from the electron positions because the wave function would then have a large amplitude for two electrons to approach one another, which is energetically unfavorable. These zeros are precisely the point charges of the equivalent plasma. Their attachment to an electron without being located precisely on it has the physical interpretation in the equivalent plasma of the presence of short-range forces (in addition to the long-range 'Coulomb' force). That is, one has for the exact wave function

$$|\Psi(z_1,\ldots,z_N)|^2 = e^{-\beta[\phi + \gamma]}, \qquad (7.5.51)$$

where $\beta = 1/m$, ϕ is given by Eq. 3.5 and γ is a multicenter 'potential' dependent on the form of the actual inter-electronic repulsions, which is *not* 'Coulombic'. The presence of a potential such as γ has *no* effect on the plasma's neutrality and only a short-range effect on its screening properties. For this reason the real ground-state charge density and quasiparticle charge, like the approximate ones, become exact as the system size increases beyond a few ℓ. One expects the error to decrease at least as fast as $\exp(-L/\lambda_D)$, where L is the system size and $\lambda_D = \ell/2^{1/2}$ is the plasma's Debye length.

Imagine now the situation, illustrated in Fig. 11, in which a macroscopically-large fractional-quantum-Hall sample, denoted A in the Figure, is embedded in a

Elementary Theory: Incompressible Quantum Fluid

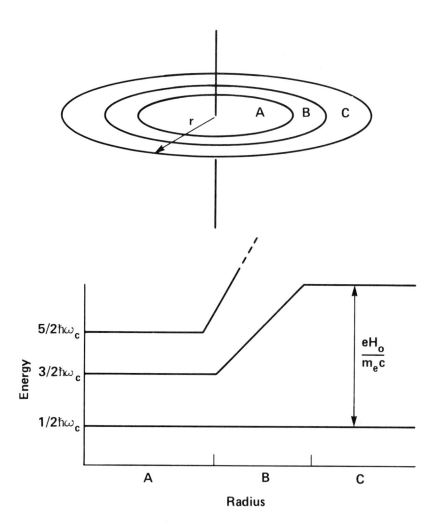

Figure 7.11 Illustration of thought experiment demonstrating exactness of quasiparticle charge. The system to be tested (A) is made contiguous with a system (C) with known properties.

second sample with a position-dependent cyclotron gap. One might accomplish this, for example, by varying the band effective mass. The gap varies sufficiently slowly over the region B that the integrity of the fractional quantum Hall state is maintained locally. In region C, the gap is as large as required to make the neglect of Landau level mixing a good approximation. This is a meaningful condition, for the Landau-level-mixing error may be made arbitrarily small by making the cyclotron gap arbitrarily large. Now imagine region A to be pierced at its center with an infinitely thin magnetic solenoid, through which is adiabatically inserted a flux quantum $\phi_0 = hc/e$ (cf. Eq. 5.1). So long as the entire system's ground state is nondegenerate, or, equivalently, so long as the manybody excitation gap is maintained in all three regions and at the interfaces between them, the operation shifts all electrons and creates a quasiparticle at the solenoid in A. The charge of this quasiparticle is *exactly* $\pm 1/m$ because the charge drawn through the surface of a Gauss's law box surrounding the solenoid *but cutting region C only* is exactly $\pm 1/m$. This reasoning is valid so long as region A is sufficiently large to contain the whole quasiparticle. This is the sense in which exact quantization is a macroscopic phenomenon.

It is also true that the charge *density* in region A is exactly quantized, so long as the region is sufficiently large, for a second Gauss's law box, placed inside the first so as to cut region A only, also transports charge $\pm 1/m$ through its surface.

The validity of these two results, the exact quantization of quasiparticle charge and the exact quantization of ground state charge density, implies that fractional quantum Hall phenomena have a physical meaning transcending the microscopic considerations in Secs. 5.3 and 5.5. Even at low magnetic fields when Landau level mixing is extreme, quasiparticles appear to maintain their identity *exactly* so long as an energy gap of *any size* exists.

7.7 More Than One Quasiparticle

Quasiparticles of like or opposite sign interact quantum-mechanically much the way electrons in the lowest Landau level do. This interaction is now believed to cause the hierarchy of fractional quantum Hall states and to provide an excitonic interpretation of the 'magnetoroton', the critical low-lying collective excitation of the fractional quantum Hall ground state (see Chap. IX). Because these behaviors are important I would like to show how they follow from the properties of S_{z0} and S_{z0}^+.

Elementary Theory: Incompressible Quantum Fluid

Quasiparticles of like sign have quantized separations just as electrons in the lowest Landau level do. As was the case with electrons, this is most easily understood physically by considering a quasiparticle pair. Suppose one wishes to find the superposition of states of the form

$$S_{z_A} S_{z_B} |m\rangle \quad \text{or} \quad S^\dagger_{z_A} S^\dagger_{z_B} |m\rangle$$

that 'diagonalizes' the many-body Hamiltonian, projected onto the set of such states. This superposition is determined solely by *symmetry* (Laughlin 1985c), just as the two-electron problem is diagonalized by symmetry with Ψ_k in Eq. 2.7. In fact, this is the two electron problem when $m=1$. The symmetry being exploited, translational and rotational invariance, follows from the uniformity of the equivalent plasma. It becomes more apparent if one takes the basis states to be of the form

$$|z_A, z_B\rangle = \exp\left\{-\frac{1}{4m}[|z_A|^2 + |z_B|^2]\right\} S_{z_A} S_{z_B} |m\rangle ,$$

(7.7.1)

for holes, and similarly for particles, so that the *diagonal* matrix elements of overlap and energy become functions of $|z_A - z_B|$ solely, in the manner

$$\langle z_A, z_B | z_A, z_B \rangle = F[|z_A - z_B|^2] \quad (7.7.2)$$

$$\langle z_A, z_B | \mathcal{H} | z_A, z_B \rangle = E[|z_A - z_B|^2] . \quad (7.7.3)$$

The functions E and F are easy to calculate. For example, in the notation of Sec. 7.5:

$$F[|z|^2] = C \frac{g_{22}[|z|^2]}{|z|^{2/m}} , \quad (7.7.4)$$

where C is a constant. These diagonal matrix elements analytically continue in the manner

$$\langle z_{A'}, z_{B'} | z_A, z_B \rangle = F[(z_{A'}^* - z_{B'}^*)(z_A - z_B)]$$

$$\times \exp\left\{-\frac{1}{4m}\left[|z_{A'}|^2 + |z_{B'}|^2 + |z_A|^2 + |z_B|^2\right]\right\}$$

$$\times \exp\left\{\frac{1}{2m}\left[z_{A'}^* z_A + z_{B'}^* z_B\right]\right\}, \qquad (7.7.5)$$

and similarly for $\mathcal{H}$, to yield matrices *necessarily* diagonalized by states of the form

$$|k\rangle = \frac{1}{(2\pi m)^2} \iint d^2 z_A d^2 z_B \, (z_A^* - z_B^*)^{2k}$$

$$\times \exp\left\{-\frac{1}{4m}\left[|z_A|^2 + |z_B|^2\right]\right\} |z_A, z_B\rangle. \qquad (7.7.6)$$

Using Eqs. 5.29 and 7.75 one obtains

$$\langle z_{A'}, z_{B'} | k \rangle = F_k \, (z_{A'} - z_{B'})^{2k} \exp\left\{-\frac{1}{4m}\left[|z_{A'}|^2 + |z_{B'}|^2\right]\right\}$$
$$(7.7.7)$$

and

$$\langle z_{A'}, z_{B'} | \mathcal{H} | k \rangle = E_k \, (z_{A'} - z_{B'})^{2k}$$

$$\times \exp\left\{-\frac{1}{4m}\left[|z_{A'}|^2 + |z_{B'}|^2\right]\right\}, \qquad (7.7.8)$$

where

$$F_k = \frac{1}{(4m)^{2k+1} \pi (2k)!} \int |z|^{4k} e^{-\frac{1}{4m}|z|^2} F^e[|z|^2] \, d^2 z, \qquad (7.7.9)$$

with

$$F[|z|^2] = \frac{1}{4\pi m} \int e^{-\frac{1}{4m}|z-z'|^2} F^e[|z'|^2] \, d^2 z'. \qquad (7.7.10)$$

Elementary Theory: Incompressible Quantum Fluid

and similarly for E. The energy eigenvalue is E_k/F_k. Note that, as in the discussion of Eq. 5.45, the moments F_k and E_k are well-defined even though the 'enhanced' functions F^e and E^e might not be. Note also that the exponent $2k$ is *even* regardless of the value of m.

Generalizations of this result to more than two quasiparticles is most conveniently done via a transformation to the 'fractional statistics representation', the existence of which was first suggested by Halperin (1984). Through this transformation one may represent arbitrary superpositions of the M-quasiparticle states

$$S_{z_1} \ldots S_{z_M} |m\rangle \quad \text{or} \quad S^\dagger_{z_1} \ldots S^\dagger_{z_M} |m\rangle$$

with a 'wave function' $\psi(\eta_1, \ldots, \eta_M)$ satisfying an equation of motion like that of the original electron problem. This wave function takes the form

$$\psi(\eta_1, \ldots, \eta_M) = \prod_{j<k}^{M} (\eta_j - \eta_k)^\alpha F(\eta_1, \ldots, \eta_M)$$

$$\times \exp\left\{-\frac{1}{4m} \sum_\ell^M |\eta_\ell|^2\right\}, \qquad (7.7.11)$$

where α is a fraction and F is symmetric and analytic in its arguments. The transformation of such a wave function to a real electron wave function $|\Psi\rangle$ is given, in the case of quasiholes, by

$$|\Psi\rangle = \frac{1}{(2\pi m)^M} \int \ldots \int \psi^*(\eta_1, \ldots, \eta_M) \prod_{j<k}^{M} (\eta_j - \eta_k)^\alpha \prod_i^M S_{\eta_i} |m\rangle$$

$$\times \exp\left\{-\frac{1}{4m} \sum_\ell^M |\eta_\ell|^2\right\} d^2\eta_1 \ldots d^2\eta_M . \qquad (7.7.12)$$

The set of transformed wave functions $|\Psi\rangle$ is *complete* because each of the original basis states

$$|z_1,\ldots,z_M\rangle = \prod_{j<k}^{M}(z_j-z_k)^\alpha \prod_{i}^{M} S_{z_i} |m\rangle \exp\left\{-\frac{1}{4m}\sum_{\ell}^{M}|z_\ell|^2\right\} ,$$

(7.7.13)

where $z_1,\ldots,z_M$ are quantum numbers, has a fractional-statistics image of the form

$$\psi_{z_1,\ldots,z_M}(\eta_1,\ldots,\eta_M) = \sum_{i_1\ldots i_M} a_{i_1\ldots i_M}$$

$$\times (z_1^*\eta_1)^{i_1}\ldots(z_M^*\eta_M)^{i_M} \prod_{j<k}^{M}[(z_j^*-z_k^*)(\eta_j-\eta_k)]^\alpha$$

$$\times \exp\left\{-\frac{1}{4m}\sum_{\ell}^{M}[|z_\ell|^2+|\eta_\ell|^2]\right\} ,$$

(7.7.14)

where

$$a_{i_1\ldots i_M} = \left[\frac{1}{(2\pi m)^M}\int\ldots\int d^2\eta_1\ldots d^2\eta_M \, |\eta_1|^{2i_1}\ldots|\eta_M|^{2i_M}\right.$$

$$\left.\times \prod_{j<k}^{M}|\eta_j-\eta_k|^{2\alpha} \exp\left\{-\frac{1}{2m}\sum_{\ell}^{M}|\eta_\ell|^2\right\}\right]^{-1} ,$$

(7.7.15)

that maps back to $|z_1,\ldots,z_M\rangle$ under the action of Eq. 7.12:

$$|z_1,\ldots,z_M\rangle = \frac{1}{(2\pi m)^M}\int\ldots\int d^2\eta_1\ldots d^2\eta_M$$

$$\times \psi^*_{z_1,\ldots,z_M}(\eta_1,\ldots,\eta_M) \prod_{j<k}^{M}(\eta_j-\eta_k)^\alpha \prod_{i}^{M} S_{\eta_i}|m\rangle$$

$$\times \exp\left\{-\frac{1}{4m}\sum_{\ell}^{M}|\eta_\ell|^2\right\} .$$

(7.7.16)

Elementary Theory: Incompressible Quantum Fluid 285

Because of this completeness, the overlap and Hamiltonian matrices may be expressed as operators acting on the fractional statistics wave function ψ. The transformed operators will be physically sensible (they are not unique) provided they are constrained to be *diagonal* in the position representation, that is, represented by functions $\mathcal{O}$ and $\mathcal{V}$ such that

$$\langle z_1, \ldots, z_M | z_1, \ldots, z_M \rangle = \int \cdots \int d^2\eta_1 \ldots d^2\eta_M$$
$$\times |\psi_{z_1, \ldots, z_M}(\eta_1, \ldots, \eta_M)|^2 \, \mathcal{O}[\eta_1, \ldots, \eta_M] \quad (7.7.17)$$

and

$$\langle z_1, \ldots, z_M | \mathcal{H} | z_1, \ldots, z_M \rangle = \int \cdots \int d^2\eta_1 \ldots d^2\eta_M$$
$$\times |\psi_{z_1, \ldots, z_M}(\eta_1, \ldots, \eta_M)|^2 \, \mathcal{V}[\eta_1, \ldots, \eta_M] \, . \quad (7.7.18)$$

Note that because of the analytic properties of $\psi_{z_1, \ldots, z_M}(\eta_1, \ldots, \eta_M)$, these expressions automatically continue properly to the off-diagonal matrix elements

$$\langle z'_1, \ldots, z'_M | z_1, \ldots, z_M \rangle \text{ and } \langle z'_1, \ldots, z'_M | \mathcal{H} | z_1, \ldots, z_M \rangle.$$

In terms of $\mathcal{O}$ and $\mathcal{V}$, ψ satisfies

$$\left[\sum_j^M \frac{1}{2m_e} \left[\frac{\hbar}{i} \nabla_j - \frac{e^*}{c} \mathbf{A}_j \right]^2 + \mathcal{V}(\eta_1, \ldots, \eta_M) \right] \psi(\eta_1, \ldots, \eta_M)$$
$$= E \, \mathcal{O}(\eta_1, \ldots, \eta_M) \, \psi(\eta_1, \ldots, \eta_M) \, , \quad (7.7.19)$$

in the limit of large magnetic field, where $e^* = e/m$. This will be equivalent to the original electron problem provided that $\mathcal{O}$ is approximately *constant* and

$$\mathcal{V}(\eta_1, \ldots, \eta_M) \cong \mathcal{O}(\eta_1, \ldots, \eta_M) \sum_{j<k}^M \frac{(e^*)^2}{|\eta_j - \eta_k|} \, . \quad (7.7.20)$$

This occurs for quasiholes when $\alpha = +1/m$. For quasielectrons the situation is less clear, although my current understanding is that the appropriate fraction is $-1/m$ as Halperin originally suggested.

In Fig. 12 I compare the actual overlap $\langle z_1,z_2|z_1,z_2\rangle$ and Hamiltonian $\langle z_1,z_2|\mathcal{H}|z_1,z_2\rangle$ matrix elements for $m=3$ with those estimated from the fractional statistics wave function assuming that $\mathcal{O}$ is exactly constant and that expression (7.20) is an equality. The normalization integral is the partition function of a plasma with two of its particles held fixed. It is proportional to the quantity $g_{22}(|z_1-z_2|)$ in the notation of Sec. 7.5 and may be calculated using Eqs. 5.15 – 5.17. The expected energy must be calculated using Monte Carlo methods. The wave function $\Psi_{z_1,z_2}(\eta_1,\eta_2)$ takes the simple form

$$\Psi_{z_1,z_2}(\eta_1,\eta_2) = e^{-\frac{1}{4m}\left[|z_1|^2+|z_2|^2+|\eta_1|^2+|\eta_2|^2\right]}$$

$$\times\; e^{\frac{1}{4m}(z_1^*+z_2^*)(\eta_1+\eta_2)} \sum_{n=0}^{\infty} \frac{\left[\frac{(z_1^*-z_2^*)(\eta_1-\eta_2)}{4m}\right]^{2n+1/m}}{\Gamma(2n + 1/m + 1)} \quad, \quad (7.7.21)$$

where Γ denotes the gamma function. Thus one has

Figure 7.12 Comparison of diagonal matrix elements of overlap and energy for two quasiholes at $m=3$ as given by Eqs. 7.22 and 7.23 (solid) and as calculated numerically (dots) vs. $x=|z_1-z_2|/(2m)^{1/2}$. Good agreement shows that core corrections in the fractional statistics representation are small.

Elementary Theory: Incompressible Quantum Fluid   287

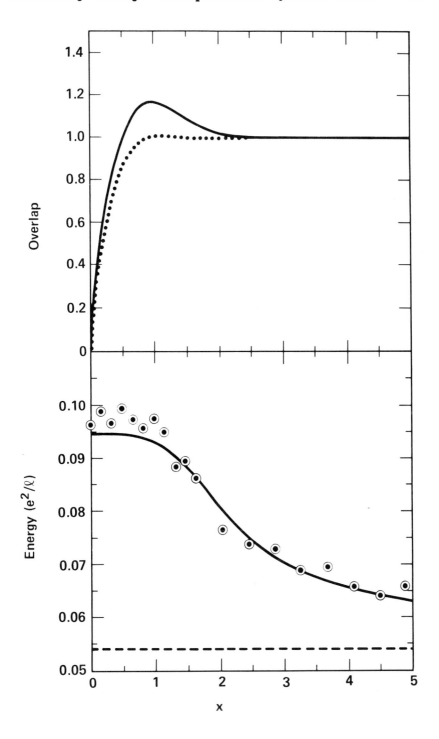

$$\iint |\Psi_{z_1,z_2}|^2 \, d^2\eta_1 d^2\eta_2 = (2\pi m)^2 \, e^{-\frac{1}{4m}|z_1-z_2|^2}$$

$$\times \sum_{n=0}^{\infty} \frac{\left[\frac{|z_1-z_2|^2}{4m}\right]^{2n+1/m}}{\Gamma(2n + 1/m + 1)} \quad , \quad (7.7.22)$$

and

$$\iint \frac{1}{|\eta_1-\eta_2|} |\Psi_{z_1,z_2}|^2 \, d^2\eta_1 d^2\eta_2 = \frac{(2\pi m)^2}{\sqrt{4m}} \, e^{-\frac{1}{4m}|z_1-z_2|^2}$$

$$\times \sum_{n=0}^{\infty} \frac{\Gamma(2n + 1/m + 1/2)}{[\Gamma(2n + 1/m + 1)]^2} \left[\frac{|z_1-z_2|^2}{4m}\right]^{2n+1/m} . \quad (7.7.23)$$

For both overlap and energy, the analytic expressions agree very well with the numerically calculated ones. The actual functions $\mathcal{O}$ and $\mathcal{V}$ obtained by solving Eqs. 7.17 and 7.18 are shown in Fig. 13. These may be seen to take on electron-like values, $\mathcal{O}=1$ and $\mathcal{V}=(e^*)^2/|z|$, except for a short-range *core correction*, as Halperin originally suggested. In general, the transformed multi-quasiparticle overlap and Hamiltonian matrices are electron-like except for small corrections whenever two or more η's come together.

Assuming the existence of a fractional statistics description of quasiparticle motion, Halperin (1984), following Haldane (1983), argued that quasiparticles should condense into a fractional quantum Hall state of the form

$$\psi(\eta_1,\ldots,\eta_M) = \prod_{j<k}^{M} (\eta_j-\eta_k)^{2k+\alpha} \exp\left\{-\frac{1}{4m}\sum_{\ell}^{M}|\eta_\ell|^2\right\} \quad (7.7.24)$$

in a manner analogous to the condensation of electrons to make $|m\rangle$. He then argued that since the wave func-

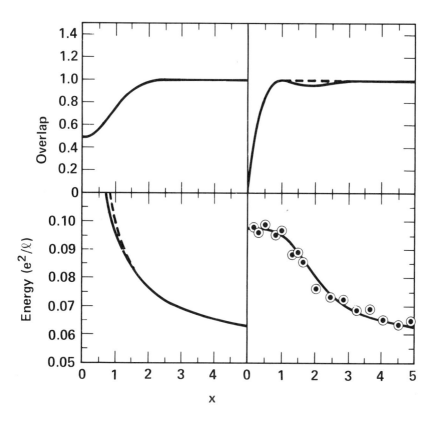

Figure 7.13 Explicit evaluation of core corrections in the fractional statistics representation. Left: The function $\mathcal{O}$ (top) and $\mathcal{V}/\mathcal{O}$ (bottom), as defined in Eqs. 7.17 and 7.18, for two quasiholes as a function of the separation $|\eta_1 - \eta_2|/(2m)^{1/2}$, for $m=3$. The dashed curve indicates a bare Coulomb potential. Right: Diagonal matrix elements reckoned from these compared with correct numerical values (dots in Fig. 12).

tion is small near the core, one could *quantitatively* estimate the energies of these states and their excitations using simple interpolation formulas such as Eq. 3.14. Assuming $m=3$ and $\alpha=1/m$, the value appropriate for quasiholes, the first 'hierarchical' state has quasiparticle density 1/7, or total electron density 2/7, the first 'new' fraction observed experimentally (Chang Chap. VI). For $\alpha=-1/3$, the value appropriate for quasielectrons, the quasiparticle density is 1/5 and the electron density 2/5. One generates the remainder of the hierarchy by iterating this reasoning, the ground state at any stage being a condensate of quasiparticles of the previous one. By quantizing ground- and excited-state energies at every stage in the hierarchy using the fractional-statistics ansatz, Halperin generated the total energy vs. electron density curve reproduced in Fig. 14. This curve has the interesting 'fractal' property of being everywhere continuous but nowhere differentiable, with the slope discontinuity at any point reflecting the energy gap of the nearest allowed fraction. Space limitations prevent me from discussing the derivation of this curve in detail. I refer the interested reader to Halperin (1984), MacDonald and Murray (1985), and MacDonald, Aers et al. (1985).

I now turn to the case of quasiparticles of opposite sign. Unlike particles of like sign, these do *not* have quantized separations, but rather form a type of itinerant exciton typical of particle motion in strong magnetic fields (Bychkov et al. 1981, Kallin and Halperin 1984, Laughlin 1985a, see also Chap. IX). The nature of these excitons may be easily understood by considering the two-particle problem

$$\mathcal{H} = \frac{1}{2m_e}\left[\frac{\hbar}{i}\nabla_1 + \frac{e}{c}A_1\right]^2 + \frac{1}{2m_e}\left[\frac{\hbar}{i}\nabla_2 - \frac{e}{c}A_1\right]^2$$
$$+ v(|r_1-r_2|) \qquad (7.7.25)$$

in a large magnetic field. The Landau-gauge ($A = By\hat{x}$) two-particle wave functions lying in the lowest Landau level are superpositions of states of the form

$$|k_1,k_2\rangle = \frac{1}{\sqrt{\pi}L} e^{i[k_1x_1+k_2x_2]} e^{-[y_1+k_1]^2/2}$$
$$\times e^{-[y_2-k_2]^2/2}, \qquad (7.7.26)$$

where L is the sample size and $\ell=1$. The Coulomb potential v, which mixes them, is translationally invariant, and thus conserves total x-momentum, $k=k_1+k_2$. For matrix elements of the Hamiltonian connecting

Elementary Theory: Incompressible Quantum Fluid   291

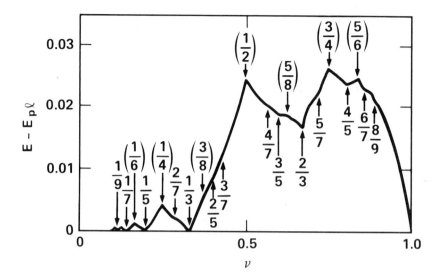

Figure 7.14 Cohesive energy per electron in excess of the 'plasma' value (dashed curve in Fig. 6) as calculated by Halperin (1984). Stable fractions appear as cusp-like minima.

states of the same k but different $\Delta \equiv k_1 - k_2$, one has

$$\langle \frac{k+\Delta'}{2}, \frac{k-\Delta'}{2} | \mathcal{H} | \frac{k+\Delta}{2}, \frac{k-\Delta}{2} \rangle$$

$$= \frac{1}{\sqrt{2\pi} \, L} \int_{-\infty}^{\infty} \int_{-\infty}^{\infty} dx \, dy \; v(\sqrt{x^2+y^2}) \; e^{i \left[\frac{\Delta - \Delta'}{2} \right] x}$$

$$\times \, e^{-\frac{1}{2}(y-k)^2} \; e^{-\frac{1}{8}(\Delta - \Delta')^2} . \qquad (7.7.27)$$

These matrix elements are a function only of the difference $\Delta - \Delta'$, and thus one may immediately write for the eigenstate

$$|k, \alpha\rangle = \frac{1}{\sqrt{8\pi}} \int d\Delta \; e^{-i\frac{\Delta \alpha}{2}} | \frac{k+\Delta}{2}, \frac{k-\Delta}{2} \rangle$$

$$= \frac{1}{\sqrt{2\pi}\, L} e^{\frac{i}{2}\left[k(x_1+x_2)-\alpha(y_1+y_2)-(x_1-x_2)(y_1+y_2)\right]}$$

$$\times\, e^{-\frac{1}{4}\left[x_1-x_2+\alpha\right]^2}\, e^{-\frac{1}{4}\left[y_1-y_2+k\right]^2}, \quad (7.7.28)$$

where k and α are the two quantum numbers. The symmetric gauge version of this wave function is

$$|z_0\rangle = \frac{1}{\sqrt{2\pi}\, L} e^{-\frac{1}{4}\left[|z_1|^2+|z_2|^2+|z_0|^2\right]}$$

$$\times\, e^{\frac{1}{2}\left[z_1 z_0^* - z_2^* z_0 + z_1 z_2^*\right]}, \quad (7.7.29)$$

with $z_0 \equiv (k+i\alpha)$. As was the case with particles of like sign, the wave function is totally constrained by symmetry to be of this form, regardless of v. The eigenvalue is

$$\langle z_0|\mathcal{H}|z_0\rangle = \frac{1}{2\pi}\int e^{-\frac{1}{2}|z-z_0|^2}\, v(|z|)\, d^2 z. \quad (7.7.30)$$

The state $|z_0\rangle$ describes particles locked in at displacement z_0 from one another but equally likely, as a pair, to be anywhere in the sample. Its momentum is iz_0, that is, proportional to z_0 but perpendicular to it, as may be seen concretely for example, through matrix elements of the charge density operator

$$\rho(z) = \delta^2(z-z_2) - \delta^2(z-z_1),$$

given by

$$\langle z_0|\rho(z)|z_0'\rangle = L^{-2}\, e^{\frac{1}{2}\left[(z_0-z_0')z^* - (z_0-z_0')^* z\right]}$$

$$\times\, e^{-\frac{1}{4}\left[|z_0|^2+|z_0'|^2\right]}\left\{e^{\frac{1}{2}z_0'^* z_0} - e^{\frac{1}{2}z_0 z_0'^*}\right\}. \quad (7.7.31)$$

Elementary Theory: Incompressible Quantum Fluid

Note in this expression that

$$\tfrac{1}{2}\left[(z_0-z_0')z^* - (z_0-z_0')^*z\right] = i\mathbf{q}\cdot\mathbf{r} .$$

Alternatively, one can postulate that $\rho(z)$ nucleates a particle-hole pair from the vacuum only when both are at z. This is close to the correct rule for 'Klein-Gordon' fermions and yields the simple result

$$\langle z_0|\rho(z)|0\rangle = \frac{1}{\sqrt{2\pi}\,L}\, e^{-\tfrac{1}{4}|z_0|^2}\, e^{\tfrac{1}{2}(zz_0^*-z^*z_0)} . \qquad (7.7.32)$$

The energy of the exciton is given by

$$\langle z_0|\mathcal{H}|z_0\rangle = \hbar\omega_c - \sqrt{\pi/2}\, e^{-\tfrac{1}{4}|z_0|^2}\, I_0\!\left[\tfrac{1}{4}|z_0|^2\right]\frac{e^2}{\ell}, \qquad (7.7.33)$$

where I_0 is a Bessel function, assuming $v = -e^2/r$. Its contribution to the static structure factor $S(q)$, is given by

$$S(q) = L^{-2}\sum_{z_0}\left|\int d^2r\,\langle z_0|\rho(r)|0\rangle\, e^{i\mathbf{q}\cdot\mathbf{r}}\right|^2 . \qquad (7.7.34)$$

The energy and contribution to the static structure factor (the latter calculated assuming Eq. 7.32) are plotted vs. $q=|z_0|$ in Fig. 15. The dashed curves illustrate the behavior of these quantities for a real charge-one magnetic exciton, as discussed for example, by Kallin and Halperin (1984).

The wave function of the exciton formed from quasiparticles is similarly determined from symmetry alone. Suppose one forces it to be a linear superposition of states of the form

$$|z_A,z_B\rangle = e^{-\tfrac{1}{4m}\left[|z_A|^2+|z_B|^2\right]}\, S^\dagger_{z_A} S_{z_B}|m\rangle . \qquad (7.7.35)$$

As with particles of like sign, one has for the *diagonal* matrix elements of overlap and energy calculable functions of $|z_A-z_B|^2$ only, in the manner

$$\langle z_A, z_B | z_A, z_B \rangle = F[|z_A - z_B|^2] \qquad (7.7.36)$$

and

$$\langle z_A, z_B | \mathcal{H} | z_A, z_B \rangle = E[|z_A - z_B|^2] . \qquad (7.7.37)$$

These, in turn, continue in the manner

$$\langle z_{A'}, z_{B'} | z_A, z_B \rangle = F[(z_A^* - z_B^*)(z_{A'} - z_{B'})]$$

$$\times e^{-\frac{1}{4m}\left[|z_{A'}|^2 + |z_{B'}|^2 + |z_A|^2 + |z_B|^2\right]}$$

$$\times e^{\frac{1}{2m}\left[z_A^* z_{A'} + z_{B'}^* z_B\right]}, \qquad (7.7.38)$$

and similarly for $\mathcal{H}$, to yield matrices *necessarily* diagonalized by states of the form

$$|z_0\rangle = \frac{1}{\sqrt{2\pi m}\,L} \iint d^2 z_A d^2 z_B \, e^{-\frac{1}{4m}\left[|z_A|^2 + |z_B|^2 + |z_0|^2\right]}$$

$$\times e^{\frac{1}{2m}\left[z_A z_0^* - z_B^* z_0 + z_A z_B^*\right]} |z_A, z_B\rangle . \qquad (7.7.39)$$

Using Eqs. 5.29 and 7.38, one obtains

Figure 7.15 Energy and contribution to structure factor $S(q)$ as given by Eqs. 7.33 and 7.34 (solid) and as expected of real inter-Landau level magneto-excitons (dashed) vs. $x = |z_0|/(2m)^{1/2}$. The dotted line is the energy of a quasiparticle pair at infinite separation.

Elementary Theory: Incompressible Quantum Fluid   295

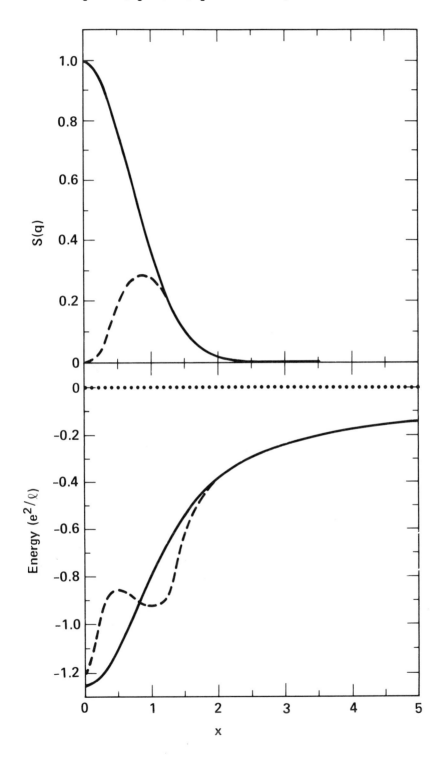

$$\langle z_{A'}, z_{B'} | z_0 \rangle = (2\pi)^2 F^e [|z_0|^2] \frac{1}{\sqrt{2\pi m}\, L}$$

$$\times e^{-\frac{1}{4m}\left[|z_{A'}|^2 + |z_{B'}|^2 + |z_0|^2\right]}$$

$$\times e^{\frac{1}{2m}\left[z_{A'} \cdot z_0^* - z_{B'}^* \cdot z_0 + z_{A'} \cdot z_{B'}^*\right]} \qquad (7.7.40)$$

and

$$\langle z_{A'}, z_{B'} | \mathcal{H} | z_0 \rangle = (2\pi)^2 E^e [|z_0|^2] \frac{1}{\sqrt{2\pi m}\, L}$$

$$\times e^{-\frac{1}{4m}\left[|z_{A'}|^2 + |z_{B'}|^2 + |z_0|^2\right]}$$

$$\times e^{\frac{1}{2m}\left[z_{A'} \cdot z_0^* - z_{B'}^* \cdot z_0 + z_{A'} \cdot z_{B'}^*\right]}, \qquad (7.7.41)$$

where

$$F[|z|^2] = \frac{1}{2\pi m} \int d^2 z \, e^{-\frac{1}{2m}|z-z'|^2} F^e[|z'|^2], \qquad (7.7.42)$$

and similarly for E. The energy eigenvalue is $E^e[|z_0|^2]/F^e[|z_0|^2]$.

The quasiexciton has the same physical interpretation as the ordinary exciton, except that the magnetic length is effectively $m^{1/2}$ times larger, reflecting the quasiparticles' fractional charge. The state $|z_0\rangle$ describes quasiparticles separated by displacement z_0 and moving as a pair, with momentum iz_0/m. The latter may be seen by evaluating the matrix element for $\rho(z)$ to nucleate a quasiexciton out of the ground state. One has

$$\langle m | S_{z_A}^\dagger S_{z_A} \rho(z) | m \rangle = e^{\frac{1}{2m}\left[|z_A|^2\right]} G[|z - z_A|^2], \qquad (7.7.43)$$

for some function G, which may be continued to yield

$$\langle z_A, z_B | \rho(z) | m \rangle = e^{-\frac{1}{4}\left[|z_A|^2 + |z_B|^2\right]}$$

$$\times e^{\frac{1}{2m} z_A z_B^*} G[(z-z_A)(z^*-z_B^*)] , \quad (7.7.44)$$

and

$$\langle z_0 | \rho(z) | m \rangle = \frac{\sqrt{2\pi m}}{L} e^{\frac{1}{4m}|z_0|^2} e^{\frac{1}{2m}[z_0 z^* - z_0^* z]}$$

$$\times \int d^2 z_B \, e^{\frac{1}{2m}\left[z_B^* z_0 - z_B z_0^*\right]} G[|z_B|^2] . \quad (7.7.45)$$

This gives a solution to $S(q)$ of

$$S(q) = 2\pi m \frac{e^{\frac{1}{2m}|z_0|^2}}{F^e(|z_0|^2)} \left| \int d^2 z_B \, e^{\frac{1}{2m}\left[z_B^* z_0 - z_B z_0^*\right]} G[|z_B|^2] \right|^2 , \quad (7.7.46)$$

with $q = |z_0|/m$.

Figures 16 and 17 compare my estimate (Laughlin 1984), based on numerical evaluations of the functions E, F, and G, for the exciton dispersion and contribution to $S(q)$ to that of Girvin, MacDonald and Platzman (1985, 1986b) and that expected of a Wigner crystal. Note that the exciton energy is raised with respect to my 1984 estimate, due to revision of the quasielectron energy. One sees that the exciton energy goes asymptotically to the sum of the quasielectron and quasihole energies, and that it exhibits an $e^{*2}/(mq)$ energy lowering, due to the attraction of the particles. This divergence is cut off at small q because the charge clouds of the particles then begin to overlap. Due to numerical difficulties, the exciton energy at small q and the overall normalization of $S(q)$ are unreliable. Fig. 16 is therefore best interpreted as indicating that a substantial fraction of the $S(q)$ sum rule is exhausted by the collective excitation (Girvin

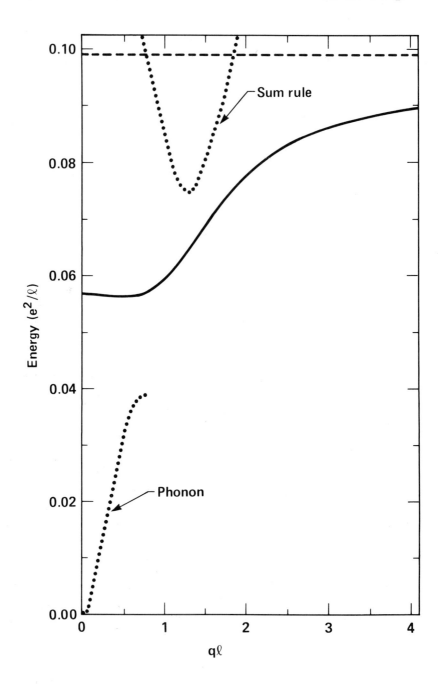

Figure 7.16 Comparison of exciton estimate of collective excitation dispersion (solid) with Wigner-crystal phonon and magnetoroton theory of Girvin, Platzman and MacDonald (sum rule). The dashed line is the energy of a quasiparticle pair at infinity. The solid curve is uncertain at small q.

Elementary Theory: Incompressible Quantum Fluid

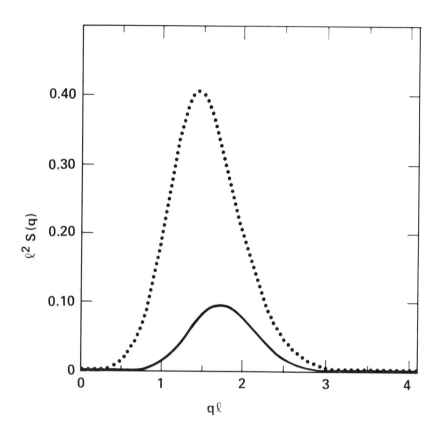

Figure 7.17 Comparison of collective excitation contribution to structure factor $S(q)$ as calculated numerically using Eq. 7.46 (solid) and as given by the ground state sum rule (dots) assumed by Girvin et al. to be exhausted. The overall normalization of the solid curve is uncertain.

et al. assume it is totally exhausted) and that the exciton and magnetoroton dispersion curves agree everywhere they are both reliable. It should be remarked that existing numerical work on small clusters supports the magnetoroton theory (see Chaps. VIII and IX).

The fact that the exciton properties so closely approximate the properties of a phonon in a Wigner crystal is not surprising, for crystals and liquids are indistinguishable on short length scales. In fact $|z_0\rangle$ actually is a transverse phonon if the Wigner crystal ground state $|W\rangle$ defined in Eq. 3.10 is substituted for $|m\rangle$ in Eqs. 7.35 and 7.39. One obtains

$$|z_0\rangle = \frac{\sqrt{2\pi m}}{L} e^{-\frac{1}{8m}|z_0|^2} \int d^2 z_A \, e^{-\frac{1}{2m}|z_A|^2}$$

$$\times \, e^{\frac{1}{4m}\left[z_A^* z_0 - z_A z_0^*\right]} S^\dagger_{z_A-z_0/2} S_{z_A+z_0/2} |W\rangle \, . \quad (7.7.47)$$

This is a phonon-like state because the operator

$$\mathcal{D} = S^\dagger_{z_A-z_0/2} S_{z_A+z_0/2}$$

effectively displaces the electron closest to z_A by z_0. When z_A is far from z_{jk}, the action of $\mathcal{D}$ on φ_{jk} (Eq. 3.11), given by

$$e^{-\frac{1}{4}\left[|z_{jk}|^2+|z|^2\right]}$$

$$\times \left[2\frac{\partial}{\partial z} - z_A^* + z_0^*/2\right]\left[z - z_A + z_0/2\right] e^{\frac{1}{2}z_{jk}^* z}$$

$$= e^{-\frac{1}{4}\left[|z_{jk}|^2+|z|^2\right]} e^{\frac{1}{2}z_{jk}^* z}$$

$$\times \left[(z_{jk}^* - z_A^* + z_0^*/2)(z - z_A + z_0/2) + 2\right] \, , \quad (7.7.48)$$

is to multiply it by the constant $|z-z_A|^2$. When $z_A = z_{jk}$, on the other hand, it multiplies it by

$$\frac{z_0^*}{2}\left[z - z_{jk} + z_0/2\right] + 2$$

which approximately equals

$$2e^{\frac{1}{4}\left[z_0^*(z-z_{jk})\right]}$$

when z_0 is small. This displaces φ_{jk} by $z_0/2$. Thus, calling D_{z_A} the operator that displaces the orbital nearest z_A by $z_0/2$, one has

$$\mathcal{D} \simeq \prod_{z_{jk} \neq z_A} |z_{jk} - z_A|^2 D_{z_A}$$

$$\simeq e^{\frac{1}{2m} D_{z_A}}, \qquad (7.7.49)$$

and

$$|z_0\rangle \simeq \int d^2 z_A \, e^{\frac{1}{4m}(z_A^* z_0 - z_A z_0^*)} D_{z_A} |W\rangle, \quad (7.7.50)$$

which is a phonon.

7.8 Conclusions

In this lecture I have tried to survey what is known about the fractional quantum Hall state from the quantum fluid point of view. I have shown how the variational ground state $|m\rangle$ and quasiparticle operators S_{z0} provide a conceptual framework for computation. I have also shown the types of calculations which have been done and, by inference, the types of calculations that still need to be done. I hope I have been helpful.

7.9 Acknowledgements

This work was performed in part under the auspices of the U. S. Department of Energy by the Lawrence Livermore National Laboratory under Contract No. W-7405-Eng-48. It was also supported by the National Science Foundation under Grant No. DMR 85-10062 and by the NSF-MRL program through the Center for Materials Research at Stanford University.

CHAPTER 8

The Hierarchy of Fractional States and Numerical Studies

F. DUNCAN M. HALDANE

8.1 Introduction

This chapter will develop the pseudopotential description of the problem of interacting electrons in the extreme quantum limit when transitions between Landau levels can be neglected, which leads to an explanation of the validity of the model wave functions introduced by Laughlin (1983a) for incompressible states of a partially filled spin-polarized Landau level with filling factor $v=1/m$, $m=1,3,5,\ldots$ In the pseudopotential approach, the Hamiltonian can be broken into two parts, H_0+H_1, where the ground state of H_0 is exactly given by the Laughlin state and its excitation spectrum has the gap characteristic of an incompressible fluid. The pseudopotential method naturally leads to the construction of both the hierarchical generalizations of Laughlin's construction (Haldane 1983, Halperin 1984, Laughlin 1984a) and extensions to multicomponent systems with internal quantum numbers such as spin, and valley indices. Once the incompressible states derived from H_0 have been identified, it must be determined whether they remain stable in the presence of the residual part of the Hamiltonian, H_1. While the truncated Hamiltonians H_0 have an exactly obtainable ground state, it has not to date proved possible to construct their excited states, which would be needed if an analytic perturbation expansion in H_1 is to be developed. Instead, a numerical approach based on exact diagonalization of the Hamiltonian describing a translationally invariant system of a small number N of electrons confined to a closed surface will be described.

In contrast to discrete lattice systems, models describing particles that can move continuously usually have an infinite-dimensional Hilbert space, so exact diagonalization is not in general possible for $N>2$. It is a remarkable fact that in the limit of extreme magnetic fields that confine the particles to a single Landau level, the Hilbert space of a finite system becomes finite in dimension, and *exact* diagonalization becomes possible. It is even more remarkable (and intimately related to the physics of the incompressible state) that numerically tractable systems (with $N \leq 10$) can be small enough to be diagonalized, yet large enough to effectively represent the physics of the thermodynamic limit. Indeed, in the most favorable case, 1/3 Landau level filling and full spin polarization, systems of as few as six electrons (involving matrices with dimensions as small as ~100) give an accurate picture of the incompressible state, and can be treated without resort to large computers.

In the incompressible state, finite-size effects fall off rapidly once the size exceeds a characteristic length scale for 'healing' of local disturbances, and properties become independent of the boundary condition geometry. In contrast, if as the model parameters are changed the incompressible state is replaced by a compressible phase, properties become *very* sensitive to variations of system size and geometry.

The conclusions reached by such studies are that, at the least, the Laughlin picture of the $v=1/3$ fractional quantum Hall effect state can be regarded as definitively confirmed, and hence that any alternative pictures that are proposed [e.g., the recent 'ring exchange' picture of Kivelson et al. (1986)] must either turn out to be equivalent to Laughlin's picture, or false.

Much of the work presented here has not been previously published, in particular, the application of the truncated pseudopotential method to multi-component systems in Sec. 8.3, and the structure-function studies of Sec. 8.8. The second-Landau-level results are also new. The pair correlations and excitation spectra in spherical systems were studied in collaboration with E. H. Rezayi (California State University, Los Angeles).

8.2 Pseudopotential Description of Interacting Particles with the Same Landau Index

The Hamiltonian describing a translationally, rotationally, and Galilean invariant system of interacting identical particles with charge q confined to the two-dimensional surface $r \cdot \hat{z} = 0$, with a uniform component of magnetic flux density in the $\hat{z}$-direction is

$$\mathcal{H} = \frac{1}{2m} \sum_i |\hat{z} \times \Pi_i|^2$$

$$+ \frac{1}{2\pi} \int d^2Q \; \tilde{V}(Q) \sum_{i<j} e^{i\mathbf{Q} \cdot (\mathbf{r}_i - \mathbf{r}_j)} , \qquad (8.2.1)$$

where $\Pi_i = -i\hbar\nabla_i - q\mathbf{A}(\mathbf{r}_i)/c$ is the dynamical momentum of the ith particle and the charge of an electron is $q=-e$. (The units will be otherwise the same as the magnetic units introduced in Chap. I.) The variables Π_i and $\mathbf{r}_i$ have commutation relations

$$[\Pi_i^\alpha, \Pi_j^\beta] = i\frac{q}{c}\delta_{ij}\epsilon^{\alpha\beta\gamma}\hat{z}^\gamma , \quad [r_i^\alpha, r_j^\beta] = 0 ,$$

$$[r_i^\alpha, \Pi_j^\beta] = i\delta_{ij}\delta^{\alpha\beta} .$$

$\tilde{V}(Q)$ is the Fourier transform of the pair interaction potential:

$$\tilde{V}(Q) = \frac{1}{2\pi} \int d^2r \; V(r) e^{i\mathbf{Q} \cdot \mathbf{r}} \qquad (8.2.2)$$

where $\mathbf{Q} \cdot \hat{z} = 0$.

In high magnetic fields, the quantum $\hbar\omega_c = \hbar|q\hat{z} \cdot \mathbf{B}|/mc$ of cyclotron orbit kinetic energy dominates the interaction energy, and at low temperatures processes that do not conserve the Landau indices $n_i = |\hat{z} \times \mathbf{\Pi}_i|^2$ of the particles are frozen out. The particles then all have $n_i = 0$, or more generally (for fermions), n Landau levels are completely filled and inert, and the residual dynamics is that of a gas of particles with Landau index n. The remaining dynamical variables of the particles once the n_i have been fixed are the 'guiding centers' (centers of the cyclotron orbital motion) $\mathbf{R}_i = \mathbf{r}_i - (q/e)\hat{z} \times \mathbf{\Pi}_i$, which commute with the $\mathbf{\Pi}_i$, and have commutation relations

$$[R_i^\alpha, R_j^\beta] = i \frac{q}{e} \epsilon^{\alpha\beta\gamma} \hat{z}^\gamma . \qquad (8.2.3)$$

The Hamiltonian describing the residual interaction is

$$\mathcal{H}' = \sum_{i<j} \mathcal{H}_{ij} ,$$

$$\mathcal{H}_{ij} = \frac{1}{2\pi}\int d^2Q |\langle n|e^{i\mathbf{Q}\cdot\mathbf{\Pi}\times\hat{z}}|n\rangle|^2 \, \tilde{V}(Q) e^{i\mathbf{Q}\cdot(\mathbf{R}_i-\mathbf{R}_j)} \qquad (8.2.4)$$

where

$$|\langle n|e^{i\mathbf{Q}\cdot\mathbf{\Pi}\times\hat{z}}|n\rangle|^2 = e^{-\frac{1}{2}Q^2}[L_n(\tfrac{1}{2}Q^2)]^2 , \qquad (8.2.5)$$

and the $L_n(x)$ are the Laguerre polynomials. $\mathcal{H}_{ij}$ commutes with the relative angular momentum operator

$$\mathcal{M}_{ij} = \tfrac{1}{4}|\mathbf{R}_i - \mathbf{R}_j|^2 - \tfrac{1}{2} . \qquad (8.2.6)$$

From (8.2.3), this is found to have integer eigenvalues $m = 0, 1, 2, \ldots$. The total z-component of angular momentum of the N-particle system is

$$M = \frac{q}{e}\left[\sum_i |\mathbf{R}_i|^2/4N + \sum_{i<j} \mathcal{M}_{ij}/N - (n-1)N\right] . \qquad (8.2.7)$$

The interaction energy of a pair of particles with the same Landau indices depends only on $\mathcal{M}_{ij}$ and n, and the interaction (8.2.4) can be written in a more fundamental form

The Hierarchy. Numerical Studies.

$$\mathcal{H}_{ij} = \sum_{m=0}^{\infty} V_m P_m(\mathcal{M}_{ij})$$

$$= \frac{1}{\pi} \int d^2Q \sum_{m=0}^{\infty} V_m L_m(Q^2) e^{-Q^2/2} e^{i\mathbf{Q}\cdot(\mathbf{R}_i - \mathbf{R}_j)} \quad (8.2.8)$$

where $P_m(\mathcal{M}_{ij})$ is the projection operator on states with $\mathcal{M}_{ij} = m$. The parameters V_m are the energies of pairs of particles with relative angular momentum m:

$$V_m = \int_0^\infty Q dQ \, \tilde{V}(Q) \left[L_n(\tfrac{1}{2}Q^2) \right]^2 L_m(Q^2) e^{-Q^2} . \quad (8.2.9)$$

These are essentially *pseudopotential* parameters: there are many inequivalent interactions $V(r)$ which have the *same* matrix elements between states where all particles have Landau index n. Thus (8.2.4) contains redundant information and the set $\{V_m\}$ — which implicitly contains the dependence on Landau index — are the only independent parameters that determine the spectrum in the high-field limit.

For a Coulomb interaction,

$$\tilde{V}(Q) = \frac{q^2}{\kappa Q} \int dz_1 \int dz_2 |\psi(z_1)|^2 |\psi(z_2)|^2 e^{-|z_1 - z_2|Q} \quad (8.2.10)$$

where $|\psi(z)|^2$ describes the density profile of the electrons *normal* to the 2D surface (Chap. I). As an example, Fig. 8.1 shows the V_m for the 'pure Coulomb' case $|\psi(z)|^2 = \delta(z)$ for Landau indices $n = 0$ and 1. If the electrons are fully spin-polarized, they are coupled by V_m with odd m only, because of Fermi statistics. If the particles were spinless bosons, only m even would appear. The dependence on the Landau index n should be noted: for $n = 0$, the V_m decrease monotonically (provided $V(r)$ does also) and $V_m \propto V(2m^{1/2})$ as $m \to \infty$. For $n = 1$, the factor $[L_n(\tfrac{1}{2}Q^2)]^2$ in (8.2.9) decreases V_0 and V_1 relative to the $n = 0$ case, and increases V_m for $m \geq 2$; the sequence is also no longer monotonic. This is a consequence of the additional structure of $n = 1$ wave functions, which have a node at the position of the guiding center.

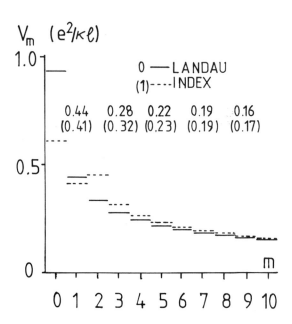

Figure 8.1 Pseudopotentials V_m (pair energies) for the pure Coulomb interaction and Landau indices $n=0$ and 1.

The effect of including a finite extent of the wave function in the normal direction is to reduce V_m, with proportionately larger reduction at small m. The decomposition of the interaction into individual terms V_0, V_1, ... may be regarded as a separation into terms describing different range components of the interaction: V_m with small m describe the short-range part, while V_m with large m describe the long-range part. This decomposition is of course not a local decomposition in real space, but the idea of splitting the interaction up into two terms, a 'short-range' piece with $m < m_0$, and a 'long range' piece with $m \geq m_0$, where choice of m_0 depends on the context, will turn out to be a very fruitful idea for understanding the FQHE.

The pair spectrum of $\mathcal{H}_{ij}$ shown in Fig 8.1 illustrates the fundamentally quantum nature of the interaction in the high field limit. Without the restriction to a given Landau index n, the operator $\mathcal{H}_{ij}$ describing the interaction between two particles has a continuous eigenvalue spectrum [$V(r_i-r_j)$ is of course diagonal in the absence of the kinetic part of the Hamiltonian]. The origin of the incompressible many-electron states that give rise to the FQHE lies in the *discrete* spectrum of the pair interaction potential $\mathcal{H}_{ij}$ after projection into the subspace of fixed Landau index. Laughlin (1983a) has suggestively called this a 'quantization of inter-particle separation'.

The reader will note that the rotational invariance of the system has been extensively invoked in the preceding discussion. However, strict rotational invariance is *not* fundamental to the FQHE. For example, it survives impurities, which break both translational *and* rotational invariance. Rotational invariance alone is removed if the effective mass tensor is anisotropic. In this case, the kinetic term in Eq. 2.1 becomes (with a suitable choice of axes)

$$\frac{1}{2m}\left[(\Lambda \Pi_i^x)^2 + (\Lambda^{-1}\Pi_i^y)^2\right] \qquad (8.2.11)$$

where $\Lambda = (m^y/m^x)^{1/4}$. The expression (8.2.5) is modified by the replacement $(Q^x, Q^y) \rightarrow (\Lambda Q^x, \Lambda^{-1} Q^y)$. Rotational invariance of $\mathcal{H}_{ij}$ is lost, and its eigenstates are no longer simply characterized by $\mathcal{M}_{ij}$. The leading effect of the anisotropy may perhaps be eliminated partially by the unitary shear transformation

$$(R_i^x, R_i^y) \rightarrow (e^\alpha R_i^x, e^{-\alpha} R_i^y),$$

generated by

$$U(\alpha) = \exp(i\alpha S), \quad S = \frac{q}{2e}\sum_i \left[R_i^x R_i^y + R_i^y R_i^x\right].$$

As will shortly be discussed, FQHE states of the pure system are characterized by an absolute gap in the density-fluctuation excitation spectrum, and for sufficiently weak effective-mass anisotropy, their character will be unchanged by breaking of rotational invariance, and controlled by the discrete spectrum of $\mathcal{H}_{ij}$. They should be well described by states $U(\alpha)\Psi_v$, where Ψ_v are the rotationally invariant incompressible states, and α is a variationally-determined anisotropy parameter.

8.3 The Principal Incompressible States: A Non-Variational Derivation

Laughlin (1983a) introduced model wave functions describing incompressible fluid states of a one-component quantum fluid at densities $1/2\pi m$ where m is odd for spinless or spin-polarized fermions, and even for spinless bosons. When introduced, the $m = 3$ and 5 wave functions in particular were given a strong variational justification, as the ground state energy evaluated with these states was substantially below the Hartree-Fock energies of previously proposed charge-density-wave states. However, the fact that a trial wave function has a good variational energy does not guarantee that its symmetry properties and long-range correlations are the same as those of the true ground state, and recently there have been some attempts (Chui et al. 1985, Kivelson et al. 1986) to introduce alternative pictures of the principal FQHE states based on the Wigner lattice. However, the justification of the Laughlin incompressible fluid wave function as a model for the $v = 1/3$ FQHE states does *not* simply rest on its variational energy: as first pointed out by Haldane (1983), it is also the exact and unique ground state of a truncated model Hamiltonian representing short-range components of the interaction.

Since the full Hamiltonian can be recovered by continuously 'turning back on' the long-range part of the interaction, alternative and qualitatively different theories of the FQHE could only be viable if a phase transition in the ground state occurred during the 'turning on' process before the full Hamiltonian is recovered. So far, it has only been possible to investigate this numerically (Haldane and Rezayi 1985a) in finite-size systems, and these results strongly suggest that the ground state of a lowest Landau level system at $n = 1/3$ with Coulomb interactions *does* have the same character as Laughlin's state.

In the experimental GaAs-Al$_x$Ga$_{1-x}$As systems, the Zeeman energy, which favors spin polarization is much smaller than the cyclotron energy, and is comparable with the Coulomb interaction energy. Various authors have considered the possibility that non-fully-spin-polarized states might occur: in particular Halperin

The Hierarchy. Numerical Studies.

(1983) listed a family of two-component generalizations of Laughlin's states that could describe such a situation.

I will here give the non-variational derivation of the principal incompressible fluid states of a spin-½ electron system. The filling factor v is defined so the density is $v/2\pi$: two families of fundamental incompressible fluid states are found – Laughlin's fully spin-polarized family with $v = 1, 1/3, 1/5, \ldots$ and an *unpolarized* family with $v = 2, 2/5, 2/9, \ldots$. This family is contained within the set of states proposed by Halperin, but are the only members of this set that are viable many-body wave functions for spin-½ electrons with spin-independent interactions.

The electronic spin enters the Hamiltonian only through the Zeeman term

$$\mathcal{H}_Z = - g\mu_B \, \mathbf{B} \cdot \sum_i \mathbf{s}_i \; . \qquad (8.3.1)$$

In the absence of spin orbit coupling, eigenstates factorize into independent spatial and spin terms. In addition to S^z, states can be classified by a total spin quantum number S, and the Fermi statistics then requires that the spatial wave function has the permutation symmetry type of the Young graph $(2^{\frac{1}{2}N-S}, 1^{2S})$. It is the requirement that, in the absence of 'accidental' degeneracy, each eigenstate must have a definite symmetry type characterized by a Young graph, that rules out most of the Halperin two-component wave functions, provided the interactions are spin-independent.

In the absence of the Zeeman term, the fully-polarized states are still stable, and have spontaneous ferromagnetic order, and both fully polarized and unpolarized incompressible states can occur. In addition, *partially* polarized states corresponding to polarized fluids of excitations of the unpolarized states are possible. However, as the Zeeman term grows stronger, the fully polarized states are favored at the expense of less polarized states, and eventually the ground state is fully polarized at all filling factors.

To construct many-particle states at a given density, it is necessary to impose some boundary condition.

In this Section, I shall use the disk geometry scheme used by Laughlin (1983a), and discuss translationally invariant boundary conditions in the context of numerical studies.

The particles will be confined to a disk-shaped region through which N_Φ quanta of magnetic flux pass. In keeping with the idea that the formulation of the problem does not depend on which Landau level is partially filled, I will use an abstract basis of states acted on by the guiding center coordinates rather than explicit wave functions. This is a gauge-independent formalism, and no gauge need in fact be picked. The basis of single-particle states will be the eigenstates of the generator of guiding center rotations $M_i = \frac{1}{2}(R_i^2 - 1)$:

$$M_i |m\rangle = m|m\rangle , \quad m = 0,1,\ldots , \quad (8.3.2)$$

$$|m\rangle = (m!)^{-\frac{1}{2}} (Z_i^+)^m |0\rangle , \quad (8.3.3)$$

$$Z_i^\pm = [R_i^x \mp i\,sgn(q/e)R_i^y]/2^{\frac{1}{2}} . \quad (8.3.4)$$

Here Z_i^+ and Z_i^-, with $[Z_i^-, Z_i^+] = 1$, are the raising and lowering operators for the angular momentum M_i, and

$$Z_i^- |0\rangle = 0.$$

A disk boundary condition is essentially equivalent to imposing the restriction that the allowed values of m in (8.3.2) are restricted to the set $m = 0,1,\ldots,N_\Phi$. This is not a rigid boundary condition in real space of the kind usually imposed, but is very convenient in this context.

The general single-particle state in this restricted Hilbert space can be written as

$$P_{N_\Phi}(Z_i^+)|0\rangle , \quad (8.3.5)$$

where $P_N(x)$ is an arbitrary polynomial of degree N. This state is not in general normalized. Two-particle states can be classified by the *relative* angular momentum $M_{ij} = \frac{1}{4}|R_i - R_j|^2 - \frac{1}{2} = 0,1,\ldots,N_\Phi$: the general

The Hierarchy. Numerical Studies.

(unnormalized) two-particle state with relative angular momentum m can be written

$$P_{(N_\Phi - m)}(Z_i^+ + Z_j^+)(Z_i^+ - Z_j^+)^m |0\rangle . \qquad (8.3.6)$$

In the finite-size disk geometry, only the pseudopotentials V_m, with $m \leq N_\Phi$ appear in the Hamiltonian. From (8.3.6), it can be seen that the general two-particle state with relative angular momentum at least m can be written

$$P_{(N_\Phi - m)}(Z_i^+, Z_j^+)(Z_i^+ - Z_j^+)^m |0\rangle , \qquad (8.3.7)$$

where $P_N(x, x')$ is a polynomial function of degree N in both its variables. If the pseudopotentials V_n vanish for $n \geq m$, the states (8.3.7) have zero energy. This property is shared by the N-particle states

$$P_{(N_\Phi - m(N-1))}(Z_1^+, Z_2^+, \ldots, Z_N^+) \prod_{i<j} (Z_i^+ - Z_j^+)^m |0\rangle , \qquad (8.3.8)$$

which can be constructed for $N_\Phi \geq m(N-1)$, i.e., for low enough filling factor $v = N/N_\Phi$.

This immediately suggests the following idea, which was essentially first given by Haldane (1983) in the pseudopotential formulation [and has been restated in an equivalent form in terms of interactions like $V(r) = \nabla^2 \delta^2(r)$ by Trugman and Kivelson (1985) and Pokrovsky and Talopov (1985)]. If $V_m = 0$ for $m \geq m_0$, and $V_m \geq 0$ for $m < m_0$, then for filling factors $v < v_c \leq 1/m_0$, the ground state energy is zero, and there is a degenerate manifold of ground states of the form (8.3.8). v_c is not necessarily equal to $1/m_0$, as permutation symmetry requirements on the state may require a factor corresponding to the polynomial prefactor in Eq. 3.8 to be present. As the magnetic field is decreased, increasing the filling factor, there is thus a critical filling factor v_c at which the zero-energy state is a unique, non-degenerate state, and above which the ground state energy is greater than zero.

The state $|0\rangle$ in (8.3.8) which is annihilated by all the Z^-_i is a fully symmetric state under particle interchange. If the particles are bosons, then the polynomial factors on (8.3.8) acting on $|0\rangle$ must be symmetric under particle interchange. If m is even, the prefactor $P_M(z_1,z_2,...z_N)$ of (8.3.8) must also be symmetric, and its lowest-degree form is $P = 1$. If m is odd, P must be antisymmetric, and its lowest-degree form is $\Pi(Z^+{}_i - Z^+{}_j)$ with $i < j$. Thus (8.3.8) with m even is the nondegenerate limiting state with $v = v_c = 1/m$ for both $m_0 = m$ and $m_0 = m-1$. Similarly, for a system of spinless fermions P must be antisymmetric if m is even, and symmetric if m is odd, and (8.3.8) with m odd are the limiting nondegenerate states for $m_0 = m$ and $m_0 = m-1$. As a nondegenerate state, the limiting state must be an angular momentum eigenstate, and the states (8.3.8) with $P = 1$ have $M = mN(N-1)/2$.

Considering the truncated problems where pseudopotentials V_m, $m \geq m_0$, have been removed thus naturally leads to a set of special states for spinless bosons at filling factors $1/m$, m even and for spinless fermions for filling factors $1/m$, m odd. If the Landau index $n = 0$, the wave functions of these states are of course Laughlin's wave functions $\psi_{1/m}$ (Laughlin 1983a), and their n-point correlation functions in the thermodynamic limit are identical to those of the classical 2D one-component plasma (OCP) with density $1/2\pi m$ and dimensionless inverse temperature $\Gamma = 2m$. (For $\Gamma > 140$, the 2DOCP is believed to have an orientationally ordered '2D solid' phase with algebraic translational order, and in this regime, the correspondence is to the rotationally-averaged correlation functions.) Two-particle correlation in a partially-filled Landau level in the high-field limit may be described in terms of the guiding-center coordinate analog of the static structure function:

$$S_0(Q) = \frac{1}{N} \sum_{i,j} \langle e^{i\mathbf{Q}\cdot(\mathbf{R}_i - \mathbf{R}_j)} \rangle \; . \quad (8.3.9)$$

If the Landau index is $n = 0$, the true static structure factor which coincides with that of the OCP is $S(Q) = 1 + \exp(-\frac{1}{2}Q^2)(S_0(Q)-1)$, and is rotationally invariant. From the OCP sum rules (Girvin 1984b, Girvin et al 1986b, Laughlin 1985a) it follows that $S_0(Q) \approx (m-1)Q^4/8$ as $Q \to 0$ for the Laughlin states $\psi_{1/m}$. This result applies to Eq. 3.8 with $P = 1$ independent of the Landau index.

For a spin-½ electron system with total spin quantum number $S \leq \frac{1}{2}N$, and $S^z = S$, the state must be anti-

The Hierarchy. Numerical Studies. 315

symmetric separately in the up-spin electron and down-spin electron coordinates. For odd m_0, (8.3.8) with $P = 1$ and $m = m_0$ satisfies this requirement, since it is antisymmetric in all coordinates. This state belongs to a multiplet of states with maximum spin polarization, $S = N/2$. The Laughlin states with $v = 1$, $1/3$, $1/5$ thus remain stable, maximum-density zero-energy ground states of short-range interactions even in the absence of a Zeeman term, when the spin polarization axis becomes arbitrary, and they are ferromagnetically ordered states.

When m_0 is even, the polynomial prefactor of the maximum-density state of the form (8.3.8) must antisymmetrize particles with the same spin direction:

$$\prod_{i<j} (z_i^+ - z_j^+)^{m(\sigma_i,\sigma_j)} |0\rangle ,$$

$$m(\sigma_i,\sigma_j) = m_0 + \delta_{\sigma_i,\sigma_j} . \qquad (8.3.10)$$

Since the polynomial must be of the same degree in all particle coordinates, the number of up and down-spin particles are equal, and this is a $S = 0$ state of Young type $(2^{\frac{1}{2}N})$, with $N_s = m_0(N-1) + \frac{1}{2}N - 1$, i.e., filling factors $2/(2m_0+1)$, m_0 even, $= 2, 2/5, 2/9, \ldots$. Just as the Laughlin states at filling factor $1/3$, $1/5$, ... can be regarded as correlated generalizations of the spin-polarized filled Landau level $v = 1$ state, these are generalizations of the non-polarized $v = 2$ filled-Landau-level state.

Because they are by construction nondegenerate eigenstates of a permutation symmetric Hamiltonian, the fully polarized- and non-spin-polarized maximum density states constructed above are guaranteed to belong to a definite Young symmetry type. In the variational spirit of Laughlin's introduction of his wave function, Halperin (1983) has catalogued many possible generalizations of the Laughlin states, including (8.3.10) for spin-½ electrons, but with $m(\sigma_i,\sigma_j)$ an arbitrary symmetric quantity. In fact, *only the two cases* $m(\sigma,\sigma') = m_0$ (odd) and $m(\sigma,\sigma') = \delta_{\sigma\sigma'} + m_0$ (even) derived above are *physically-acceptable* states of spin-½ electrons with spin-independent forces. The reason is that states must not only be eigenstates of S^z, but also of the total spin operator. This is the same as requiring states to belong to a definite Young symmetry type: not only must states be antisymmetric under exchange of particles whose labels are in the same column of the Young graph, but they must also be annihilated by the attempt to antisymmetrize any further [the Fock cyclic condition, (Hammermesh 1962)]. The majority of the Halperin states (e.g., the unpolarized $v = 2/7$ state he proposed) fail this test. This example reveals the power of the non-variational truncated-pseudopotential method of deriving FQHE states: it automatically produces states satisfying the various symmetry requirements, without resorting to guesswork.

As noted by Girvin (1984b), the lowest Landau level states of the form (8.3.10) have correlation functions identical to those of a generalized two-component plasma. Using the various sum rules, it is straightforward to show that again $S_0(Q) \propto Q^4$ as $Q \to 0$.

To summarize: the truncated pseudopotential models lead to a family of states related to classical plasmas at filling factors 1/2, 1/4, ... (bosons), 1, 1/3, 1/5, ... (spinless fermions) and 2, 1, 2/5, 1/3, 2/9, 1/5 ... (alternately non-polarized and ferromagnetically polarized) for spin-½ fermions. These are uniform density states (as seen from the plasma correspondence), in the thermodynamic limit. This can also be established from their quantum numbers when constructed in finite-size geometries without boundary conditions that break translational invariance. In the disk geometry, the angular momentum quantum number $M = \frac{1}{2}mN(N-1) = \frac{1}{2}NN_s$ is exactly what is expected for a uniform density state, where the mean occupation of each single-particle angular momentum basis state approaches a constant at large M_i.

At these filling factors, the ground state energy per particle of the truncated pseudopotential models is nonanalytic: at lower fillings all its derivatives with respect to v vanish. Above v_c, the ground state energy is *linear* in $v-v_c$, $E/N \propto |v-v_c|$ for small $|v-v_c|$. This cusp in ground state energy as a function of filling factor is the signal of *incompressibility*, and arises because increases of filling factor above v_c can only be accommodated by the creation of quantized charged defects or 'quasiparticles', with a finite energy cost per defect, and maintaining the background density $v/2\pi$ of the incompressible state far from the defects. A rigorous proof of this, in the form of a lower bound for the quasiparticle energy of the truncated pseudopotential model, has not been obtained, but both the structure of the model, and finite-size numerical studies described later strongly support this picture.

8.4 Quasiparticle and Quasihole Excitations

For the truncated pseudopotential interactions, the 'quasihole' excitation made by increasing N_Φ by one flux quantum from the critical value giving the incompressible fluid are exactly given by the construction of Laughlin (1983a):

$$\psi(Z_0) = \prod_i (Z_i^+ - z_0)\Psi_v, \qquad v = 1/m. \qquad (8.4.1)$$

By construction, these are zero-energy eigenstates. Since removing one *particle* from the Laughlin states gives the state

$$\prod_i (Z_i^+ - z_0)^m \Psi_v, \qquad v = 1/m \qquad (8.4.2)$$

where m 'quasiholes' are present at the location of the removed particle, the excess charge localized around r_0, where $z_0 = x_0 + i\,\mathrm{sgn}(q/e)y_0$, must be $-1/m$ times the particle charge q.

In the case of spin-½ particles, increase of N_Φ by one flux quantum generally produces a two-parameter state:

$$\psi(z_0(1), z_0(2)) = \prod_i (Z_i^+ - z_0(\sigma_i))\Psi_v, \quad \sigma_i = 1,2. \quad (8.4.3)$$

This can be interpreted as a state with two elementary excitations. In the case of unpolarized states, $v = 2/(2m_0+1)$, m_0 even, the elementary quasihole is a spin-½, charge-$1/(2m_0+1)$ object:

$$\psi(z_0,\sigma) = \prod_i (Z_i^+ - z_0)^{\delta_{\sigma_i,-\sigma}} \Psi_v. \qquad (8.4.4)$$

(Equality of the polynomial degree for spin-up and spin-down particles shows that $N_\sigma = N_{-\sigma} + 1$.) Thus, despite the exotic fractional charge quantum numbers (possible because the incompressible fluid may be regarded as a condensate that can accept or release charge in fractional amounts as the magnetic field is varied), the excitations of the unpolarized state carry the usual spin quantum numbers of a spin-rotationally-invariant system.

In the case of the *ferromagnetic* polarized states at $v = 1/m$, m odd, fractionalization of the quasihole as in (8.4.3) occurs at the expense of the Zeeman energy cost of producing spin-reversal of some of the particles. An 'elementary' state such as (8.4.4) cannot be constructed, and still have equal-degree polynomials for up and down spins. However, an interpretation of (8.4.3) in terms of two charge-$1/2m_0$ objects, confined by the Zeeman energy, can be developed (Haldane, unpublished).

Expansion of the general quasihole states in powers of z_0 gives a set of linearly independent angular momentum eigenstates:

$$\psi(z_0) = \sum_{m=0}^{N'} z_0^m \, \psi(m) \qquad (8.4.5)$$

where $N' = N$ for (8.4.1) and $\frac{1}{2}(N-1)$ for (8.4.4). This is the number of independent one-quasihole states, and corresponds precisely to the number of independent states expected for a fractionally charged particle confined to a region with N_ϕ flux quanta.

Just as decreasing the filling factor produces 'quasiholes', it is expected that increasing it produces 'quasiparticles' of opposite charge. While the quasihole states constructed above are *exact* eigenstates of the truncated pseudopotential model, there is no known exact construction of the corresponding quasiparticle states. Laughlin (1983a) made the natural *ansatz* corresponding to the replacement

$$(Z_i^+ - z_0) \longrightarrow (Z_i^- - z_0^*)$$

in (8.4.1) and (8.4.4). Numerical studies (Haldane and Rezayi, 1985a) show that this state has properties close to the exact quasiparticle state in finite-size systems.

The Hierarchy. Numerical Studies. 319

Crucial to the derivation of the incompressible states from the truncated pseudopotential models is the assumption that the energy of the exact quasiparticle state remains finite in the thermodynamic limit, i.e., that it can be bounded below by an amount $\alpha(m_0)$ times $\min\{V_m: m < m_0\}$, where $\alpha(m_0)$ is positive definite. A naive argument might suggest that $\alpha = 1$: since v_c is the highest filling factor at which a state with no pairs of particles having $m_{ij} < m_0$ can be constructed, states at higher filling factors 'must' have at least one such pair, giving a lower energy bound with $\alpha = 1$. This argument is not valid, as m_{ij} is not a good quantum number. In fact, numerical studies described later suggest $\alpha \simeq 0.5$ for the $v = 1/3$ spinless fermion state. For this state, Pokrovsky and Talapov (1985) state as 'obvious' that adding a *particle* (i.e., introducing three quasiparticles) must produce a state with at least one $m=1$ pair (so $\alpha = 1/3$ gives a lower bound). However, in the absence of any detailed justification, it is not clear that this argument is any different from the incorrect 'naive' argument given above.

It is interesting to speculate on the significance of the crystallization of the OCP at $\Gamma > 140$ (i.e., $m_0 > 70$). In the crystalline phase, algebraic divergence of the structure factor at reciprocal lattice vectors will survive rotational averaging. Girvin et al. (1985) have derived a variational upper bound to the collective excitation spectrum in terms of the structure factor that shows gapless excitations are present if it diverges. This implies that for $m_0 > 70$, the truncated pseudopotential model at the critical filling becomes gapless in the thermodynamic limit, with an orientationally ordered, and presumably compressible, ground state. If this is the case, any rigorous derivation of $\alpha(m_0)$ would show $\alpha = 0$ for $m_0 > 70$, implying that α must also be derived in terms of some correlation function such as the structure factor.

The two-quasihole generalizations of (8.4.1) can be written

$$\psi(z_0^1, z_0^2) = \prod_i (Z_i^+ - z_0^1)(Z_i^+ - z_0^2) \psi_v . \qquad (8.4.6)$$

These states are symmetric in the variables z_0^1 and z_0^2. If they are expanded in powers of $(z_0^1 + z_0^2)$ and

$(z^1{}_0 - z^2{}_0)$, a set of orthogonal eigenstates of total angular momentum and center-of-mass angular momentum is generated. For the truncated pseudopotential interaction, these are of course degenerate zero-energy eigenstates, but if a weak $m = m_0$ pseudopotential is added to the interaction to lift the degeneracy, the eigenstates are precisely those components of (8.4.6) with fixed powers of $(z^1{}_0 + z^2{}_0)$ and $(z^1{}_0 - z^2{}_0)$. In the thermodynamic limit, translational invariance is restored, and the energy depends only on the relative angular momentum m of the quasiholes, where the eigenstate is the term in the expansion of (8.4.6) proportional to $(z^1{}_0 - z^2{}_0)^m$, $m = 0,2,4,\ldots$. A similar analysis for the two-quasihole state (8.4.3) of the unpolarized spin-½ electron states shows that even values of m correspond to states with total spin S=0 and odd m to S=1. (For example, if the two quasihole coordinates are equal, the factor in (8.4.3) acting on Ψ_v is fully symmetric in particle coordinates, and does not change the Young symmetry type which determines S.) If the relative angular momentum of quasiholes is defined so that creating two quasiholes at the same point creates them in an m=0 relative angular momentum state, the quasiholes of the spinless fermion and spinless boson Laughlin states may be regarded as combining with the rules of Bose statistics, while the spin-½ quasiholes of the unpolarized spin-½ fermion Laughlin-type state combine as spin-½ fermions. However, this assignment of 'statistics' follows from the arbitrary definition of the relative angular momentum of the two-hole state as the exponent of $(z^1{}_0 - z^2{}_0)$ in the expansion of (8.4.6) or (8.4.3). Other assignments lead to other nominal 'statistics'. Under the replacement

$$(Z^+_i - z_0) \rightarrow (Z^-_i - z_0^*) ,$$

the angular momentum analysis of the two-quasiparticle states is identical to the above.

8.5 The Hierarchy of Quasiparticle-Quasihole Fluids

The energies of a pair of quasiparticles or quasiholes in a state of relative angular momentum m define quasiparticle interaction pseudopotentials $\mathcal{V}_m$. There

The Hierarchy. Numerical Studies.

will of course be effective three-body forces, etc., between the quasiparticles even though the underlying interacting particle model has only pairwise interactions. However the hierarchical scheme for constructing general incompressible fluid states rests on the idea that the dominant interaction between the quasiparticles is the short-range repulsive part of the pair interaction. If this is the case, and if the quasiparticle interaction energies are small compared to the energy gap for collective excitations of the parent incompressible fluid state, it is valid to apply the truncated pseudopotential model to the interacting gas of quasiparticles present at filling factors close to the filling factor of the parent incompressible state.

There are three independent versions of this idea, due to Haldane (1983), Laughlin (1984a), and Halperin (1984). These differ in language, particularly in assigning different 'statistics' to the quasiparticles which can be traced to different definitions of the relative angular momentum of the quasiparticles. The latter two derivations involve the explicit construction of Schrödinger wave functions, which may differ in detail. However, just as the derivation of states of particles confined to a given Landau level can be carried out algebraically, in terms of pseudopotentials, and without direct construction of Schrödinger wave functions, an algebraic derivation (Haldane 1983) of the quasiparticle hierarchy is possible.

It must be emphasized that the construction of the hierarchy is valid only to the extent that (a) short-range pair interaction pseudopotentials dominate the quasiparticle interactions, and (b) these energies are themselves dominated by the collective excitation gap of the parent fluid. This is precisely analogous to the corresponding condition on the parent fluid state: that its interactions are dominated by short-range pseudopotentials, and that the interaction energies are small compared to the gap for excitations to higher Landau levels. Whether these conditions are satisfied is in the end a quantitative question to be settled by direct calculation. At least in the case of the principal (2/5 and 2/7) daughter fluids of the 1/3 state, finite-size numerical studies (Haldane and Rezayi 1985a) confirm that these conditions are satisfied.

If the parent state is one of the spinless-particle Laughlin states with $v = 1/m$, the first level hierarchy equation is

$$N_\phi = m(N-1) + \alpha N_{qp} , \quad \alpha = \pm 1 \quad (8.5.1)$$

$$N = p(N_{qp} - 1) , \quad p \text{ even} . \quad (8.5.2)$$

In the first equation, the number N_{qp} of quasiparticles ($\alpha = -1$) or quasiholes ($\alpha = +1$) is given by the difference of the polynomial degree N_ϕ and the polynomial degree $m(N-1)$ of the pure Laughlin state. The second equation is analogous to the first, except that (a) N_ϕ, the number of independent single-electron states, has been replaced by N, the number of independent quasiparticle states, and (b) p must be even, because of the nominal Bose statistics of quasiparticles in this scheme. The second equation describes the quasiparticle fluid at its maximum density for avoiding pair states with relative angular momentum less than p. From (8.5.2), $N_{qp} = N/p$ in the thermodynamic limit, and the filling factor N/N_ϕ is $1/(m+\alpha/p) = p/(pm+\alpha)$. For the 1/3 state the two most stable daughter fluids will correspond to $p = 2$, assuming that the contact repulsion between quasiparticles is the dominant pseudopotential. These states are the 2/5 and 2/7 states.

Adding a particle to the system, so $N \to N^0 + 1$ produces the changes

$$N_\phi = m(N^0 - 1) + \alpha(N_{qp}^0 + \alpha m) ,$$

$$N^0 + 1 = p(N_{qp}^0 + \alpha m - 1) - \alpha(mp - \alpha) . \quad (8.5.3)$$

Thus addition of one electron to the two-fluid system produces $(mp - \alpha)$ quasiparticle or quasihole excitations of the daughter incompressible fluid. These therefore have charge $\pm 1/(mp-\alpha)$, i.e., the reciprocal of the denominator of the filling fraction.

Clearly the hierarchy construction can be iterated, if interactions between the quasiparticles of the daughter fluid are also predominately described by a truncated pseudopotential interaction. The general

The Hierarchy. Numerical Studies.

filling factor for a multi-component incompressible fluid of spinless particles is expressed as a continued fraction (Haldane 1983):

$$v = \cfrac{1}{m + \cfrac{\alpha_1}{p_1 - \cfrac{\alpha_2}{p_2 - \cfrac{\alpha_3}{p_3 - \cdots}}}} \quad (8.5.4)$$

For a Fermi system (m odd) these are all rationals with an odd denominator while for a Bose system (m even) these are all of the form (even)/(odd) or (odd)/(even). In fact, the fluids derived from the Laughlin fluids with $m = 3,5,7,\ldots$ are in one-to-one correspondence with all rationals of the appropriate type (i.e., with an odd denominator) in the range $0 < v < \frac{1}{2}$, and the equivalent statement for the Bose case ($m=2,4,6,\ldots$) holds for $0 < v < 1$. Fermion fluids with $\frac{1}{2} < v < 1$, can be obtained by particle-hole symmetry, however, it is interesting to note that the construction (8.5.3) with $m=1$ generates all odd-denominator rationals in the range $\frac{1}{2} < v < \infty$. Those based on 'quasihole' excitations of the $v=1$ state generate the particle-hole symmetric partners of the states with $v < \frac{1}{2}$. If the 'quasiparticle' excitation of the filled Landau level $v = 1$ state is identified as a particle in the next Landau level, then the $v = 2$ state with two filled Landau levels can be considered a two-component fluid daughter state of the simple fluid $v = 1$ state, just as the 2/5 state is a daughter of the 1/3 state.

The equations (8.5.1), (8.5.2) also express the 'fractional statistics' aspect of the quasiparticles. A 'fractional statistics' object (Wilczek 1982) is modelled by an object that is both charged and carries magnetic flux. When two such objects exchange, additional anomalous phase changes can be produced by the Bohm-Aharonov effect. Eq. 5.2 shows that quasiparticles experience particle density of the parent fluid just as the particles experience magnetic flux density. Unlike the original particles, the quasiparticles carry 'flux' as there is a depletion of the parent fluid density in their vicinity. Eqs. 5.1-2 relate total magnetic flux, particle number and quasiparticle number. Arovas et al. (1984) have shown in a simple way that these relations hold locally as well, and that the phase change accompanying motion of a quasiparticle around a closed path counts the total number of particles in the area enclosed by the path, just as the phase change accompanying motion of a particle around a path counts the enclosed flux.

The charge of the quasiparticle in the general state can be obtained as before by adding a particle, and seeing how many quasiparticles are produced at the lowest level of the hierarchy. The general result is that if the filling factor is p/q (with p,q relatively prime) the quasiparticle charge is $\pm 1/q$ times the particle charge in spinless particle systems.

Tao (1985) has found a simple argument that seems to explain the odd denominator selection rule for fermions and its equivalent for bosons, in simple fluids. If the incompressible fluid has filling factor p/q, then if a solenoid pierces the fluid, and its flux is increased by q flux quanta, the equivalent of p particles must accumulate around it from the radial Hall current induced by the transient electric field. Exchange of the two solenoids (with their charge accumulations) produces a Bohm–Aharonov phase factor $\exp(2\pi i pq)$. Since p particles are exchanged this must be equal to $(-1)^{pp} = (-1)^p$ for fermions, and 1 for bosons. This rules out even q for fermions and p and q which are both odd for bosons. The crucial assumption made implicitly in this argument is that the state produced by the action of the solenoids has the same quantum numbers as the state produced by the addition of particles. This is not in general true in more complex fluid systems where the particles carry internal quantum numbers that are not all the same. The assumption is correct in the spinless or spin-polarized fluids of the hierarchy scheme. However it is not even clear whether or not it applies as a general theorem in the spin-polarized fluid case. It is not implausible that some interaction that is attractive at short distances and repulsive at larger distances could favor a state where electrons pair into charge-2 bosons, which condense into an even denominator Laughlin state (Halperin 1983). For this case, the state (with no broken pairs) produced by passing a flux quantum through each solenoid is not at all equivalent to the state produced by addition of particles.

The Hierarchy. Numerical Studies. 325

The hierarchy construction can of course be applied to the unpolarized fluids of spin-½ electrons: since the quasiparticles now carry spin, the daughter fluids can themselves be of either the unpolarized or fully polarized types. A unpolarized parent fluid with a fully polarized daughter fluid is of course a partially polarized incompressible fluid state. Since the principles of constructing such possible states are straightforward generalizations of those in the single component case, and such states are not relevant to present experimental systems, they will not be given here.

The hierarchy picture of the origin of general incompressible fluid states is a falsifiable theory in the sense that it uniquely assigns a hierarchical structure to a given observed quantum Hall effect plateau and implies that if a given state is observed, its parent states must also be observed. This condition has been satisfied by all the experimentally observed plateaus to date.

8.6 Translationally Invariant Geometries for Numerical Studies

To date, the FQHE problem has proved curiously intractable to the usual second-quantized theoretical techniques. However, numerical diagonalization of systems of small numbers of particles has proved to give a remarkably clear picture of the principal FQHE states such as the 1/3 spin-polarized electron state. The success of this method seems to be due to the physics of the incompressible state: provided the dimensions of the system are larger than the characteristic length scale (the inter-particle spacing for simple 'one-fluid' FQHE states), the ground state becomes very insensitive to the details of the boundary conditions.

In the absence of a substrate potential, the Hamiltonian in the thermodynamic limit is invariant under translations and the 2D parity operation (rotation by π about any point). Assuming central forces, it is also invariant under the combination of time-reversal and reflection about lines in the surface parallel to a principal axis of the effective mass tensor. If the effective mass is isotropic, there is full rotation symmetry in the surface.

In order to maximize the possible symmetry reduction of the finite system, and to interpret its quantum numbers, it is desirable to choose boundary conditions that retain as many as possible of the symmetries of the infinite system. There are only two types of boundary conditions that preserve translational invariance: spherical and periodic.

If rotational invariance is present, all the symmetries of the infinite system are retained by 'compactifying' the 2D surface onto the surface of a sphere (Haldane 1983). Topologically, this is the simplest closed surface. The price paid is that the finite system is a surface of non-zero curvature, and the surface densities at which particular incompressible states occur depend somewhat on the size of the system, only becoming equal to the planar surface values in the thermodynamic limit. In this geometry, states are characterized by an angular momentum quantum number L, and fall into $2L+1$-degenerate multiplets. Translationally and rotationally invariant states are characterized by the quantum number $L=0$. If the ground state is such a $L=0$ state, the dispersion of its neutral collective excitations is given by the L-dependence of the excitation energy: the normal modes on a spherical surface are spherical harmonics, with quantum numbers L related to wave numbers k by $kR = L$, where R is the radius of the sphere.

Since the spherical surface is merely a technical device to impose spatial boundary conditions, it is convenient (in the absence of spin-orbit coupling) to keep the magnetic field that enters in the Zeeman term as a fixed direction field singling out a quantization axis in spin space. If the radial field is also used for the Zeeman term, the curvature of the surface induces a spin-orbit coupling that only vanishes in the thermodynamic limit. A similar (but unavoidable) effect prevents separation of the degrees of freedom of a finite spherical system into cyclotron motion and 'guiding center' variables.

The other choice is the more conventional one of periodic boundary conditions. These were used in the initial studies by Yoshioka et al. (1983), but a fully two-dimensional treatment of the translational symmetry using these boundary conditions has only recently been given (Haldane 1985). There are really a family of periodic boundary conditions: physical (gauge-invariant) quantities are required to be invariant under a set of discrete translations defining a 2D Bravais lattice with a quantized unit-cell area (the surface area of the sphere is also quantized). The family of 2D Bravais lattices form a continuous two-parameter set parameterized by the unit cell aspect ratio and the angle between basis vectors, and provided that the shortest PBC lattice vector does not become smaller than the characteristic length scale of the problem, it is useful to study various PBC geometries. Periodic boundary conditions break the rotational invariance of the Hamiltonian, and the surface acquires the topology of the torus, but remains flat so the densities of the various incompressible states are size-independent and can be unambiguously identified. The highest degree of symmetry is achieved if the PBC is based on a hexagonal Bravais lattice (or if the effective mass is anisotropic, on one of the rectangular lattices – simple or centered – oriented along the principal axes of the effective mass tensor).

In this geometry it is not possible to construct a perfectly translationally-invariant state of a finite system: if the 'unit cell' contains N_s flux quanta, and the filling factor is p/q, the most nearly translationally-invariant states that can be constructed have physical properties that vary periodically with a unit cell area q/N_s^2 times the PBC unit cell area. This irreducible spatial 'ripple' of physical quantities only vanishes when the thermodynamic limit is taken at a rational filling factor. The fundamental quantum number (Haldane 1985) is a 2D 'Bloch vector' k defined in a large Brillouin zone related to the minute unit cell of the irreducible 'ripple': this Brillouin zone expands to cover all reciprocal space in the thermodynamic limit at rational filling factor. One particular value of the quantum number k is uniquely singled out by the absence of point rotation degeneracy (generically the finite-size Hamiltonian has two-fold rotational invariance, and for the high-symmetry square and hexagonal PBC Bravais lattices, four-and six-fold rotation symmetries occur). This value provides the origin $k = 0$ about which reciprocal space expands as the thermodynamic limit is taken and defines states that become intrinsically rotationally invariant in this limit.

A feature of translationally-invariant Hamiltonians in this geometry is that every eigenstate belongs to a q-fold degenerate multiplet of states with the same k related by center-of-mass translations. This degeneracy has suggested to some authors that a broken symmetry may occur in the FQHE, but is a general feature of periodic boundary conditions, unrelated to the physical nature of the ground state (Haldane 1985). The wave function of a translationally-invariant Hamiltonian can (in the absence of curvature of the surface) be factorized into uncoupled terms describing the center-of-mass and relative motion degrees of freedom. The degeneracy is a general property of the center-of-mass wave function in the periodic geometry.

If rotational invariance is present, but translational invariance is broken, as when one studies the response of the FQHE state to an isolated impurity, the disc boundary conditions used in the previous sections may be useful, but the spherical geometry is still the geometry of choice (Zhang et al. 1985, Rezayi and Haldane 1985).

In the case of a general substrate potential, periodic boundary conditions are probably the most convenient, as they allow the Hall conductance of the non-uniform system to be described in terms of the Chern number (Niu et al. 1985, Avron and Seiler 1985, Chap. IV). The wave function itself satisfies not simple periodic boundary conditions, but a gauge-dependent quasiperiodic boundary condition characterized by two additional phase parameters which evolve linearly in time if a uniform electric field is applied parallel to the surface. The Chern number is a topological invariant characterizing how the wave function behaves as these two phases are varied, and is well-defined in the 'generic' case when all degeneracies are lifted by an inhomogeneous substrate potential. In the absence of a substrate potential, a change in these phases is equivalent to a translation of the center-of-mass, which leaves the eigenvalue spectrum invariant.

The Laughlin wave functions, originally given in the inhomogeneous disc geometry, have now been explicitly constructed in both spherical (Haldane 1983) and periodic (Haldane and Rezayi 1985b) geometries, and are very conveniently obtained as reference states in finite-size studies by numerically diagonalizing the Hamiltonian with the appropriate truncated pseudopotential potential. In the coordinate representation on the sphere, factors (z_i-z_j) become factors

($u_i v_j - v_i u_j$) where [u,v] are spinor coordinates [$\cos(\theta/2)\exp(i\varphi/2)$, $\sin(\theta/2)\exp(-i\varphi/2)$]. In periodic boundary conditions, it essentially becomes $\theta_1(\pi(z_i-z_j)/L_1|(L_2/L_1))$ where $\theta_1(z|\tau)$ is an elliptic theta function, and L_1 and L_2 are complex representations of the PBC translations. In both geometries, the Laughlin states have the quantum numbers ($L=0$, $k=0$) that signal intrinsic translational and rotational invariance in the thermodynamic limit.

The moduli of these wave functions are again equivalent to the Boltzmann factor of an OCP defined in the two respective geometries. In the periodic case, this correspondence requires an average over the center-of-mass degree of freedom (parameterized by the q-fold degeneracy parameter and the two phase parameters of the quasiperiodic boundary condition on the wave function), which restores full translational symmetry to the finite periodic system.

8.7 Ground-State Energy Studies

The earliest numerical studies of a homogeneous system in the high-field limit (Yoshioka et al. 1983) were principally aimed at studying ground-state energies and comparing them with predictions of the Hartree-Fock charge-density-wave state. At 1/3 filling, the ground-state energy per particle was found to be significantly lower than the Hartree-Fock prediction, and the various numerical estimates of around -0.412 ± 0.003 e^2/ℓ per particle are in good agreement with estimates of the Laughlin state energy obtained from OCP studies. Cusps in the ground-state energy as a function of filling factor were also found.

A similar study, this time using the truncated pseudopotential interaction and a spherical geometry, is shown in Fig. 8.2. By changing the magnetic field, the Landau level degeneracy of a fully spin-polarized six-particle system is gradually increased from its minimum value of six, thus decreasing the filling factor. Ground-state degeneracies are shown: nondegenerate states, correspond to filling factors (identified from the hierarchy scheme) of 1, 2/3, 2/5, and 1/3. At and below 1/3 filling, the ground state has zero energy, and rapidly becomes highly degenerate as the number of zero-energy multiplets proliferates. Cusps

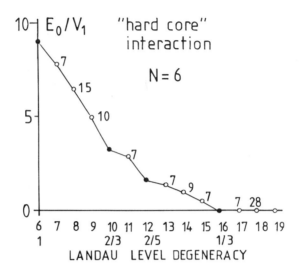

Figure 8.2 Ground-state energy and degeneracy for a six-particle, spherical geometry, spin-polarized electron system with the 'hard core' truncated pseudopotential interaction $V_m=0$ for $m > 1$. Open points indicate ground states with the indicated degeneracy. Closed points indicate a non-degenerate ground state.

of the ground-state energy function are evident at these special filling factors.

The study by Yoshioka et al. (1983) used a simple rectangular boundary condition lattice, and found small sensitivity of the ground-state energy to the aspect ratio. The insensitivity of the ground-state properties to area-preserving changes is characteristic of the incompressible state. Figure 8.3 shows a general exploration of the effect of changes of periodic boundary condition geometry on a five-electron system with the lowest-Landau-level Coulomb interaction pseudopotentials. If the two basis vectors of the PBC Bravais lattice have lengths L_1 and L_2, and the angle between them is θ, the family of 2D Bravais lattices can be parameterized in the triangular region $0 \leq 2\cos(\theta) \leq L_1/L_2 \leq 1$ with the physical condition that L_1 should not be less than the mean inter-particle spacing. The oblique lattices are in the interior of this region bounded by the simple and centered rectangular lattices. The two physical corners represent

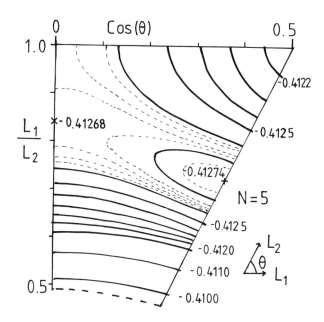

Figure 8.3 Ground state energy per particle of the five-particle, periodic-geometry, spin-polarized system with lowest-Landau-level Coulomb interactions as a function of boundary condition geometry. Variation of less than 0.15% shows the extreme insensitivity to boundary conditions of the incompressible state, even at small particle numbers.

the higher symmetry square and hexagonal lattices. The ground-state energy in the physical region varies by less than 0.15% as the geometry is varied. The lowest energy is not in fact achieved at the simple rectangular lattice point reported by Yoshioka et al. (1983). This is a saddle point, and the actual minimum corresponds to a centered rectangular lattice. When L_1/L_2 becomes pathologically small, rapid variation of the ground-state energy occurs.

The ground-state energy can be generally expressed in terms of the interaction pseudopotentials and the pair amplitudes

$$A_m = \langle \xi^\dagger_{m,n} \xi_{m,n} \rangle$$

where $\xi^\dagger_{m,n}$ is an operator that creates a pair of particles in a state of relative angular momentum m, and center-of-mass momentum n. (For a translationally-invariant ground state, A_m is independent of n). The ground-state energy is the substrate potential energy plus the sum of $V_m A_m N_m$ where N_m is the number of independent pair states with relative angular momentum m. The pair amplitudes for the Laughlin state and the lowest-Landau-level Coulomb-interaction ground state of a six-particle, 1/3 filling, spherical system are shown in Table 8.1, together with the Coulomb pseudopotentials of the finite system. Even-m pair amplitudes of the spin-polarized system vanish because of the Pauli principal. The Laughlin state is distinguished by an exactly vanishing $m=1$ pair amplitude, and is the exact ground state provided V_1 is the only non-zero pseudopotential. The perturbative effect of the other V_m is responsible for a very small but finite $m=1$ pair amplitude in the Coulomb case. However, the variational Coulomb energy of -0.449954 e^2/ℓ per particle of the Laughlin state of this finite system is practically indistinguishable from the true Coulomb ground-state energy of -0.450172. [These values are

Table 8.1 Pair amplitudes for the $N = 6$, 1/3 filling Laughlin state and lowest-Landau-level Coulomb interaction ground state (spherical geometry, Landau level degeneracy 16). Also shown: lowest-Landau-level Coulomb pseudopotentials $V(m)$, based on chord length as the inter-particle separation. These are size dependent, and differ from the infinite plane values given in Fig. 8.1 because of the curvature of the surface.

m	Laughlin	Coulomb	$V(m)$
1	0.0000000	0.0004048	0.4781002
3	0.2456732	0.2388983	0.3092084
5	0.1439109	0.1598708	0.2526069
7	0.1094809	0.1075856	0.2237067
9	0.1378118	0.1187430	0.2067833
11	0.1429911	0.1384482	0.1965555
13	0.1484027	0.1697052	0.1907857
15	0.1543784	0.1913457	0.1884637

somewhat lower than the extrapolated values for an infinite system because of curvature effects (Haldane and Rezayi, 1985a).]

8.8 Pair Correlations and the Structure Factors

At 1/3 filling, the Laughlin state is the only state with a pair correlation function $g(r)$ vanishing as r^6 as the particles come together. All other spin-polarized electron states have $g(r) \sim r^2$. Yoshioka (1984a) has studied this behavior in the case of the lowest-Landau-level Coulomb interaction as a function of filling factor. The r^2 term is very small at and below 1/3 filling, and linear in the excess filling above 1/3. In the truncated pseudopotential model, it vanishes at and below 1/3 filling. The actual pair-correlation function depicted by Yoshioka is somewhat anisotropic due to boundary condition effects. Haldane and Rezayi (1985a) have exhibited the correlation function of the six-particle system on the sphere. Further results, for sizes of four to eight particles, are shown in Fig. 8.4, as a function of the geometric (chord) distance between the particles. The background density to which $g(r)$ must asymptote is somewhat size-dependent, a consequence of curvature. As the sphere surface area is increased and the surface flattens, there is systematic relaxation of the correlation function making quantitative extrapolation to the flat limit difficult. However, the general structure of the oscillatory behavior of $g(r)$ is apparent, and the eight-particle system is large enough to show a secondary maximum. The short-distance r^6 behavior is clearly seen. The residual r^2 component (not visible on the scale of the main Figure) is shown as a highly magnified inset, since the projection of the Laughlin state on the Coulomb ground state is close to 100%.

The direct pair correlation function $g(r)$ turns out to be much more sensitive to finite-size effects than its Fourier transform, the static structure factor. In the case of periodic boundary conditions, this is probably because it is a discrete sum over allowed Q-values of the structure factor, while real-space curvature is to blame in the spherical case.

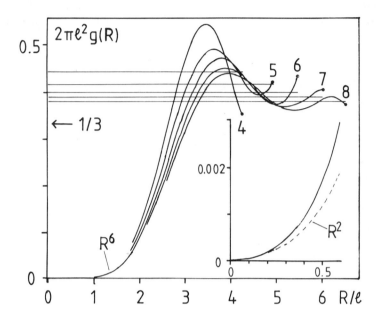

Figure 8.4 Pair correlation functions $g(R)$ in the ground state of $N = 4$- to 8-particle lowest-Landau-level spherical-geometry systems with $\nu=1/3$ and Coulomb interactions. R is the geometric (chord) distance. The uniform background density for the different sizes is shown. Its size-dependence is a consequence of surface curvature in this geometry. The maximum separation possible is the sphere diameter.

The fundamental quantity is the dynamic structure factor, given at zero temperature by

$$S(\mathbf{Q},\omega) = \frac{1}{N}\sum_{\nu} |\langle 0|\sum_{i} e^{i\mathbf{Q}\cdot\mathbf{r}_i}|\nu\rangle|^2 \, \delta(E_\nu - E_0 - \omega). \quad (8.8.1)$$

On the periodic surface, this is only defined at wave vectors compatible with the PBC. On the spherical surface, spherical harmonics are the analogs of Fourier components:

$$S_L(\omega) = \frac{4\pi}{N}\sum_{\nu} |\langle 0|\sum_{i} Y_{LM}(\hat{\Omega}_i)|\nu\rangle|^2 \, \delta(E_\nu - E_0 - \omega) . \quad (8.8.2)$$

The Hierarchy. Numerical Studies.

In the thermodynamic limit, if orientational order is absent, $S_L(\omega)$ coincides with $S(Q,\omega)$ if one takes $L=QR$, where R is the radius of the sphere.

The usual static structure factor $S(Q)$ is the integral of the dynamic structure factor over all frequencies. In the high-field limit when the cyclotron energy gap becomes effectively infinite, it is useful to follow Girvin et al. (1985, 1986b, Chap. IX) and study the 'projected static structure factor' $\bar{s}(Q)$ defined by taking the limit of infinite cyclotron energy *before* integrating over frequency, with the effect that only intermediate states not involving inter-Landau-level transitions are kept. This projected 'static' structure factor gives the sum rule for the spectral weight of the interesting low-energy excitations of the system, and is related directly to the conventional one: $S(Q)$ is $\bar{s}(Q)$ plus a weighted average of the (trivially-calculable) structure factors that would be obtained by changing the electron density to (a) empty and (b) completely fill the partially-filled Landau level. In the case that the lowest Landau level is partially filled,

$$S(Q) = 1 - e^{-Q^2/2} + \bar{s}(Q) . \qquad (8.8.3)$$

In turn, $\bar{s}(Q)$ can be related to the 'intrinsic' or 'guiding center' structure factor $S_0(Q)$ given by expression (8.3.9), and which is independent of Landau level form factors:

$$\bar{s}(Q) = \frac{v}{2n+v} e^{-Q^2/2} L_n(Q^2/2) S_0(Q) . \qquad (8.8.4)$$

This quantity is positive definite, and quite generally vanishes as $Q \to 0$ (neglecting the delta-function contribution at $Q=0$). The quantity $2n+v$ in (8.8.4) is the total filling factor (assuming only spin-degeneracy of the Landau levels), while v is the filling factor of the Landau level with index n. If the ground state is not orientationally ordered, it can be shown (Girvin et al. 1985, 1986b, Chap. IX) that

$$\lim_{Q \to 0} S_0(Q)/Q^2 = 0 , \qquad (8.8.5a)$$

and [Haldane unpublished, see Girvin et al. (1986b) for an equivalent expression]:

$$\lim_{Q \to \infty} S_0(Q) = 1 + v\langle P \rangle , \qquad (8.8.5b)$$

where $\langle P \rangle$ is the expectation value of the pair exchange operator P_{ij} averaged over all distinct pairs and v is the filling factor.

$S_0(Q)$ can also be defined on the sphere using the expansion of the exchange operator to define the tensor operator analogs of $\exp(i\mathbf{Q}\cdot\mathbf{R})$.

Fig. 8.5 shows a compendium of results for $\bar{s}(Q)$ on the sphere for the 1/3 states of 4, 5, 6, 7, and 8 particles. In contrast to the direct-space results of Fig. 8.4, there is remarkable size-independence. These results were for the Coulomb interaction in the lowest Landau level, but because of the dominance of the Laughlin state, should be very similar to the pure Laughlin state (truncated pseudopotential) results. For comparison, the results of a plasma analogy calculation for the Laughlin state using the modified HNC and Monte Carlo techniques (A. H. MacDonald, private communication) are shown. While at the time of writing the corresponding finite-size spherical geometry data for the Laughlin state was not available for a more direct comparison with the plasma theory calculations, similar data from periodic geometry calculations suggests that the slight discrepancy in peak heights is a real effect of the small difference between the Laughlin state and the Coulomb interaction ground state, rather than being due to the difference in computational techniques. Curvature effects seem to be absent in the discrete reciprocal space constructed through the relation $L = QR$.

Also shown in Fig. 8.5 is the fraction of the spectral weight of $\bar{s}(Q)$ that is exhausted by the collective excitation found in the numerical study by Haldane and Rezayi (1985a). Using the 'single mode approximation' (SMA) (related to the Feynman-Bijl *ansatz* for the ^{4}He phonon), Girvin et al. (1985, 1986b, Chap. IX) were able to account quantitatively for the prominent 'roton-like' minimum in the numerical results for the collective-mode dispersion. The SMA is a variational approximation that becomes exact if the collective excitation exhausts the full spec-

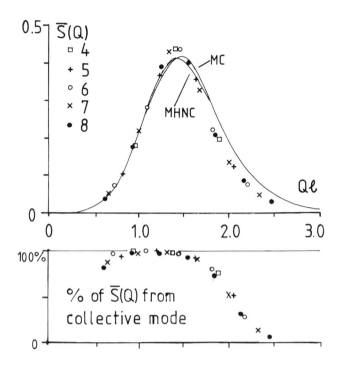

Figure 8.5 (Top): projected structure function $\bar{s}(Q)$ of spherical-geometry $v=1/3$ states for 4 - 8 particle systems, showing a remarkable size-independence. The lowest-Landau-level Coulomb interaction was used. Also shown: Monte-Carlo and modified hypernetted chain plasma theory calculations for the projected structure factor of the 1/3 Laughlin state (courtesy of A. H. McDonald and co-workers). (Bottom): fraction of the spectral weight of $\bar{s}(Q)$ contributed by the collective mode. The SMA assumes this to be 100%, and is evidently a very good approximation at intermediate Q.

tral weight of $\bar{s}(Q)$. From Fig. 8.5, it can be seen that this is indeed the case in the region of Q around 1.4 inverse magnetic lengths that dominates $\bar{s}(Q)$. The SMA is clearly not valid at large Q, and also appears to lose its validity at small Q, though since the projected structure factor vanishes identically for $L=1$, finite size limitations make this region of $\bar{s}(Q)$ difficult to explore. I will return to the SMA in Sec. 8.10.

In Fig. 8.6 I show a similar structure function study carried out in the periodic geometry. The choice of an oblique lattice PBC usefully increases the number of distinct values of $|Q|$ that are sampled. The Figure shows results for the lowest-Landau-level Laughlin state, and despite the small size (six particles) the results are very isotropic and similar to the spherical-geometry (and plasma) results of Fig. 8.5.

In the periodic geometry, the Landau-index-independent quantity $S_0(Q)$ can also be obtained directly and is shown in Fig. 8.7, again for the Laughlin state. Because the quantity $\exp(iQ \cdot R_i)$ translates particle i by $\hat{z} \times Q$, $S_0(Q)$ for a *finite* periodic system is periodic in reciprocal space. The data in Fig. 8.7 is only plotted for Q in the corresponding 'Brillouin zone,' which becomes larger as the size of the system is increased. The surprising fact that a system of as few as six particles can be a 'large' system is illustrated by Fig. 8.7: the function $S_0(Q)$ already appears to reach its asymptotic, large-Q behavior within the 'window' of reciprocal space accessible at this size! Fig. 8.8 shows the results for the same calculation repeated for the lowest-Landau-level Coulomb-interaction ground state. The results are very similar, as they must be, given that the projection of the Laughlin state onto the Coulomb interaction ground state is close to 100%. A striking feature of Fig. 8.7 and Fig. 8.8 is the isotropy of the results. This is *not* true of the very different results obtained by repeating the calculation for the *second*-Landau-level Coulomb interaction (Fig. 8.9). As will be seen in Sec. 8.10, the ground state of this interaction is quite different from the Laughlin state, and represents a *gapless*, *compressible* state that is very sensitive to the boundary condition geometry. Evidently the data of Fig. 8.9 is not well-represented by a plot against $|Q|$: Fig. 8.10 contrasts $S_0(Q)$ as a two-dimensional function of Q for lowest- and second-Landau-level Coulomb interactions with hexagonal PBC geometry. The compressible system has adapted to the anisotropy introduced by the boundary condition, while the incompressible system remains highly isotropic. The details of the sharp peaks of $S_0(Q)$ in the compressible state should not be taken too seriously, as these features are very sensitive to system size and geometry. However they do seem to suggest a tendency to charge density wave formation.

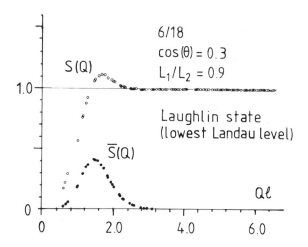

Figure 8.6 Structure factor $S(Q)$ and projected structure factor $\bar{S}(Q)$ for the 1/3 Laughlin state in the lowest Landau level (six particles in an oblique lattice PBC geometry).

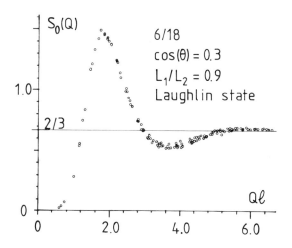

Figure 8.7 The Laughlin state data of Fig. 8.6, converted to the Landau-level-independent form $S_0(Q)$ (see Eq. 8.3.9). The asymptotic large-Q value $1+v\langle P\rangle = 2/3$, where P is the pair-exchange operator, is attained within the 'window' of reciprocal space that can be studied in a six-particle system.

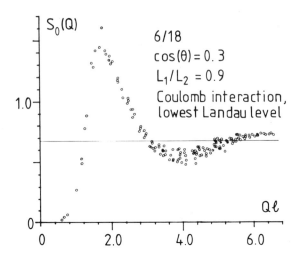

Figure 8.8 Same quantities as Fig. 8.7, but calculated for the true ground state of the lowest-Landau-level Coulomb interaction rather than the Laughlin state.

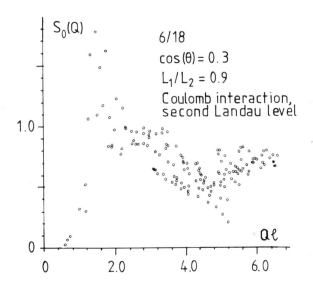

Figure 8.9 Same quantities as Figs. 8.7-8.8 but recalculated using the *second*-Landau-level Coulomb-interaction ground state -- a gapless, compressible state. Unlike the data of Figs. 8.7-8.8, this data is very sensitive to changes in boundary conditions.

The Hierarchy. Numerical Studies. 341

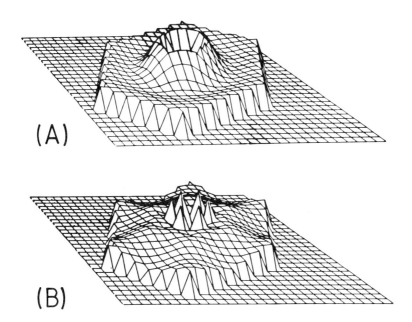

Figure 8.10 Two-dimensional plots (with linear interpolation) of $S_0(Q)$ for (A) lowest- and (B) second-Landau-level Coulomb-interaction ground states, with hexagonal lattice PBC geometry. The corners of the hexagonal 'window' of reciprocal space that can be studied at this size (6 particles, 18 flux quanta) correspond to $|Q\ell| = 6.60$. The almost perfect rotational invariance of the incompressible state (A) is in marked contrast to the adaptation to the boundary condition orientation exhibited by the compressible state (B). The peak structure in (B) is probably a finite-size artifact, but may reflect a tendency towards charge density wave formation in the compressible state.

8.9. Quasiparticle Excitations

Having studied the properties of the incompressible ground state, the natural next step is to examine the elementary quasiparticle and quasihole states and compare them to model states such as the Laughlin construction discussed in Sec. 8.4. The basic quantities of interest are the charge carried by the excitation and the energy of the excitation, or rather the sum of the excitation energies of the elementary quasiparticle plus the elementary quasihole, since at fixed total charge and magnetic flux, these excitations always occur in neutral pairs.

The state corresponding to an incompressible system with a single charged quasiparticle or quasihole excitation is easily recognized in a numerical study as an isolated ground-state multiplet with a degeneracy proportional to the size of the system. The charge carried on such a quasiparticle can be determined from this degeneracy, and its magnitude is always the electronic charge divided by the denominator of the rational v. Quite general arguments show that this latter quantity is an integer multiple of the quasiparticle charge: addition of an electron to the system must produce an integer number of elementary quasiparticles, as must removal of a quantum of magnetic flux. By Faraday's law, coupled with the assumption of ideal Hall conductivity, the latter operation causes charge v to flow into the system. It thus follows that the ratio of electronic charge to quasiparticle charge is an integer which is a multiple of the denominator of v.

To prove that the multiplying factor is unity if an isolated quasiparticle multiplet can be constructed, it is convenient to consider periodic boundary conditions, and $v = p/q$. The incompressible state then has a q-fold center-of-mass degeneracy, as discussed by Haldane (1985). After modification of the total charge and flux to produce the elementary quasiparticle state there will generally be no common divisor of the charge N and flux N_ϕ, and every state will be N_ϕ-fold degenerate. This must approximately derive from the q-fold ground-state degeneracy of the incompressible state times an (N_ϕ/q)-fold positional degeneracy of the charged quasiparticle, from which it follows that the ratio of quasiparticle charge to electronic charge is just $1/q$.

In the more general case where this ratio is $1/nq$ (e.g., the ferromagnetically ordered state alluded to in Sec. 8.4, which appears when the Zeeman energy vanishes), quasiparticles can only be produced in groups of n, and the total multiplicity of an isolated group of quasi-degenerate low-lying multiplets corresponding to n-quasiparticle states allows n to be identified.

These exotic $n > 1$ examples only occur in a limit that a continuous broken symmetry involving polarization of an internal degree of freedom of the electrons occurs, and the usual charge-$1/q$ quasiparticle fractionalizes. While it is clearly impossible to 'switch off' the Zeeman energy while maintaining a fixed Landau level filling factor, Rasolt et al. (1985) have suggested such a situation might be realizable if 'spin' were reinterpreted as a valley index.

The above discussion appears to make the 'hidden' assumption that the system is an incompressible fluid exhibiting the QHE through the assertion that if the net flux through a region is changed, a net charge accumulates as a consequence of the Hall current generated by the Faraday EMF. This assumption is not valid if the system is in a *gapless, compressible* state such as a Wigner lattice, in which screening prevents local charge accumulation.

Numerical studies of the behavior of the multiplicities of elementary quasiparticle and quasihole states as a function of size in the spherical geometry (Haldane and Rezayi 1985a) confirm the fractional charge. For the 1/3 state with lowest-Landau-level Coulomb interactions, the coherent states of the excitations, centered at a point, were prepared. The charge-density profiles for, e.g., the six-particle system are extremely close to those of the model states proposed by Laughlin.

The energy gap for making an infinitely separated quasiparticle plus quasihole pair for the 1/3 state with lowest-Landau-level Coulomb interactions has been estimated by extrapolation of finite-size data as 0.105 ± 0.005 $e^2/\kappa\ell$ by Haldane and Rezayi (1985a). However, the comparison of states with different charges requires some rather delicate Madelung-type subtractions if long-range Coulomb interactions are present, and a somewhat more direct estimate of this quantity can be obtained from the *neutral* excitation spectrum, as described in the next section.

8.10 Collective Excitations

The full energy-level spectrum of a seven-electron system with the lowest-Landau-level pseudopotential corresponding to filling factor $v=1/3$ in a spherical geometry is shown in Fig. 11. The ground state is a

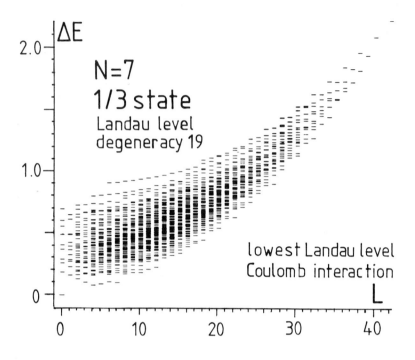

Figure 8.11 Complete excitation spectrum for the seven-particle system in spherical geometry corresponding to $v=1/3$, with the lowest-Landau-level Coulomb interaction.

non-degenerate state with angular momentum number $L=0$. It is well separated from other energy levels and has the quantum number of an isotropic state, and an essentially 100% projection on the model Laughlin wave function in this geometry. The second feature of note is the set of low-lying excitations with $L=2,3,\ldots,N$, which may be identified as a branch of *neutral elementary excitations* or collective modes. The states of a neutral particle in a spherical geometry are described by spherical harmonic wave functions. Wave number and angular momentum are related by $kR=L$, where R is the radius of the spherical surface. [This is more appropriate than another plausible relation, $kR=(L(L+1))^{1/2}$, since kR is the number of wavelengths in a distance $2\pi R$, the circumference of a great circle, and L has a similar significance for spherical harmonics.] In the hierarchy picture applied to spherical systems (Haldane 1983), the elementary excitations carry angular momentum $L=N'/2$, where N' is the number of particles of the parent fluid. The large-L collective-mode states can be interpreted as a quasiparticle-quasihole combination, with $L \leq N'$. For $v=1/3$, $N'=N$, the number of electrons.

Above the branch of collective excitations lies a quasi-continuum of non-elementary states with multiple excitations. The maximum energy state is a highly degenerate $L=42$ multiplet which can be identified as a state where the seven particles are compressed into a droplet with the density of the filled Landau level state.

Figure 12 shows a similar spectrum at $v=1/5$. The feature to note is the banded structure of the quasi-continuum. At $v=1/5$, there are two distinct energy scales controlling excitation energies: V_1-V_5 and V_3-V_5. If $V_1-V_5 \gg V_3-V_5$, and N is finite, there will be a separation of the excitation energies into a low-energy set not involving $m=1$ pairs, and those with one two, three, etc., $m=1$ pairs. This can be seen in Fig. 12 and was noted by Trugman and Kivelson (1985) who considered a model Gaussian interaction for which $V_1 \gg V_3 \gg V_5$, etc.

Figure 13 shows details of the low-lying energy levels of lowest-Landau-level Coulomb pseudopotential systems corresponding to $v=1/3$, $1/5$, $1/7$, and $2/7$. The collective excitation with $L=2,3,\ldots,N$ can be identified at $v=1/5$ and $v=1/7$, although it is clearest at $v=1/3$. The characteristic excitation energy scale drops from of order $0.08\ e^2/\kappa\ell$ at $v=1/3$ to $0.02\ e^2/\kappa\ell$ at $v=1/5$ and $0.008\ e^2/\kappa\ell$ at $v=1/7$. The spectrum at $v=2/7$ illustrates the hierarchy picture: The low-lying group of levels can be identified as those of a gas of four quasihole excitations of an underlying $v=1/3$ state. The lowest $L=2$, 3 and 4 levels may be identified as the collective mode of the quasihole gas, which has a characteristic energy scale of order $0.02\ e^2/\kappa\ell$, while the underlying parent fluid has a much larger gap.

In Fig. 13, the conversion of L into collective-mode wave number is shown. The results are quite size and boundary-condition independent, as seen from the equivalent calculation at $v=1/3$ using periodic boundary conditions with a hexagonal cell (Fig. 14). The spectrum in Fig. 14 is shown as a function of the absolute value of the two-dimensional quantum number k

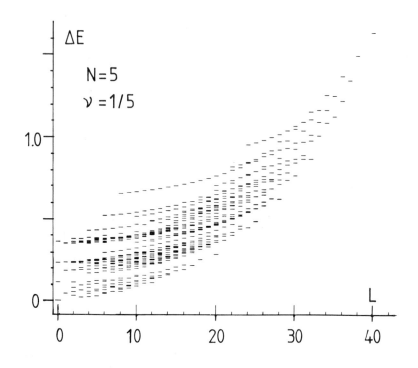

Figure 8.12 Five-particle $v=1/5$ system, otherwise same as Fig. 8.11.

which classifies states at rational filling factors (Haldane 1985). The ground state again has the quantum number of a homogeneous state ($k=0$) and an almost 100% projection on the Laughlin state in this geometry.

A striking feature of the $v=1/3$ collective excitation is that the minimum gap lies at a non-zero wave number $k\ell \simeq 1.4$. This feature approximately coincides with the peak in the projected structure factor $\bar{s}(Q)$ (Fig. 8.5). As described in Chap. IX, the 'single mode approximation' (SMA) *ansatz* of Girvin *et al.* (1985, 1986b) relates the spectrum to $\bar{s}(Q)$, with the assumption that the full weight of the dynamic projected structure factor $\bar{s}(Q,\omega)$ [the integral of which over frequency is $\bar{s}(Q)$] comes from the collective mode alone. The SMA gives a rigorous upper bound to the true collective mode energy, and Girvin *et al.* found that in the region of the gap minimum, there was extremely good agreement between the SMA and the exact finite-size results. A direct test of the SMA is

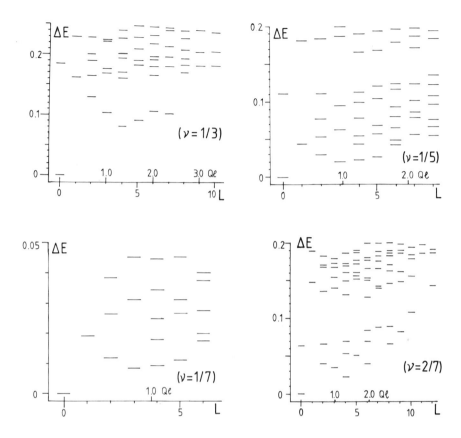

Figure 8.13 Details of low-lying excitations vs. dimensionless wave number $Q\ell$, for $v=1/3$ ($N=7$), $v=1/5$ ($N=5$), $v=1/7$ ($N=5$) and $v=2/7$ ($N=6$), again for a spherical lowest-Landau-level Coulomb system.

shown in the lower part of Fig. 5, which shows the fraction of the total weight of $\bar{s}(Q)$ that comes from the collective mode. For $Q\ell$ in the range 0.8 to 1.5, the SMA is seen to be essentially exact. This is also the region that dominates the response of the system to probes that couple to the electrons. (The response becomes negligible when $\bar{s}(Q)$ becomes small.) The SMA assumes that the collective mode involves excitation of a *single* electron. At large wave number, the collective mode is instead best described as a quasiparticle-quasihole pair (Kallin and Halperin 1984, Laughlin 1985, Chap. VII, Girvin et al. 1986b, Chap. IX) which is not a simple excitation in terms of electron coordinates, and hence the SMA breaks down, as

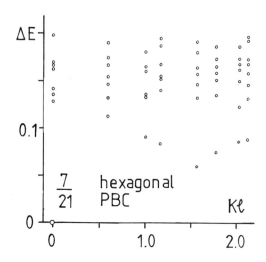

Figure 8.14 Low-lying energy levels of the $N=7$, $v=1/3$ lowest-Landau-level Coulomb system with periodic boundary conditions and a hexagonal cell.

seen directly in Fig. 5. However this region gives vanishingly small contribution to the response functions, which should be well-described by the SMA. As noted previously, Fig. 5 also suggests that the SMA loses validity at *long* wavelengths, though the finite size of the system precludes study at very long wavelengths. It is not clear whether the projection of the SMA *ansatz* onto the exact collective mode remains finite or vanishes in the long wavelength limit. The SMA *ansatz* is modelled on the Feynman-Bijl *ansatz* for the collective mode of liquid ^{4}He. For quantitative agreement with experiment in ^{4}He, it is necessary to include two-particle 'back-flow' corrections, but evidently this is *not* the case in the FQHE. Girvin *et al.* (1986b) have discussed the possible explanation for why the simple SMA scheme appears to work so well (see Chap. IX for a fuller discussion).

In Sec. 8.3 the idea of a perturbation treatment of the difference between the truncated pseudopotential and the true interaction was discussed. Fig. 15 shows the numerical implementation of this idea for the lowest-Landau-level Coulomb interaction at $v=1/3$. The Hamiltonian may be written as $H_0 + \lambda H_1$ where $\lambda = 0$ represents the truncated pseudopotential where only $V_1 \neq 0$

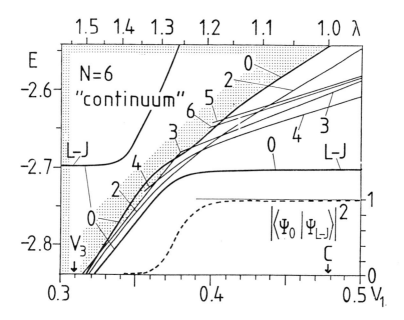

Figure 8.15 Effect of varying V_1 for a spherical $v=1/3$, $N=6$ lowest-Landau-level Coulomb system. 'C' marks the pure Coulomb value of V_1 [which is size-dependent in this geometry since the geometric (chord) distance for inter-particle separation is used]. Low-lying energy levels (labelled by angular momentum L) are shown, as is the region of the quasi-continuum. 'L-J' marks the state which is very close to the pure Laughlin(-Jastrow) state. The projection of the Laughlin state on the ground state is also shown. For $V_1 \gtrsim 0.37$ the system is incompressible, but is compressible for smaller values. The perturbation parameter λ of the decomposition $H=H_0+\lambda H_1$ into a truncated pseudopotential model H_0 plus residual Coulomb interactions is also shown.

and $\lambda=1$ the full Coulomb interaction in the lowest Landau level. Equivalently, V_1 can be varied, keeping V_m at their Coulomb values, and this is depicted in Fig. 15. As $V_1 \to +\infty$, the Laughlin state is the only state for which $\langle \Psi|H|\Psi \rangle$ is independent of V_1, and becomes the exact ground state. All other states have energies proportional to V_1 in this limit, and there is a large energy gap. As V_1 is lowered, it is seen that as long as the gap persists ($\lambda \lesssim 1.25$) the Laughlin state has essentially 100% projection on the true ground state. This regime includes the pure lowest-Landau-level Coulomb-interaction ground state, which is thus well-described by the Laughlin state. For $\lambda \simeq 1.25$, there is a rapid change of ground-state character, and the system becomes gapless. Studies of the effect of changing v for $\lambda > 1.25$ indicate it also becomes compressible (Haldane and Rezayi 1985, and unpublished). While the ground-state symmetry in the spherical geometry does not change at the transition, the transition appears to be first order. The collective mode gap appears to remain finite up to the transition, and the Laughlin state can be found as an *excited* state (with an excitation energy $\propto N$) in the gapless regime. The first order nature of the transition is clearer in periodic-boundary-condition systems with high symmetry (square or hexagonal) cells, as the *point symmetry* of the ground state changes.

In the *second* Landau level, it turns out that the $v=1/3$ Coulomb pseudopotential system has a *gapless* spectrum (Fig. 16) and is compressible. In Fig. 16, which shows the spectrum of a periodic six-electron system with a hexagonal unit cell, there are two almost degenerate $k=0$ ground states with different point

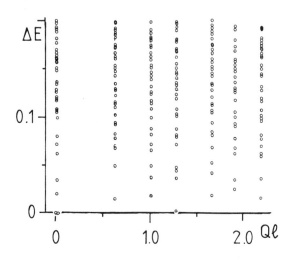

Figure 8.16 Excitation spectrum *vs.* dimensionless wave number $Q\ell$ of the *gapless* $v=1/3$, *second*-Landau-level Coulomb-interaction state ($N=6$, periodic boundary conditions with an hexagonal cell).

The Hierarchy. Numerical Studies.

symmetry under $2\pi/6$ rotations and a multiplet of very low-energy $k{\neq}0$ levels. The spectrum is clarified by applying the technique of varying V_1 away from the second-Landau-level Coulomb-interaction value, keeping the other pseudopotential components fixed (Fig. 17). This shows that the pure Coulomb system is just *at* the first-order transition from the Laughlin state to the gapless state, and that incompressibility is restored by slightly increasing V_1. The absence of incompressibility at 1/3 filling in the second Landau level (i.e., $v=7/3$) may be understood as a consequence of the reduction in V_1-V_3 due to the node of the wave function at the center of the cyclotron orbit when the Landau index is 1 (see Fig. 1).

The numerical 'perturbation' studies of Figs. 15 and 17 clarify what was an initially puzzling feature of the Laughlin state. While introduced as a 'variational' trial wave function, it has no free parameters

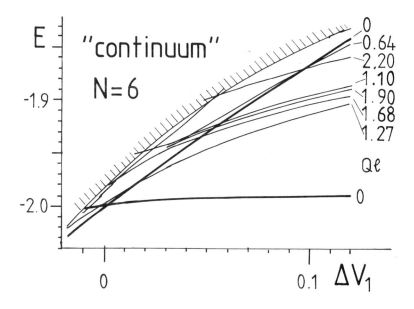

Figure 8.17 Effect of varying V_1 on the $v=1/3$ *second*-Landau-level Coulomb system excitation spectrum vs. dimensionless wave number $Q\ell$. $\Delta V_1=0$ marks the pure Coulomb interaction. This Figure is similar to Fig. 8.15, but uses the periodic geometry with a hexagonal cell. The pure Coulomb system is on the borderline between gapless and incompressible states.

at fixed filling factor and has no apparent dependence on the actual inter-particle interaction. This contrasts with other model states, such as those of the BCS model of superconductivity, which describes a family of states that vary with coupling strength. The Laughlin state is an essentially unique and rigid state at $v=1/m$, and is favored by strong short-range pseudopotential components. If longer-range components become important, a different type of ground state is favored, but the transition between the two regimes is essentially discontinuous. So long as the system is in the incompressible regime, and there is a gap in the excitation spectrum, the ground-state properties are essentially insensitive to the details of the potential.

8.11 Acknowledgements

The author is an Alfred P. Sloan Fellow, and thanks the Alfred P. Sloan Foundation for its support. Some of the work presented here was carried out at the University of Southern California and was supported in part by NSF grant DMR-8405347.

CHAPTER 9

Collective Excitations

STEVEN M. GIRVIN

9.1 Introduction

Many condensed-matter systems exhibit collective excitation modes involving coherent oscillations of the medium. For instance classical liquids and gases support sound waves. On quite general grounds one expects a well-defined sound wave to exist for all wavelengths much greater than the typical particle spacing. These modes are hydrodynamical in the sense that during one period of oscillation, many collisions occur between particles thereby maintaining local quasi-equilibrium. The particular details of the molecular collisions are largely irrelevant so long as they maintain the local quasi-equilibrium. The restoring force for the oscillation is the local pressure change induced by the local density change, and to a good approximation, one can compute the local pressure vs. density relationship using equilibrium thermodynamics.

Other condensed matter systems exhibit collective modes which are *not* hydrodynamic in nature. A good example is the case of phonon excitations in solids. Phonons, which are quantized lattice vibrations, are rather different from hydrodynamic modes because they represent true *elementary excitations* of the system. That is, a quantum state consisting of a single phonon is an *eigenstate* of the system Hamiltonian (at least within the harmonic approximation). Nevertheless because phonon modes are essentially independent harmonic oscillators, one can construct coherent states (Schulman 1981) of phonons which behave very much like the density oscillations (sound waves) in classical liquids and gases. An additional point to bear in mind in the following discussion is that because solids can support shear stresses, they exhibit transverse as well as longitudinal phonons.

Superfluid helium provides another interesting example which will yield several useful analogies in the discussion of the fractional quantum Hall effect (FQHE). Superfluid ^{4}He is a remarkable quantum liquid which exhibits phonons as elementary excitations. Unlike the hydrodynamic sound-wave modes in say, liquid argon, the phonons in superfluid helium are true elementary excitations of the quantum system. Feynman's theory of helium (Feynman 1972) shows that these elementary excitations remain well-defined even for microscopic wavelengths approaching the inter-atomic spacing--much as the phonons in a solid are well defined for all wave vectors out to the Brillouin zone boundary. This contrasts with the hydrodynamic modes in classical fluids which, as noted earlier, are well-defined only for wavelengths much larger than the particle spacing. We shall see shortly that the essential property of superfluid helium is the lack of low-lying single-particle excitations. This allows the phonons to be *collisionless* elementary excitations rather than collision-controlled collective modes as classical sound waves are.

It is possible for a fluid to support transverse modes provided the fluid is charged and is subjected to a magnetic field. The Lorentz force perpendicular to the particle velocity can supply the necessary transverse restoring force. Plasma physicists refer to the longitudinal mode as the 'upper-hybrid mode' and to the transverse mode as the 'lower-hybrid mode' (Platzman and Wolff 1973). Strictly speaking, the magnetic field mixes the longitudinal and transverse motion so that neither mode is purely longitudinal or transverse. We shall see shortly what the analogous modes are in the two-dimensional electron gas in a quantizing magnetic field.

9.2 Density Waves as Elementary Excitations

Feynman's theory of helium (Feynman 1972) asks the following question: Given some knowledge of the quantum ground state, what can one learn about the nature of the low-lying excited states of the system? This is precisely the question we are interested in addressing here for the fractional quantum Hall effect. The ground state of the system appears to be rather well described by Laughlin's variational wave function or the related hierarchical wave functions (Haldane 1983, Laughlin 1984a, Halperin 1984, Girvin 1984a). Given this information one would like to develop a physical picture of and a quantitative theory for the low-lying excited states.

Because Feynman's methods will prove to be very useful and because the analogies with superfluidity turn out to be of considerable importance, it is worthwhile to review briefly Feynman's theory of ^{4}He.

Feynman argues that because ^{4}He is a boson system there can be no low-lying single-particle excitations and that the only low-energy excitations are collective density waves in which the energy is minimized by coherently distributing the momentum amongst a large number of atoms. Supposing that the exact ground state ψ is known, Feynman writes down the following variational wave function for the density-wave excited state at wave vector k:

$$\Phi_k = N^{-1/2} \rho_k \psi \qquad (9.2.1)$$

where N is the number of particles and ρ_k is the Fourier transform of the density operator:

$$\rho_k \equiv \sum_{j=1}^{N} e^{i k \cdot r_j} . \qquad (9.2.2)$$

In order to develop a physical picture of the state described by this variational wave function, consider the following points. Because Φ_k contains the ground-state wave function ψ as a factor, good correlations are automatically built in. At the same time it is straightforward to show that Φ_k is orthogonal to the ground state, as required [Problem 1]. Consider now some particular configuration of the particles. If this configuration is likely in the ground state (i.e., ψ is not too small for this configuration) and if ρ_k is fairly large, then Φ_k will be relatively large. This means that the particular configuration will be likely in the excited state. But ρ_k will be large only if the configuration has a significant density modulation at wave vector k. Hence the excited-state wave function describes a density wave (Feynman 1972, Girvin et al. 1985, 1986a,b).

Having found a form for the variational wave function which we can justify on physical grounds, we now need to evaluate the expectation value of the excitation energy. This is given by the usual variational expression:

$$\Delta(k) = f(k)/s(k) , \qquad (9.2.3)$$

where

$$f(k) \equiv N^{-1} \langle \psi | \rho_k^\dagger (\mathcal{H} - E_0) \rho_k | \psi \rangle , \qquad (9.2.4)$$

$\mathcal{H}$ is the Hamiltonian, E_0 is the ground-state energy and the norm of the excited state is

$$s(k) \equiv N^{-1} \langle \psi | \rho_k^\dagger \rho_k | \psi \rangle . \qquad (9.2.5)$$

It turns out to be remarkably easy to evaluate Eq. 2.3. The quantity $f(k)$ is nothing more than the oscillator strength and is given by the usual sum rule (Mahan 1981)

$$f(k) \equiv \frac{\hbar^2 k^2}{2m} . \qquad (9.2.6)$$

Furthermore, the quantity $s(k)$ is precisely the static structure factor which can be directly determined from neutron scattering experiments. Eq. 2.3 is the famous Bijl-Feynman formula (Feynman 1972, Mahan 1981). The great advantage of this expression is that it yields a dynamical quantity, namely the collective-mode energy, in terms of purely static properties of the ground state. We can interpret Eq. 2.3 as saying that the collective-mode energy is the single-particle energy at the appropriate momentum divided by a factor which is something like a mass renormalization associated with the correlations among the particles.

An alternative interpretation of Eq. 2.3 is the following. The assumption that Φ_k is an eigenfunction of $\mathcal{H}$ is equivalent to the assumption that a single mode (the density wave) saturates 100% of the oscillator strength sum. I shall refer to this assumption as the single-mode approximation (SMA). This interpretation of Feynman's theory will prove useful in the study of the modes in the FQHE.

9.3 Collective Modes in the FQHE

We are now in a position to consider the nature of the collective modes in the FQHE (Kallin and Halperin 1984, Laughlin 1985a, Girvin et al. 1985, 1986a,b). Since we are dealing with a charged system in a magnetic field, we expect that there will be some similarities to the classical plasma modes discussed above. At the same time the system bears a strong resemblance to superfluid ^{4}He in that it is highly quantum-mechanical and exhibits dissipationless current flow. One might therefore expect the collective modes to be non-hydrodynamic elementary excitations as in ^{4}He but having both transverse and longitudinal components as in the classical plasma. As we shall see, this expectation turns out to be correct.

One difficulty which we must now face before developing a quantitative theory is the fact that Feynman's method would seem to rely crucially on the Bose statistics in ^{4}He. Recall that it was the Bose statistics which eliminated the possibility of low-lying single-particle excitations and therefore made the SMA valid. As is well known, Fermi systems exhibit a high density of single-particle excitations in the form of a continuum of particle-hole excitations across the Fermi surface (Pines 1964, Mahan 1981). It seems clear on phenomenological grounds however that in the FQHE, the single-particle continuum must somehow be suppressed because the current flow is nearly dissipationless rather like it is in a superfluid. [There are some important distinctions, however, between ordinary superfluidity and the FQHE. These will be discussed in more detail in Sec. 9.5.] The origin of this effect is the destruction of the Fermi surface by the magnetic field. Indeed in two dimensions the continuum of kinetic energy is completely quenched by the magnetic field. As noted in Chap. I, the single-particle quantum states are highly degenerate Landau levels separated by gaps in the kinetic energy of $\hbar\omega_c$. In the limit of large fields it is sensible to classify the elementary excitations according to their kinetic energy. Excitations involving transfer of an electron from one Landau level to the next higher level have an energy of order $\hbar\omega_c$ and correspond to the upper-hybrid mode in plasmas. Intra-Landau-level excitations have much smaller energies on the order of the Coulomb energy $e^2/\kappa\ell$ (the natural unit of energy which I will use throughout this discussion) since the kinetic energy is zero. These latter excitations are the analogs of the quasi-transverse lower-hybrid mode in plasmas. I emphasize that this picture is valid in the high field limit where the particle interactions do not lead to much Landau level mixing (see the further discussion in Chap. X).

In view of the above remarks it seems plausible that one could construct a theory of the excitations in the FQHE which is similar in spirit to Feynman's ^{4}He theory but which takes into account the existence of two important modes rather than just a single mode. For simplicity let us assume that the lowest Landau level has fractional filling v and let us ignore for the moment any spin and valley quantum numbers. Then a natural pair of variational excited states to consider is:

$$\Psi_k = P_1 \rho_k \psi \qquad (9.3.1)$$

and

$$\Phi_k = P_0 \rho_k \psi , \qquad (9.3.2)$$

where P_n is the projection operator for the nth Landau level and ψ is the exact ground state of the system (and is therefore largely in the lowest Landau level). It is clear from the construction that Ψ_k and Φ_k are eigenstates of the kinetic energy operator T:

$$T \Phi_k = 0 \Phi_k , \qquad (9.3.3)$$

Collective Excitations

$$T \Psi_k = \hbar\omega_c \Psi_k \ . \qquad (9.3.4)$$

(Note that I have dropped the $\hbar\omega_c/2$ zero-point energy from T). If these are in fact good approximations to eigenfunctions of the full Hamiltonian, then Ψ_k will correspond to the upper-hybrid or magnetoplasmon mode and Φ_k will correspond to the lower-hybrid or magnetophonon (or as we shall see, magnetoroton) mode.

It is not yet clear however that these are sensible wave functions. In particular it is not even clear that the elementary excitations can be labelled by a conserved wave vector. Recall from the discussion in Chap. I that the system is not translation-invariant in the presence of a magnetic field, due to the vector potential appearing in the Hamiltonian. However translations of the vector potential are equivalent to gauge transformations (Problem 2). Hence there *can* be a conserved momentum but the quantity which is conserved depends on which gauge is chosen. For example in the Landau gauge used in Chap. I the *x*-component of the *linear* momentum is conserved and the single-particle states are labelled by a one-dimensional wave vector. On the other hand in the symmetric gauge the *angular* momentum is a good quantum number and the single-particle states are labelled by the integer azimuthal quantum number.

We must now consider how this affects the nature of the collective modes. The density operator creates a *neutral* excitation; i.e., a particle-hole pair. Imagine that we construct an effective Hamiltonian for this particle-hole pair. How does the magnetic field enter? Using relative and center-of-mass coordinates the problem simplifies in the usual way provided that one recognizes that the vector potential causes the relative separation coordinate to make a contribution to the center-of-mass momentum. Because the pair is neutral the center of mass coordinate does not couple to the magnetic field and so the center-of-mass momentum (i.e., the total momentum of the excitation) is conserved independent of the gauge. These points are presented in more detail by Kallin and Halperin (1984).

A more physical way to picture this is to note that if two particles have the *same* charge, their mutual Lorentz force will cause them to circle around each other. This semi-classical periodic motion leads to discretely quantized bound states. On the other hand, if the particles have *opposite* charge, their mutual Lorentz force causes them to drift uniformly in a direction normal to their separation. This uniform semiclassical motion is associated with the conserved momentum quantum number (see Chap. VII for a discussion of the states of two quasiparticles with like or opposite charges).

The result of all this is that the neutral collective excitations *can* be labelled by a conserved wave vector. This is just what we have done in Eqs. 3.1-2 and so this form for the variational wave functions seems quite reasonable. I will discuss the evaluation of the energy for these two states in Secs. 9.4 and 9.7.

9.4 Magnetophonons and Magnetorotons

As noted earlier the intra-Landau-level mode given by Eq. 3.2 has zero kinetic energy and a potential energy on the order of $e^2/\kappa\ell$. This low-energy mode is the primary one of interest to the FQHE and we would like to evaluate its energy using methods analogous to those in the Feynman theory of ^{4}He. Following Eqs. 2.2-3 we therefore define the projected oscillator strength and the projected structure factor by:

$$\bar{f}(k) \equiv N^{-1}\langle\psi|\Lambda_k^\dagger\,(\mathcal{H}-E_0)\,\Lambda_k|\psi\rangle\,, \quad (9.4.1)$$

$$\bar{s}(k) \equiv N^{-1}\langle\psi|\Lambda_k^\dagger\,\Lambda_k|\psi\rangle\,, \quad (9.4.2)$$

where

$$\Lambda_k \equiv P_0\,\rho_k\,P_0, \quad (9.4.3)$$

is the projected density operator. In analogy with Eq. 2.3 the collective mode energy is then:

$$\Delta(k) = \bar{f}(k)/\bar{s}(k)\,. \quad (9.4.4)$$

Fortunately we can take advantage of the analytic simplicity of the single-particle eigenstates in the lowest Landau level to explicitly compute the projected density operator Λ_k and hence compute $\bar{f}$ and $\bar{s}$. The technical details can be found in Girvin and Jach (1984), Itzykson (1985, 1986) and Girvin, MacDonald and Platzman (1985, 1986a,b). The results are:

$$\bar{f}(k) = \frac{v}{2\pi}\int \frac{d^2q}{(2\pi)^2} v(q) \int d^2r\, [g(r) - 1]$$

$$\times [e^{-|k|^2/2} e^{iq\cdot r}(e^{(kq^* - k^*q)/2} - 1)$$

$$+ e^{i(k+q)\cdot r}(e^{k\cdot q} - e^{k^*q})] \qquad (9.4.5)$$

and

$$\bar{s}(k) = s(k) - (1 - e^{-|k|^2/2}) . \qquad (9.4.6)$$

where q and k are complex number representations of $\mathbf{q}$ and $\mathbf{k}$. In Eq. 4.5 $g(r)$ is the two-point correlation function related to the Fourier transform of $s(k)$, the (unprojected) static structure factor of the ground state (Mahan 1981, Girvin et al. 1986b). The quantity $v(q)$ is the Fourier transform of the interaction potential. Notice that the scale of $\bar{f}(k)$, and hence the scale of the collective-mode energy, is set solely by the scale of the Coulomb interaction $v(q)$ since the kinetic energy has been completely quenched. Clearly the evaluation of Eq. 4.4 is somewhat more complex than for the case of Eq. 2.3, but the spirit of the result is still the same as the Bijl-Feynman formula, namely that one can express the collective mode energy in terms of static properties of the ground state.

Unlike the case of ^{4}He we do not have the advantage of knowing $s(k)$ from neutron scattering measurements so we must rely on a specific model of the ground state. Laughlin's ground-state wave function discussed in Chap. VII is a good choice because it appears to be quite accurate (Yoshioka 1984a, Haldane and Rezayi 1985a, Haldane Chap. VIII) and because the static structure factor is available through a variety of numerical and analytical means (Laughlin 1983a, Levesque et al. 1984, Girvin 1984b, Girvin et al. 1986b). The results of evaluating Eq. 4.4 for the collective-excitation energy vs. wave vector are plotted in Fig. 9.1 for filling factors $v = 1/3$, $1/5$, and

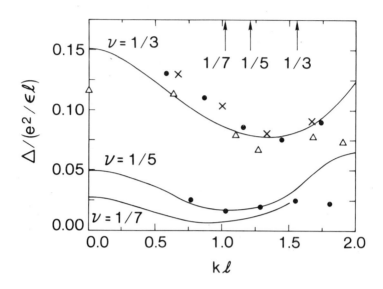

Figure 9.1 Comparison of SMA prediction of collective mode energy for $v = 1/3$, $1/5$, and $1/7$ (solid lines) with small-system numerical results. Crosses indicate ($N=7$, $v = 1/3$) spherical system, triangles ($N = 6$, $v = 1/3$) hexagonal unit cell (Haldane and Rezayi 1985a). Solid dots are from ($N = 9$, $v = 1/3$) and ($N = 7$, $v = 1/5$) spherical system calculations of Fano et al. (1986). Arrows at the top indicate the magnitude of the reciprocal lattice vector of the Wigner crystal at the corresponding density.

1/7. This Figure displays several striking features which are discussed below.

The first thing to notice is that the collective mode is *not* a massless Goldstone mode. That is, $\Delta(k)$ does not vanish linearly for small k as it does in ^{4}He. The magnetophonon has a finite gap ('mass'). This is *not* due to the fact that we are dealing with a charged system, since the Coulomb force is not sufficiently long-ranged to give a finite plasma frequency in two dimensions (Ando et al. 1982). Furthermore we are not discussing the magnetoplasmon but rather the magnetophonon, for which the long-range of the interaction is not relevant, since this is the quasi-transverse (lower-hybrid) mode. The origin of the finite gap is the incompressibility of the system at rational fractional filling factors. Because the system is incompressible, long-wavelength density fluctuations

Collective Excitations

are strongly suppressed and $\bar{s}(k)$ vanishes rapidly ($\approx k^4$) for small k. One can show from quite general oscillator-strength arguments [Note 1] that $\bar{f}(k) \approx k^4$ for small k so that $\Delta(k)$ in Eq. 4.4 has a *finite* value for small k. These points are discussed in more detail by Girvin, MacDonald and Platzman (1985, 1986a,b), but the essential thing to keep in mind is that the two-dimensional one-component plasma (2DOCP) analog system derived from Laughlin's wave function (see Chap. VII) has long-ranged interactions ($\approx \log r$) which for small k strongly suppress $\bar{s}(k)$ in the ground state. The incompressibility of the quantum system is thus intimately related to the incompressibility of the 2DOCP. This is a somewhat subtle point which bears rephrasing. Recall that I noted above that the excitation gap at small wave vectors is *not* due to the long range of the (true) Coulomb interaction, but we can interpret the 2DOCP discussion above as saying that the origin of the gap is the long range of the two-dimensional (*pseudo-*) Coulomb potential in the 2DOCP analog problem. One should not, however, make the mistake of assuming that the *dynamics* of the 2DOCP has any direct connection with the *dynamics* (i.e., the excited-state spectrum) of the quantum system. Recall that the 2DOCP gives a description of the particle distribution in the *ground* state of the quantum system and *not* the excited states. Fortunately linear response theory (the fluctuation-dissipation theorem) tells us that the compressibility can be computed knowing only the static fluctuations in the ground state. This is precisely the same information used above to compute the collective mode energy. The existence of a finite gap in the excitation spectrum follows from the incompressibility of the system.

The second feature to notice in Fig. 9.1 is the deep minimum in $\Delta(k)$ at finite wave vector. This 'magnetoroton' minimum is precisely analogous to the roton minimum in ^{4}He (Feynman 1972, Mahan 1981) and is due to a peak in the structure factor at the wave vector associated with the mean particle spacing (recall that the structure factor appears in the denominator of the Bijl-Feynman expression for the mode energy). Note that as the filling factor v is reduced, the minimum value of the excitation gap rapidly collapses towards zero. This gap collapse is a precursor to the Wigner crystal instability which is believed to occur in the vicinity of $v = 1/7$–$1/10$ (Lam and Girvin 1984,

1985, Levesque, Weis and MacDonald 1984). The magnitude of the wave vector at the magnetoroton minimum lies close to the magnitude of the reciprocal lattice vector of the corresponding Wigner crystal as indicated by the arrows in Fig. 9.1. In mean-field theory the magnetoroton does not go completely soft prior to the transition because the transition is (weakly) first-order (there exist cubic invariants of the form $\rho_p \rho_q \rho_{-p-q}$ in the Landau-Ginsburg free energy functional). The actual nature of this purely quantum (zero-temperature) phase transition is however not known at present and remains one of the important unanswered questions in the field. It would be extremely interesting to examine possible connections with classical defect-mediated two-dimensional melting (Nelson and Halperin 1979). A step in this direction was taken by Fisher (1982) but it was prior to the discovery of the FQHE and Laughlin's liquid-state wave function.

Having provided a fairly simple physical picture of the low-lying collective excitations based on the analogy with superfluid ^{4}He, we must now consider the quantitative accuracy of the SMA on which the calculations are based. Fig. 9.1 shows the very good agreement with the exact small-system numerical results for $v = 1/3$ of Haldane and Rezayi (1985a), and for $v = 1/3$ and $1/5$ of Fano et al. (1986). Note that there is a small amount of scatter in the points because of finite-size effects in the small-system calculations. This problem can be alleviated by substituting for Laughlin's wave function in Eq. 4.5-6, the exact numerical ground state. Haldane discussed this in Chap. VIII and showed that the SMA is quite accurate, particularly near the magnetoroton minimum where the lowest mode absorbs 98% of the oscillator strength.

Having in our possession rather accurate eigenfunctions and eigenvalues for the low-lying collective modes we can compute various quantities such as the microwave conductivity (Platzman et al. 1985) and the static susceptibility (Girvin et al. 1986b). The former shows a square-root singularity at the frequency corresponding to the threshold for magnetoroton production. This singularity ought in principle to be experimentally observable, provided it is not swamped by conductivity associated with quasiparticles which are nucleated and localized by the inevitable disorder which is present.

Collective Excitations

The static susceptibility is given by

$$\chi(q) = -2\alpha(q) \qquad (9.4.7)$$

where

$$\alpha(q) = \frac{\bar{s}(q)^2}{\bar{f}(q)} = \frac{\bar{s}(q)}{\Delta(q)}. \qquad (9.4.8)$$

Figure 9.2 shows the quantity α as a function of dimensionless wave vector $q\ell$. Notice that in addition to being sharply peaked at the roton wave vector, the peak-height of the susceptibility rapidly diverges as the filling factor is reduced. This is a further indication of the Wigner crystal instability and suggests that perturbations such as disorder can easily destroy the liquid state at low densities even before the crystal instability is actually reached.

As noted above, the SMA is essentially exact in the vicinity of the roton minimum. But as can be seen from Fig. 9.2, this is precisely the region which makes the most contribution to the linear response of the system. Hence the SMA should give an excellent description of the susceptibility and related response functions.

As discussed in Chap. VIII the lowest mode's fraction of the oscillator strength drops to about 80% for small wave vectors. This may be an indication that a low-momentum state consisting of a bound pair of rotons may have significant weight in the small-wave-vector excited state. A bound roton-pair state, being a two-particle excitation, would have no oscillator strength. Notice from Fig. 9.1 that for $\nu = 1/3$, it happens that there is an accidental (?) degeneracy $\Delta(0) \approx 2\Delta(k_{min})$ so that one might expect strong mixing between the magnetophonon at small wave vectors and a pair of rotons of opposite momenta. This is an area that deserves further investigation both numerically and analytically.

In the opposite limit of very large wave vectors ($k\ell \gg 1$) the SMA breaks down because it is not sensible to have a density wave with wavelength much shorter than the particle spacing. In this regime the natural low-lying collective mode is the quasi-exciton discussed in Sec. 9.6.

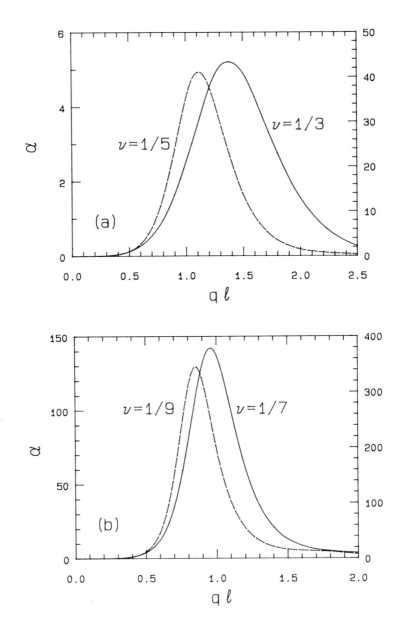

Figure 9.2 Susceptibility parameter α vs. dimensionless wave vector $q\ell$. (a) $v = 1/3$ (scale on left), $v = 1/5$ (scale on right); (b) $v = 1/7$ (scale on left), $v = 1/9$ (scale on right).

9.5 Further Superfluid Analogies: Vortices

In addition to having phonons and rotons as elementary excitations, superfluid helium exhibits quantized vortices (Feynman 1972). It will prove useful to review the physics of these excitations in order to see what can be said about analogous objects in the FQHE.

We can begin by considering the microscopic wave functions which describe macroscopic superfluid hydrodynamics. If ψ is the exact ground state then the wave function

$$\Phi_k = \exp\left[i \sum_{j=1}^{N} k \cdot r_j\right] \psi \qquad (9.5.1)$$

describes a state with a uniform current density (the center of mass momentum has been boosted). Note the difference between Eq. 5.1 and Eqs. 2.1-2. Eq. 2.1 describes a microscopic phonon excitation. Eq. 5.1 on the other hand describes a macroscopic current flow. We can generalize Eq. 5.1 to describe nonuniform flows by writing

$$\Phi = \exp\left[i \sum_{j=1}^{N} \varphi(r_j)\right] \psi \qquad (9.5.2)$$

where φ is a spatially varying phase factor. Eq. 5.2 can also be rewritten as

$$\Phi = \exp\left[i \int d^2r \, \varphi(r) \rho(r)\right] \psi \qquad (9.5.3)$$

where $\rho(r)$ is the density operator. It is straightforward to show that the current density for this state satisfies (Feynman 1972)

$$\langle J(r) \rangle = \frac{\hbar}{m} n \, \nabla\varphi(r) \qquad (9.5.4)$$

where $n = \langle \rho(r) \rangle$ is the particle density (assumed uniform). At this point one sees that there is a close connection between the *microscopic* wave function (9.5.3) and the *macroscopic* Landau-Ginsburg order parameter (Tinkham 1975).

Eq. 5.2 is a good approximation to an eigenfunction only if the steady-state continuity equation $\nabla \cdot \langle J \rangle = 0$ is satisfied (Feynman 1972). Hence the phase must satisfy $\nabla^2 \varphi = 0$. It also follows from Eq. 5.4 that the current flow is irrotational: $\nabla \times \langle J \rangle = 0$. Thus the circulation

$$\Gamma \equiv \oint_\Lambda \langle J(r) \rangle \cdot dr = \int_\Lambda d^2 r \; \nabla \times \langle J(r) \rangle \qquad (9.5.5)$$

must vanish in any simply-connected region [Note 2].

A famous paradox now arises (Feynman 1972). What happens if we take a rotating bucket of helium and cool it below the superfluid transition point? How can the conserved angular momentum (i.e., the circulation) be supported by the superfluid? The answer is that the superfluid must become effectively multiply-connected by developing point (in two dimensions) defects where the wave function (more precisely, the order parameter) vanishes. These point defects are vortices which carry the circulation of the fluid and can be described by giving the phase function in Eqs. 5.2-3 a branch cut. For example take

$$\varphi(r) = \lambda \theta \qquad (9.5.6)$$

where λ is a constant and θ is the azimuthal angle (between r and the x-axis). Note from Eq. 5.4 that the current is purely azimuthal

$$\langle J(r) \rangle = n \frac{\hbar}{m} \frac{\hat{\theta}}{r} \qquad (9.5.7)$$

as is appropriate for a vortex. The angle θ has a discontinuity (branch cut) of 2π across the positive real axis. If Φ is to be single-valued then the constant λ must be an integer. This requirement leads to the quantization of the circulation around the vortex

$$\Gamma = n \frac{h}{m} \lambda \; . \qquad (9.5.8)$$

Collective Excitations

Putting Eq. 5.6 into Eq. 5.1 results in a good approximation to an eigenfunction except near the origin where the kinetic energy density diverges due to the centrifugal barrier. This is remedied by giving the vortex a 'normal core' so that Φ vanishes (linearly) near the origin (Feynman 1972). This gives the multiple-connectedness discussed above.

The energy density of the vortex is proportional to $|J|^2$ so that the total energy

$$E \sim \int d^2r \; |\langle J(r) \rangle|^2 \sim \ln L \qquad (9.5.9)$$

diverges logarithmically with the size L of the system [Note 3]. This very important result means that isolated vortices can not exist in equilibrium (i.e. in an 'unstirred' fluid). Vortices can only appear in bound pairs of zero net vorticity. As the density of vortex pairs rises with temperature, the pairs begin to screen each other and eventually can unbind. This is (in a very small nutshell!) the essence of the Kosterlitz-Thouless phase transition (José et al. 1977).

After this brief review of vortices in superfluid helium we are now ready to consider the analogous objects in the FQHE. Because of the applied magnetic field, the current density operator is modified by the vector potential:

$$J = \frac{\hbar}{m} \sum_{j=1}^{N} \left\{ p_j + \frac{e}{c} A(r_j) \right\}. \qquad (9.5.10)$$

With this in mind let us consider the energy of an isolated vortex. It is possible for the system to avoid the usual logarithmic divergence in energy by shifting the particles closer to or further from the core so that the change in vector potential seen by each particle cancels out the change in the momentum and hence removes the kinetic energy. In other words while δp is forced to be non-zero by the vortex, $\delta p + (e/c)\delta A$ can still vanish (outside the small core region). The kinetic energy of the vortex is essentially zero (there is a small potential energy contribution from the core region). This makes sense when we recall that all the eigenstates in the lowest Landau level are degenerate in kinetic energy.

In addition to producing a finite vortex energy, the vector potential gives the vortices another even more remarkable property: they carry fractional charge

$$e^* = -\lambda v e \qquad (9.5.11)$$

where λ is the circulation quantum number of the vortex and v is the filling factor. This is readily demonstrated as follows. Consider a ring Λ of radius r centered on the vortex. From the single-valuedness of the wave function we have:

$$\oint_\Lambda n^{-1} \langle p \rangle \cdot dr = h\lambda . \qquad (9.5.12)$$

As discussed above the vector potential causes the ring to move in or out to minimize the kinetic energy. The change in the vector potential seen by the ring must satisfy

$$\oint_\Lambda n^{-1} \frac{e}{c} \langle \delta A \rangle \cdot dr = -h\lambda . \qquad (9.5.13).$$

Hence the *change* in the flux enclosed by the ring (due to the change in the ring size, *not* due to any change in the flux itself) is $-\lambda$ flux quanta. Another way to say this is that there are λ flux quanta in the annulus between the original and shifted radii of the ring. At filling factor v, the density is v particles per flux quantum, so that there are λv particles in the annulus and the vortex has acquired a charge given by (9.5.11) [Note 4].

By now it should be clear that these quantized vortices are in fact Laughlin's fractionally charged quasiparticles (Laughlin 1983a, Chap. VII). Indeed we can write the following modified version of Eq. 5.2 which keeps the state in the lowest Landau level

$$\Phi = \exp\left[\sum_{j=1}^{N} i\theta(r_j) + \ln(|r_j|) \right] \psi \qquad (9.5.14)$$

which can be rewritten in complex notation as

$$\Phi = \left[\prod_{j=1}^{N} z_j\right] \psi . \qquad (9.5.15)$$

This is precisely Laughlin's variational quasihole wave function.

The energy of the Laughlin quasiparticle states (as well as the exact vortex state for small systems) has been evaluated by a variety of means (Laughlin 1983a, Su 1984, Haldane 1985, Chakraborty 1985, Girvin et al. 1986b, Morf and Halperin 1986, MacDonald and Girvin 1986). The more recent of these results suggest that the sum of the quasielectron and quasihole energies ($\Delta(\infty) \sim 0.99 - 0.106 \, e^2/\kappa\ell$ for $v = 1/3$) is considerably larger than the original hypernetted-chain estimates of Laughlin (with most of the increase being in the quasielectron energy). MacDonald and Girvin (1986) have recently proposed a vortex wave function which differs slightly from Laughlin's in the core region in such a way that the energy can be easily computed directly from the two-point correlation function of the ground state. The results are in good agreement with the Monte Carlo and small-system calculations mentioned above.

In summary, the results of this and the previous sections point out the following facts: There is a close analogy between the FQHE and superfluidity, the common factor being the absence of low-lying single-particle excitations. Both systems exhibit elementary excitations consisting of phonons, rotons and quantized vortices. There are however three vital differences:

a) Phonons in the FQHE have a 'mass'.
b) Isolated vortices in the FQHE have a *finite* energy.
c) Vortices in the FQHE bind a fractional charge.

Each of these points deserves discussion from several perspectives. The finite phonon mass is, as mentioned previously, connected with the incompressibility of the system at rational fractional filling factors. The fact that isolated vortices have finite energy precludes the possibility of a Kosterlitz-Thouless phase transition but at the same time allows one to condense a group of vortices into a correlated

Laughlin state of their own and thereby generate a hierarchy of incompressible states at arbitrary rational filling factors (Haldane 1983, Chap. VIII, Laughlin 1984a, Chap. VII, Halperin 1984, Girvin 1984a). Another very important consequence of the finite energy and charge of the vortices is that unlike the case of superfluid helium, the dissipation is *finite* (thermally activated) at finite temperatures. This dissipation is similar in some ways to flux-flow dissipation in type-II superconductors (Tinkham 1975). There are, however, important differences. In type-II superconductors the vortices carry no charge and cause dissipation through the following mechanism. The normal core region contains low-lying, essentially gapless excitations which can be excited when the vortex moves past impurities. This causes a viscous damping of the vortex motion. In the FQHE the 'normal core' is microscopic in radius ($\sim \ell$). It is not known for certain at present but seems likely that there do no exist quasi-gapless excitations internal to such vortices (see the numerical work of Haldane and Rezayi 1985a). On the other hand the vortices in the FQHE carry a charge and so can directly conduct electricity. Dissipation by motion of such objects would be analogous to the dissipation due to electrons in a partially filled Landau level (in the absence of the FQHE!) [Note 5]. Another possible source of dissipation involving scattering of thermally activated rotons is discussed in Platzman et al. (1985).

It is interesting to note that properties (a) and (b) listed above occur in superconductors via the Anderson-Higgs mechanism (Anderson 1958, 1963, Higgs 1964a,b, 1966, Moriyasu 1983) due to the coupling of the charged Cooper pairs to the (self-generated) vector potential. Despite some suggestions to the contrary (Friedman et al. 1984a,b) it must be emphasized that this is *not* what is literally occurring here since the magnetic fields generated by the currents in the inversion layer are irrelevant (Haldane and Chen 1984) to the FQHE. However properties (a-c) strongly suggest that the dynamics of interacting electrons in Landau levels can be described by an *effective* field theory in which the Anderson-Higgs mechanism is operative (along with a PT violating term which gives the vortices charge). This intriguing possibility is mentioned again in Chap. X where the first steps toward a Landau-Ginsburg theory of the FQHE will be taken.

9.6 Quasi-Excitons

One can demonstrate that in the limit of large wave vectors ($k \to \infty$) the excited state given in Eq. 3.2 represents a particle-hole pair separated by a distance $k\ell^2$ (Kallin and Halperin 1984, Girvin et al. 1986b). Recall from the discussion in Chap. I that particles in the lowest Landau level do not change momentum by changing their *velocity* (which is always essentially zero in the absence of drift forces) but rather by changing their *position* (and hence their vector potential). The density wave acquires its momentum $\hbar k$ by moving a particle a distance $k\ell^2$ from its original position (in the direction $k \times \hat{z}$). For $k\ell \gg 1$ the particle has been moved a large distance and is uncorrelated with its surroundings so that the expectation value of the energy is large (I say expectation value because the state is not a good eigenfunction of $\mathcal{H}$). This is another interpretation of the fact that Eq. 3.2 is a poor representation of the lowest excited state for $k\ell \gg 1$.

Kallin and Halperin (1984) suggested that there should exist a collective excitation at wave vector k consisting of a bound pair of Laughlin's quasiparticles of opposite charge ($\pm v$). In order to carry the correct momentum these would have to be separated by a distance $d = k\ell^2/v$ rather than $d = k\ell^2$ since the effective magnetic length is larger for the fractionally charged particles. One therefore expects the dispersion relation for this quasiparticle exciton to be:

$$\Delta(k) = \Delta(\infty) - v^3/k\ell . \qquad (9.6.1)$$

Here $\Delta(\infty)$ represents the (dimensionless) energy to create a widely separated quasihole-quasielectron pair and the second term represents the Coulomb attraction between the fractional charges. It is interesting to note that the group velocity $v \equiv \nabla_k \Delta(k)/\hbar$ for the quasi-exciton mode is precisely what one would compute semiclassically from the **E**x**B** drift velocity of the two quasiparticles under the influence of their mutual Coulomb interaction.

This physical picture of the quasi-exciton and the expression in Eq. 5.1 for the dispersion relation are both sensible in the limit $k\ell \gg 1$ where the quasihole and quasielectron are separated by a distance d much greater than their characteristic size ($d \gg \ell$). I am of the opinion (Girvin et al. 1985, 1986b) that for moderate and small values of k the low-lying collective mode crosses over continuously from being a quasi-exciton to being the density-wave magnetoroton-magnetophonon described earlier (with a possible two-roton bound state for very small k). The density-wave is distinct from the quasiexciton because the former is a particle-hole pair excitation while the latter is a quasiparticle-quasihole (vortex-antivortex) excitation.

Laughlin (1985a) holds to a different picture which is based on an explicit variational wave function for the quasi-exciton. He has investigated its energy numerically and the results are consistent with Eq. 5.1 at large k but do not show a roton minimum at finite k (The minimum is at $k = 0$) [see Fig. 7.16]. This is inconsistent with the magnetophonon energy shown in Fig. 9.1 and with the numerical results of Haldane and Rezayi (1985a) discussed earlier. This disagreement is not fully understood at present but appears to be due to numerical difficulties in evaluating the variational energy of Laughlin's quasi-exciton wave function at small wave vectors (see the discussion in Chap. VII).

9.7 Magnetoplasmons and Cyclotron Resonance

As mentioned previously the magnetoplasmon mode is the analog of the classical plasma upper-hybrid mode. In the limit $k \to 0$ Kohn's theorem (Kohn 1961) guarantees that the magnetoplasmon mode will have energy $\Delta^1(0) = \hbar\omega_c$ independent of the particle interactions [Note 1]. For finite wave vectors interactions cause a dispersion in $\Delta^1(k)$ which has been computed in various standard approximations (Chiu and Quinn 1974, Horing and Yildiz 1976, Lerner and Lozovik 1978, Theis 1980, Bychkov et al. 1981, Kallin and Halperin 1984, Hawrylak and Quinn 1985). Microwave and far-infrared experiments (Kennedy et al. 1977, Wagner et al. 1980, Wilson et al. 1981) nominally probe only the $k = 0$ point but in fact yield an indirect measure of the dispersion because weak impurity scattering breaks momentum conservation and allows the photons to couple to modes at all wave vectors (Schlesinger et al. 1984, Kallin and Halperin 1985).

If one uses the SMA excited-state wave function given by Eq. 3.1, the magnetoplasmon dispersion can be computed by methods similar to those employed above in the study of the magnetoroton (MacDonald, Oji and Girvin 1985). If one approximates the required static structure factor by its Hartree-Fock value (which is essentially exact in the extreme quantum limit for $v = 1$) the SMA result for the mode dispersion is identical to the expression obtained by summing ladder diagrams (Lerner and Lozovik 1978, Kallin and Halperin 1984). For the case $v < 1$ perturbative and mean-field methods fail because of the ground-state degeneracy. Ground-state correlations are zeroth order in the interaction and must be explicitly built in by hand. This has been done using the static structure factor from Laughlin's wave function for $v = 1/3$, $1/5$, and $1/7$ (MacDonald, Oji and Girvin 1985). The net effect of correlations is to increase the upward dispersion of $\Delta^1(k)$ in a manner which seems to be consistent with the measured narrowness of the resonance line shape found by Wilson, et al. (1981) in Si inversion layers. Recent work by Rikken et al. (1985a,b) in GaAs suggests that there is a narrowing of the resonance line width at those fractional filling factors at which the FQHE is observed. This lends tentative support to the theory.

An alternative explanation of the data of Wilson, et al. which was developed before the discovery of the Laughlin liquid state, but which may still be relevant, is the theory by Fukuyama and Lee (1978) for the cyclotron resonance of a pinned Wigner crystal. It may well be that disorder (which is particularly significant in Si devices) can stabilize the Wigner crystal in a glassy state which is lower in energy than the Laughlin liquid (Girvin et al. 1986b). This will be discussed further below in the Sec. 9.9.

There are also other puzzles concerning anomalous line shapes. Schlesinger et al. (1984) have suggested that if the magnetoplasmon energy has a minimum at finite k which lies near or slightly below $\hbar\omega_c$ then one could explain the two-peaked structure which they found in the resonance line shape. The SMA theory indicates however that an improved description of the dispersion at fractional v causes $\Delta^1(k)$ to be a monotone increasing function of k eliminating the minimum at finite k found in simpler approximations. In summary, it is safe to say that there are still several interesting unsolved questions in this aspect of the collective-mode problem.

One can of course generalize this discussion to allow for direct cyclotron absorption at harmonics of $\hbar\omega_c$ by substituting P_n ($n > 1$) for P_1 in Eq. 3.1. Kohn's theorem tells us that such transitions will have zero oscillator strength at $k = 0$ but they can be observed at finite wave vectors.

9.8 Other Collective Modes

Up to this point I have neglected spin and possible valley quantum numbers. Including these brings in several interesting possibilities for spin-density and valley-polarization waves (Giuliani and Quinn 1985a,b,c, Rasolt et al. 1985). The $k = 0$ gap for spin waves depends on the Zeeman splitting. The effective g-factor (which is generally small because of the small effective mass) can in principle be varied by tilting the applied field relative to the plane of the inversion layer. The spins can follow the tilt but the orbital motion is confined to the plane so that $\hbar\omega_c$ depends only on the normal component of B. Hence one has in principle some control over the nature of the spin waves.

Unlike GaAs, Si inversion layer states are characterized by an additional valley quantum number (Ando et al. 1982, see also Chaps. I, II). If two valleys are degenerate (which may happen for appropriate crystallographic orientation of the inversion layer, assuming strain effects can be eliminated) and if the ground state is valley-polarized, then gapless valley-polarization waves are expected to exist (Rasolt et al. 1985). Although it has been very difficult to obtain samples of sufficiently high mobility, recent evidence now suggests a very weak FQHE in Si devices (Pudalov et al. 1984a, Gavrilov et al. 1984, 1985, Furneaux et al. 1985). All of these samples have a 100 crystallographic orientation which lifts the valley degeneracy. The full valley degeneracy is expected for the 111 orientation, but unfortunately

Collective Excitations 377

this has never been observed except in heavily annealed, low-mobility samples (see for example, Tsui and Kaminsky 1979). This is another area that deserves further study.

9.9 Role of Disorder

Up to now I have assumed an ideal system with no disorder and have developed an approximate diagonalization of the Coulomb interaction part of the Hamiltonian. Given these results it is now possible to consider the effects of disorder (Laughlin et al. 1985, MacDonald, Liu et al. 1986). It is known that the fractional quantum Hall effect can be observed only in extremely high-mobility, low-disorder samples (see Chap. VI). Even though the disorder is small in some absolute sense it is still significant compared to the very small scale of the experimental excitation gap ($\lesssim 5$ K).

There are two pieces of evidence for the importance of disorder. The first is that the observed activation energy for the dissipation is much lower than the naive theoretical predictions. Assuming that the dissipation is due to thermally activated quasiparticles, the law of mass action (Ashcroft and Mermin 1976) tells us that the dissipation will obey:

$$\rho_{xx} \approx \exp(-\beta\Delta(\infty)/2) \qquad (9.9.1)$$

where β is the inverse temperature and $\Delta(\infty)$ is, as discussed in Sec. 9.6, the energy required to create a widely separated quasielectron and quasihole pair. For $v = 1/3$ we have $\Delta(\infty) \approx 0.1\ e^2/\kappa\ell$ (Haldane and Rezayi 1985, Girvin et al. 1985, 1986b, Morf and Halperin 1986). At a field of $B = 20$ T this corresponds in GaAs to $\Delta(\infty) \approx 22$ K (in temperature units). This is much larger than the experimental value reported in Chap. VI (≈ 5 K). Part of the discrepancy is due to the finite thickness of the inversion layer which softens the Coulomb repulsion between electrons at small distances (Girvin et al. 1986b, Zhang and Das Sarma 1986). This reduces the gap by approximately a factor of two. The remaining discrepancy is still large however and must be presumed to be due to the finite disorder in the system.

The second piece of experimental evidence for the importance of disorder is the existence of finite plateau widths in the FQHE. In analogy with the picture of the integral effect, one presumes that the finite plateau width arises from localization of the quasiparticles which appear in the system when the magnetic field is varied about the center position of the plateau (Laughlin 1983a, Laughlin et al. 1985, Chap. VII). Localization implies that disorder is playing a nontrivial role.

Disorder represents an extremely difficult complication which is only beginning to be addressed. Laughlin, et al. (1985) have proposed that as disorder is increased the quasiparticle gap is renormalized downwards and eventually disappears thereby destroying the FQHE. These authors have developed a scaling picture of this effect based on an analogy with the scaling theory of localization in the integral effect (Pruisken Chap. V, Levine et al. 1983). The physical picture involves a background liquid state plus localized quasiparticles. An alternative possibility which may apply to some or all of the fractional Hall steps is that the FQHE disappears because the gap collapses at a finite wave vector near the roton minimum (Girvin et al. 1986b, MacDonald, Liu et al. 1986). In this case the physical picture of the transition would be rather different. Here one imagines that the Wigner crystal is stabilized in a glassy state by the disorder. Because the Wigner crystal has gapless excitations it is more susceptible to the perturbation caused by disorder than the liquid, and hence could be more stable than the liquid. As mentioned in Sec. 9.7, such a glassy state was discussed some time ago in connection with cyclotron resonance anomalies in Si. I suspect that it will be difficult to make significant progress on these disorder questions without further experimental developments.

9.10 Acknowledgments

The material presented here is based on work done in close collaboration with my friends and colleagues A. H. MacDonald (NRC, Ottawa) and P. M. Platzman (AT&T Bell Laboratories, Murray Hill).

9.11 Notes

Note 1. The total oscillator strength (summed over all Landau levels) is $\hbar^2 k^2/2m$. The arguments used to prove Kohn's theorem (Kohn 1961) show that for small k all of the oscillator strength is absorbed by the transition to the next higher Landau level. Hence $\bar{f}(k)$, the intra-Landau level contribution to the oscillator strength, must vanish faster than k^2 and it is straightforward to show that in fact $\bar{f}(k) \approx |k|^4$.

Kohn's theorem implies that the magnetoplasmon (cyclotron resonance) mode at $k=0$ has an energy of precisely $\hbar\omega_c$ unless there are long-range forces present. The Coulomb interaction in two dimensions is not sufficiently long-ranged to perturb the magnetoplasmon because the plasma frequency in the absence of a B field vanishes for $k \to 0$.

Note 2. The first integral in Eq. 5.5 is along the perimeter of the region Λ. The second uses Stokes theorem to relate this to the integral of the curl of $\langle J(r) \rangle$ over the interior of Λ.

Note 3. I emphasize that this divergence has nothing to do with the core region discussed above but rather with the very slow r^{-1} falloff of the current density at large distances from the core.

Note 4. The density of particles is $n = \nu/2\pi\ell^2$. The density of flux quanta is $n_B = B/(hc/e) = 1/2\pi\ell^2$. Hence $n/n_B = \nu$. If B is in the $-\hat{z}$ direction, the sign of the vortex charge will be as in Eq. 5.11.

Note 5. In a superconductor there are also quasi-particles (broken Cooper pairs, as opposed to vortices) which have a *finite* activation energy (given by the superconducting gap). These do not cause dissipation since they are 'shorted out' by the infinite conductivity associated with the supercurrent. This does *not* occur in the FQHE because the supercurrent produces a finite Hall electric field which can pump energy into the (charged) quasiparticles. A slicker way to say this is that even though the supercurrent gives zero dissipation ($\rho_{xx}=0$), the conductivity σ_{xx} is zero not infinite. Hence the supercurrent does *not* short out the quasiparticles. Assuming that the total conductivity tensor is the sum of the supercurrent and quasiparticle conductivities gives ρ_{xx} proportional to the σ_{xx} contributed by the quasiparticles.

9.12 Problems

Problem 1. Show that if the ground state has uniform density then the variational excited state given in Eq. 2.1 is in fact orthogonal to the ground state.

Problem 2. Show that in the presence of a uniform magnetic field, translations of the system are equivalent to gauge transformations. Prove this generally without relying on any particular choice for the initial gauge.

Part C
The Quantum Hall Effect

CHAPTER 10

Summary, Omissions and Unanswered Questions

STEVEN M. GIRVIN

10.1 Integer Effect at Zero Temperature

We have seen in the previous chapters that the IQHE is, in its essence, a single-particle localization phenomenon associated with the finite mobility gap between extended states in different Landau levels. The gap is a single-particle effect in that it is associated with the kinetic energy. The only many-particle effect which is essential to include is the Pauli exclusion principle which allows the mobility gap to produce dissipationless current flow at zero temperature [Problem 1]. It is this lack of dissipation which is the key to the effect [Note 1]. Theoretical work on this problem has largely centered on proving the fundamental quantization theorem:

$$\sigma_{xx} = 0 \implies \sigma_{xy} = ie^2/h, \qquad (10.1.1)$$

where σ is the macroscopic conductance. The original approach taken by Laughlin (1981) has provided a new language with which to address this problem and has led ultimately to the very beautiful picture that the macroscopic conductance is quantized because it is a topological invariant (Thouless Chap. IV).

The importance of topological features of the problem also appears in the field-theoretic description of localization (Pruisken Chap. V). The localization problem is rather different in the presence of a quantizing magnetic field because of the existence of nonperturbative topological terms in the action of the effective field theory. The physical origins of these differences in the localization problem have been made clear in the work of Prange (1981, Chap. III) and Joynt and Prange (1984).

Our basic theoretical understanding of the effect at zero temperature in the presence of disorder thus appears to be well established. There remain however some unsolved problems. In particular there does not exist at present a good understanding of the width and shape of the Hall steps (see Chap. II). This in turn is related to a poor understanding of the density of localized states between the Landau levels. Recent heat-capacity (Gornik et al. 1985a,b), magnetization (Eisenstein et al. 1985a,b) and magnetocapacitance (Kaplit and Zemel 1968, Goodall et al. 1985, Smith and Stiles 1984, Smith et al. 1985a,b) measurements indicate a significant density of localized states. The heat capacity and magnetization measurements have been interpreted assuming a constant (magnetic-field-independent) carrier number but the validity of this approximation has not been firmly established. In addition macroscopic inhomogeneities may well be important. I believe that these are potentially serious sources of error which need to be investigated. Recent work by Das Sarma and Stern (1985) indicates that the transport scattering time may be very much larger than that deduced from the Landau level broadening. This complicates the connection which exists (in principle) between mobility measurements and level broadening measurements.

10.2 IQHE at Finite Temperatures

In contrast to the zero-temperature limit, the situation at finite temperatures is not so well understood. Several finite temperature effects were discussed in Chap. II. Among the most important of these is the thermal activation of the dissipation. Assuming that the dissipation is due to activation of carriers

Summary, Omissions and Unanswered Questions

across the mobility gap, one expects

$$\sigma_{xx} \sim e^{-\beta\hbar\omega_c/2} \qquad (10.2.1)$$

at the center of the Hall step. The predicted value of the activation energy is in fairly good, but not excellent, quantitative agreement with experiment (Cage Chap. II). The origin of the discrepancies is not known and I believe this may bode ill for quantitative prediction of the activation energy in the more difficult case of the FQHE.

As discussed in Chap. II, the activated conduction freezes out at low temperatures and is replaced by what appears to be variable-range hopping. There has been only limited theoretical work on this phenomenon (Ono 1982, Wysokinski and Brenig 1983) and more needs to be done. Despite the fact that the quantization of σ_{xy} at zero temperature is now well understood, we have very little understanding of the corrections at finite temperature. It was noted in Chap. II that one often sees very wide, flat Hall steps for which the Hall resistance is in error by an amount *linearly* proportional to the dissipation. This is a much larger and more important source of error than the original estimates (Laughlin 1981) which suggested that the error in quantization should be *quadratic* in the dissipation. The linear relationship (2.10.1) violates time-reversal symmetry unless the coefficient s is odd under time-reversal. As we shall see below this gives a hint as to the origin of the effect. Consider a model in which the conductivity tensor contains two contributions:

$$\sigma = \sigma_0 + \delta\sigma . \qquad (10.2.2)$$

The first is the ideal conductivity tensor:

$$\sigma_0 = \frac{ie^2}{h} \begin{bmatrix} 0 & 1 \\ -1 & 0 \end{bmatrix}, \qquad (10.2.3)$$

and the second is the conductivity due to an additional mechanism which is assumed to act in parallel with the ideal. This alternate channel could represent variable-range hopping or a physically separate process such as conduction on the sample surface. Inverting the conductivity tensor and expanding to lowest order in $\delta\sigma$ yields:

$$\delta\rho_{xy} \sim - s\, \delta\rho_{xx}, \qquad (10.2.4)$$

with

$$s = - \frac{\delta\sigma_{xy}}{\delta\sigma_{yy}}. \qquad (10.2.5)$$

Thus s is the tangent of the Hall angle in the alternate channel. As noted in Chap. II, s varies between roughly 0.01 and 0.5. If one neglects the possibility of a finite Hall angle in the alternate channel, then the error in the Hall resistivity becomes quadratic in the dissipation. Wysokinski and Brenig (1983) have concluded that the contribution to s from variable range hopping is in fact negligibly small, but this is clearly an area which requires further work both experimentally and theoretically. It would be particularly useful to have a quantitatively accurate realization of the scaling theory (Chap. V) which could be applied by assuming a cutoff of the scaling at the inelastic scattering length (if one had available a quantitative theory for computing that length!).

10.3 Metrology

The IQHE represents a revolutionary development in electrical metrology. Many fundamental constant determinations directly or indirectly involve high-precision electrical measurements and so the IQHE is of great importance in this area. In addition to giving a direct determination of the fine structure constant (Chap. II), the IQHE makes an indirect improvement to fundamental constant determinations in general.

As noted in Chap. II there is excellent evidence that, at sufficiently low temperatures, the Hall resistance of a given plateau is device-independent to within the experimental precision ($\sim 2\times 10^{-8}$). This strongly suggests that the quantization condition (10.1.1) is in fact realized at zero temperature. To prove this however requires more than just relative

Summary, Omissions and Unanswered Questions

agreement between different devices. One must make an *absolute* determination of the Hall resistance in units of the SI ohm. This is a much more difficult procedure but present results indicate (Taylor 1985) that the value of the fine structure constant obtained with the IQHE is consistent with previous values to within the experimental uncertainty of ~0.3 ppm. It is expected that this uncertainty will be significantly reduced within the next few years.

10.4 Basic Picture of the Fractional Effect

There now exists a rather well-established 'standard picture' of the FQHE based on Laughlin's ground-state and quasiparticle wave functions and their hierarchical generalizations (Chaps. VII and VIII). While there has been as yet no direct experimental confirmation for the hierarchy, the sequences of fractional states which have been found so far do not violate the predictions of the hierarchical model. Haldane (Chap. VIII) and Trugman and Kivelson (1985) have pointed out that the Laughlin state is an exact eigenstate for short-range repulsive potentials. One expects that the long-range part of the potential can be treated in a perturbation expansion which has a *finite* radius of convergence due to the excitation gap generated by the short-range potential. Numerical evidence from exact diagonalizations of the Hamiltonian for small systems strongly suggest that the radius of convergence is sufficiently large that the Laughlin state is a good approximation to the actual ground-state eigenfunction of the full potential.

One area in which the basic picture of the FQHE is incomplete is in the understanding of the role of disorder. Localization of quasiparticles by the disorder is believed to account for the finite step-width in the FQHE in a manner analogous to the IQHE. The observed activation energy for dissipation is approximately a factor of 4 smaller than the value predicted by the simplest theoretical models. About half of the discrepancy is due to finite thickness effects which soften the Coulomb repulsion at short distances. As discussed in Chap. IX the remaining factor of 2 is presumably due to disorder. Very little is understood at this point, however.

Some interesting alternatives to the standard picture have recently been proposed. Chui, et al. (1985) have proposed a ground-state wave function which gives a particle distribution corresponding to a finite-temperature two-dimensional classical solid with a log r interaction. The ground-state energy is comparable to that of the Laughlin liquid, and because of the lack of true long-range spatial order, the phonons are able to circumvent Goldstone's theorem and have a *finite* gap. Another very interesting proposal is that of Kivelson, et al. (1985) who suggest that large ring exchanges in the Wigner crystal can produce the required cusp in the free energy at rational fractional filling factors (see also Chui 1985a, Baskaran 1986). They map the problem onto a Coulomb gas which resembles the 2DOCP associated with the Laughlin wave function.

While the full implications of these proposals are not yet clear, I believe that there are two reasonable interpretations. The first is that the true ground state *does* indeed have some nontrivial solid-like symmetry which could easily be missed in the numerical calculations because of finite size effects. The second possibility is that the suggested alternatives are approximations to states which are not significantly different from Laughlin's. In either case these suggestions provide a new and useful perspective from which to view the problem.

It would be very helpful to have numerical calculations on larger systems. Chui (1985b) has used a real-space-renormalization-group method which looks promising but needs further investigation to prove its validity. It seems likely that present computer technology will not allow an exact solution of the ground state for N much larger than about ten. One could however attempt a restricted-basis-set calculation. In view of the success of the single-mode approximation discussed in Chap. IX, basis states consisting of multiple phonons and magnetorotons on top of Laughlin's ground state should prove useful. This would be equivalent to adding variational freedom to Laughlin's wave function ψ by writing states of the form:

Summary, Omissions and Unanswered Questions

$$\Psi = \left\{ 1 + \sum_q \alpha(q)\, \Lambda_q^\dagger \Lambda_q + \sum_k \sum_q \beta(k,q)\, \Lambda_{k+q}^\dagger \Lambda_k \Lambda_q \right.$$

$$\left. + \sum_p \sum_k \sum_q + \ldots \right\} \psi \qquad (10.4.1)$$

where the notation of Chap. IX has been used. This basis set is complete (actually over-complete) as can be shown by the following (rather imprecise) argument. The Hamiltonian can be expressed solely in terms of the projected density operator as (Girvin et al. 1985, 1986b):

$$\mathcal{H} = \sum_q v(q)\, \Lambda_q^\dagger \Lambda_q + \text{constant} . \qquad (10.4.2)$$

The time evolution of the projected density operator depends on the commutator $[\mathcal{H}, \Lambda_q]$. We can now take advantage of the fact that the Λ operators form a closed algebra under the commutation relation (Girvin et al. 1986b):

$$[\Lambda_k, \Lambda_q] = F_{kq}\, \Lambda_{k+q} , \qquad (10.4.3)$$

where F_{kq} is a complex number given by

$$F_{kq} = 2i e^{k \cdot q} \sin(k \times q) \qquad (10.4.4)$$

(where the component of the cross-product normal to the plane is implicitly assumed). This implies that multi-phonon and -roton states can scatter or decay *only* into other multi-phonon and -roton states. The basis given by states of the form in (10.4.1) is therefore complete. This should in fact be a very efficient basis in which to work. A good first project would be the two-roton bound-state problem discussed in Chap. IX.

10.5 Remarks on the Neglect of Landau Level Mixing

Almost all theoretical calculations to date have neglected the mixing between different Landau levels. At first sight this seems to be reasonable in the high field limit where the scale of the Coulomb energy $e^2/\kappa\ell$ is much smaller than the Landau level spacing $\hbar\omega_c$ (a limit not frequently realized in practice). However in making a perturbation expansion in the small parameter $r \equiv (e^2/\kappa\ell)/\hbar\omega_c \sim B^{-1/2}$ one must not forget that as the *spacing* between Landau levels rises so does the *degeneracy* (each going linearly with B).

As an extreme example consider the single-particle eigenfunctions in the presence of a delta-function potential. We know that the wave functions must have a cusp at the location of the delta function. This can be achieved only by significant mixing of infinitely many Landau levels. A smooth potential on the other hand, would not have this problem.

MacDonald (1984b) has shown that for the case of the Hartree-Fock Wigner crystal, Landau-level mixing has little effect on the ground-state energy. Yoshioka (1984b, 1985a,b) has investigated the failure of particle-hole symmetry between the FQHE states at filling factor v and $1 - v$ due to Landau-level mixing. He finds a weak (~10%) effect which is qualitatively inconsistent with the large asymmetry ($\Delta_{1/3}/\Delta_{2/3} \sim 3$) found experimentally by Wakabayashi, et al. (1985). It seems likely that finite thickness and B-field-dependent disorder effects are dominating present measurements.

In sum, it appears that neglect of Landau-level mixing is a reasonable approximation within the present state-of-the-art of experiment and theory. The reader is directed to Chap. 7.6 for a lucid discussion of the exactness of the fractional quantization in the presence of Landau-level mixing.

10.6 Wanted: More Experiments

The experimental discoveries of the integral and fractional QHE were surprises which caught the theoreticians off guard. The theoretical picture has developed considerably in recent years however and further

Summary, Omissions and Unanswered Questions

progress is now dependent in part on new measurements. The basic transport measurements discussed in Chap. VI are now fairly complete (although the results are not yet totally understood). The 'wish list' for new measurements would include:
— microwave conductivity
— more cyclotron resonance and spin resonance
— Raman scattering
— thermoelectric power in the FQHE regime
— direct tunneling into the inversion layer
— direct observation of the structure factor
— more FQHE experiments in silicon devices.

It would be very nice if one or more of these techniques could directly observe the excitation gap, the liquid-solid transition, and related phenomena. The problem is that there are so few electrons in the inversion layer that signals are generally swamped by bulk effects. Hopefully some clever ideas will be forthcoming.

10.7 Towards a Landau-Ginsburg Theory of the FQHE

The FQHE is a truly remarkable many-body phenomenon. As discussed in Chap. IX, the FQHE bears some resemblance to superconductivity and superfluidity, but, as I would now like to discuss, the historical development of our theoretical understanding followed quite different routes in the two cases. Empirical Landau-Ginsburg theories of superconductivity existed for many years without any understanding of the microscopic origins of the effect. Following the development of the microscopic BCS theory, the macroscopic Landau-Ginsburg picture could be formally derived and justified. The situation is just reversed in the case of the FQHE. The explicit ground-state and excited-state wave functions displayed in Chaps. VII-IX are deceptively simple in appearance but profound in their implications. I believe that a full understanding of these implications has not yet been achieved. It is as if, prior to the development of the Landau-Ginsburg theory of superconductivity, one had been given an explicit real-space representation of the BCS wave function. Given the appropriate Hamiltonian one could find by numerical calculation that the wave function had a favorable variational energy but all of the beauty of the underlying physics would remain hidden and unrealized.

A related point is, I believe, that one ought to be able to map the problem of the partition function in the FQHE onto an effective field theory with some unusual and interesting topological properties. This has not yet been achieved.

With these thoughts in mind I would like to close this Chapter with some speculative remarks on how one might arrive at a Landau-Ginsburg theory (or perhaps one should say a London theory) of the FQHE. The best starting point is with the superfluid analogies drawn in Chap. IX. Recall that the important points were that:

(a) the phonons have a *gap*
(b) isolated vortices have *finite* energy
(c) vortices bind a fractional charge.

I remarked in Chap. IX that points (a) and (b) were reminiscent of the Anderson-Higgs mechanism in superconductors in which the charge density is coupled to the electromagnetic field. One might therefore consider a free-energy functional of the form:

$$S = \int d^2r \; |(-i\nabla + \mathbf{A})\psi(r)|^2 \; . \qquad (10.7.1)$$

Here ψ is a complex order parameter with $\langle \rho(r) \rangle = |\psi|^2$ being the local particle density. As remarked in Chap. IX, the Anderson-Higgs mechanism is not *literally* occurring since the magnetic flux generated by currents in the inversion layer is negligibly small. The fictitious vector potential $\mathbf{A}$ in (10.7.1) therefore is *not* the physical vector potential.

The next step is to give an interpretation of $\mathbf{A}$ and decide how to relate it to ψ. To do this consider the following argument. We want a result which favors a particular filling factor v_0. At $v_0 = 1/M$ the system can condense into a state with microscopic wave function ψ_M, the Mth Laughlin state. This is allowed because the Laughlin state has its filling factor precisely pinned at $v_0 = 1/M$. If the density deviates from this value, the system is *frustrated* because it must necessarily be forced out of the Laughlin state. I propose to interpret this frustration literally in the statistical mechanical sense. The way to represent frustration in a Landau-Ginsburg model is to cause **A** to have a nonzero curl (Villain 1977, Brézin et al. 1985, Halsey 1985). This is because, as Brézin, et al. note, it is impossible to make the gradient term in (10.7.1) vanish everywhere. One cannot minimize the gradient energy by imposing the 'obvious' phase relation,

$$\psi(r) = \psi(0) \exp\left\{-i\int_0^r dR \cdot A(R)\right\} , \qquad (10.7.2)$$

because the line integral is path-dependent and the phase is thus ill-defined. The energy is forced to rise in the presence of a curl of **A**; i.e., the system is frustrated by its inability to make local phase changes which globally cancel out the effect of **A**. The lattice version of this argument yields precisely the usual picture of frustrated spins on a lattice (Villain 1977, Halsey 1985).

Let us therefore determine **A** by choosing the degree of frustration to be linearly proportional to the local deviation in the density away from the ideal density $n_0 = v_0/2\pi$:

$$\nabla \times A = 2\pi\alpha(|\psi|^2 - n_0) . \qquad (10.7.3)$$

where α is a constant and where by '$\nabla \times$' I mean the component of the curl normal to the plane. Note that this somewhat peculiar looking relationship builds a handedness into the theory [Note 2].

We can now examine the self-consistent solutions of Eqs. 7.1 and 7.3. The simplest is for the case $\rho = n_0$ for which we may take:

$$\psi = \sqrt{n_0} \, e^{i\varphi} , \qquad (10.7.4)$$

where φ is an arbitrary constant phase. Here the 'flux density' $\nabla \times A$ vanishes everywhere and the free energy takes on its minimum possible value (zero). One can thus interpret this *macroscopic* unfrustrated

Landau-Ginsburg state as representing the *microscopic* Laughlin state.

Consider now the simplest states which have the form of a vortex (located at the origin):

$$\psi(r) = \sqrt{n_0}\, e^{im\theta} f(r) , \qquad (10.7.5)$$

where θ is the azimuthal angle, m is an integer quantum number, $f(r)$ is a (real) radial function which approaches unity at large r, and I have dropped the arbitrary phase factor which was displayed in Eq. 7.4. It is well known that vortex solutions of theories of the form given in Eq. 7.1 exhibit the remarkable topological feature of flux quantization. To see how this comes about, note that in the absence of the vector potential, the kinetic energy density

$$T(r) = |\nabla\psi(r)|^2 \qquad (10.7.6)$$

satisfies

$$T(r) = n_0 \left\{ \frac{m^2}{r^2} f^2(r) + \left[\frac{\partial f}{\partial r}\right]^2 \right\}. \qquad (10.7.7)$$

For large r the radial function f approaches a constant (unity) and we obtain the asymptotic result

$$T(r) \sim n_0 \frac{m^2}{r^2} . \qquad (10.7.8)$$

This slow falloff with distance causes the total kinetic energy to diverge logarithmically with the size L of the system:

$$S \sim \ln L . \qquad (10.7.9)$$

Recall that this occurs in the assumed absence of the vector potential. The system will therefore pay any *finite* energy cost required to generate a vector potential to cancel out the kinetic energy divergence. The energy will be finite only if the 'current density' vanishes (rapidly enough) for large r:

$$J(r) = \frac{1}{2}\{\psi^*(r)(-i\nabla + A)\psi(r)$$
$$+ \psi(r)(+i\nabla + A)\psi^*(r)\} \longrightarrow 0 . \qquad (10.7.10)$$

Now consider a contour Γ encircling the vortex at a large distance. We have from (10.7.5) the topological quantization condition

$$\frac{1}{n_0} \oint_\Gamma dr \cdot \{\psi^*(-i\nabla)\psi\} = 2\pi m , \qquad (10.7.11)$$

where m must be an integer due to the single-valuedness of ψ. Using this in expression (10.7.10) yields

$$\oint_\Gamma dr \cdot A(r) = -2\pi m . \qquad (10.7.12)$$

Application of Stoke's theorem produces the flux quantization condition:

$$\int_\Gamma d^2r \; \nabla \times A(r) = -2\pi m , \qquad (10.7.13)$$

where the integral is now over the (large) area enclosed by the contour Γ.

It should be noted that this argument is quite general and does not make reference to the specific relationship between A and ψ. In the case of a superconductor the flux is determined by the *current* associated with ψ. In the present case the flux density is related to the *charge* associated with ψ. Substitution of Eq. 7.3 into (10.7.13) yields

$$\int_\Gamma d^2r \; (|\psi|^2 - n_0) = -\frac{m}{\alpha} . \qquad (10.7.14)$$

Note that the integrand is the deviation in the local charge density. Hence this remarkable result tells us that an isolated vortex carries a total charge m/α whose sign depends on the sign of the vorticity. Choosing the parameter α to be $1/v_0$ yields a result consistent with interpretation given in Chap. IX in which vortices were shown to correspond to Laughlin's fractionally charged quasiparticles.

Central to the above results is the equation of motion given by (10.7.3). This could be (for example) obtained from an action of the form

$$S(\psi,\phi,\mathbf{A}) = \int d^2r \, |(-i\boldsymbol{\nabla} + \mathbf{A})\psi(r)|^2 + i\int d^2r \, \phi(\psi^*\psi - n_0)$$

$$- i\frac{\theta}{2\pi}\int d^2r \, \frac{1}{4\pi}(\phi\boldsymbol{\nabla}\times\mathbf{A} + \mathbf{A}\times\boldsymbol{\nabla}\phi) \qquad (10.7.15)$$

where the 'vacuum angle' θ is given by $\theta = 2\pi/\alpha$ and the last term on the right is a 'Chern-Simons' topological contribution to the action. We can interpret the Lagrange multiplier field ϕ as a scalar potential and note that both the charge density $\psi^*\psi$ and the 'flux density' $\boldsymbol{\nabla}\times\mathbf{A}$ couple to the potential. This is what causes flux tubes to carry charge in this model (similar models have been discussed by field theorists, see Niemi and Semenoff 1983, de Vega and Schaposnik 1986, Schonfeld 1981, Jackiw and Templeton 1981, Deser et al. 1982a,b).

In addition to having the correct fractional charge these vortices also have a finite energy. The binding of the fractional charge by the vortex is analogous to the Anderson-Higgs mechanism discussed above and causes the current density to be screened out at large distances from the vortex in such a way that conditions (b) and (c) listed above are satisfied.

We now must verify condition (a). The phonon gap is closely related to the incompressibility of the system, which in turn requires the existence of a cusp in the free energy at filling factor v_0 (Halperin 1983, Morf and Halperin 1986). Consider the case where the average density n differs from n_0 and take as a trial ground-state solution

$$\psi(r) = \sqrt{n} \, . \qquad (10.7.16)$$

Substitution of this into expression (10.7.3) for the vector potential yields

$$\boldsymbol{\nabla}\times\mathbf{A} = 2\pi\alpha(n - n_0) \, , \qquad (10.7.17)$$

which corresponds to a uniform magnetic field. The trial solution (10.7.16) is not self-consistent because of the excess kinetic energy represented by the presence of the uniform magnetic field. The system will try to minimize the frustration by collecting the flux into uniformly spaced, quantized fluxoids. An equivalent way to state this is that the system will accommodate the deviation in the density away from the ideal value, n_0 in the form of charged quasiparticles. The number of quasiparticles formed is linearly proportional to the deviation in the average density away from n_0. This means that there will be a cusp in the free energy at the density n_0 with a discontinuity in the slope of (Halperin 1983)

$$\left.\frac{dS}{dn}\right|_{n_0}^{+} - \left.\frac{dS}{dn}\right|_{n_0}^{-} = \alpha(\Delta_+ + \Delta_-) \qquad (10.7.18)$$

where Δ_+ and Δ_- are the energies of the positive and negative ($m = \pm 1$) vortices (quasiparticle energies). This is precisely the desired result and can be shown to cause condition (a) to be satisfied. One may say that the analog of flux exclusion in a superconductor is charge exclusion (incompressibility) in the FQHE.

Hence this deceptively simple looking Landau-Ginsburg theory seems to have most of the correct phenomenology built into it. The only not fully satisfactory points are that the charge of the quasiparticles ($1/\alpha$) and the critical density n_0 are built in by hand. A complete theory would automatically yield cusps in the free energy at the observed sequence of rational fractional filling factors. An additional limitation of the present simplified model is that while there is a gap for amplitude fluctuations in ψ, the phase fluctuations remain gapless. The cure for these two problems will undoubtedly require the addition of quantum fluctuations (in imaginary time) to the static action given in (10.7.15).

What could one do with such a Landau-Ginsburg theory? In addition to providing a deeper physical insight into the FQHE, such a theory would allow one to address specific questions such as the nature of the liquid-solid transition and the effects of disorder. By adding a perturbation to the action of the form

$$V = \int d^2r \, v(r) \, |\psi(r)|^2 , \qquad (10.7.19)$$

one could study the collapse of the excitation gap and the questions raised in Chap. IX concerning whether, as disorder is increased, a Wigner-glass transition occurs or whether there is a quasiparticle delocalization transition. One could also study the gap-reducing screening effects due to localized quasiparticles. A Landau-Ginsburg theory would allow much easier semiquantitative calculation than would a microscopic theory.

The question now arises of how one might provide a microscopic justification for such a Landau-Ginsburg theory. Following the discussion concerning Eq. 4.1 one could consider constructing microscopic wave functions of the form

$$\Phi = \exp\left\{ \sum_k \psi_k \Lambda_k \right\} \psi_{\text{ground}} \qquad (10.7.20)$$

where ψ_k would be identified with (the log of the Fourier transform of) the Landau-Ginsburg order parameter. An approximate expression for the partition function would then be:

$$Z = \int \mathcal{D}\psi \, \frac{\langle \Phi | e^{-\beta \mathcal{H}} | \Phi \rangle}{\langle \Phi | \Phi \rangle} , \qquad (10.7.21)$$

where the Hamiltonian $\mathcal{H}$ is given by (10.4.2) and the functional integral is over the field ψ with some appropriate measure. Integrating out the electron coordinates would yield an effective action for ψ. A related but more rigorous formal result can be obtained by applying the Hubbard-Stratonovich transformation (Schulman 1981) in the exact expression for the partition function. The mechanism by which a nontrivial action like the one in Eq. 7.1 could appear can be seen in Eqs. 4.2-4. The required handedness in the dynamics is built into the commutation relation for the projected density operators. It is quite possible that the situation is similar to that of the field theory of the IQHE discussed in Chap. V by Pruisken. The fact that different components of the dynamical

Summary, Omissions and Unanswered Questions

momentum do not commute leads to nontrivial topological terms in the effective action. I believe that these points are worth pursuing.

It is amusing in this context to consider the Schrödinger equation (without projection onto a particular Landau level) satisfied by a wave function which has had the Laughlin wave function factored out. That is if we write the full wave function as

$$\Psi = \Phi \Psi_m , \qquad (10.7.22)$$

then Φ satisfies the following Schrödinger equation:

$$-\sum_{j=1}^{N} \left\{ \tfrac{1}{2}(\nabla_j^2 \Phi) + i(\nabla_j \Phi)\cdot(-i\mathcal{E}_j - \mathcal{A}_j) \right\} + V\Phi = (E - \tfrac{1}{2})\Phi \qquad (10.7.23)$$

where E is the energy eigenvalue and V is the interaction potential. The quantity $\mathcal{A}_j$ is associated with the gradient of the phase of the Laughlin liquid wave function and satisfies

$$\nabla_j \times \mathcal{A}_j = 2\pi \left[m \sum_{k \neq j} \delta^2(r_j - r_k) - \tfrac{1}{2\pi} \right] . \qquad (10.7.24)$$

Hence $\mathcal{A}_j$ can be interpreted as a 'vector potential' computed as if each particle was bound to a flux tube carrying m flux quanta and was moving in a uniform background flux density. Likewise, the quantity $\mathcal{E}_j$ is the 'electric field' due to the gradient in the amplitude of the Laughlin liquid and is proportional to the 'electric field' in the analog plasma system (two-dimensional charges plus background) and satisfies

$$\nabla_j \cdot \mathcal{E}_j = 2\pi \left[m \sum_{k \neq j} \delta^2(r_j - r_k) - \tfrac{1}{2\pi} \right] . \qquad (10.7.25)$$

Eqs. 7.23-24 suggest the sense in which there is a flux density associated with the particle density as was assumed in the frustration *ansatz* of Eq. 7.3.

I would like to close by pointing out to the reader certain additional references in the field-theory literature which may be relevant to this problem.

Axial anomalies in 2+1-dimensional field theories are discussed by (Niemi and Semenoff 1983, Hagen 1984, 1985a,b,c). These papers also contain other useful references. Other papers of particular interest by Cardy and Rabinovici (1982) and Cardy (1982) discuss the phenomenon of *oblique confinement*. The system they study contains an electric and a magnetic condensate and is characterized by a so-called θ parameter. The system exhibits phase transitions to special vacuum states at rational fractional values of $\theta/2\pi$ and the elementary excitations consist of objects with the same ratio of electric to magnetic charge as the condensate. The similarities to the FQHE are quite striking and should be pursued.

One of the problems slowing progress is the rather different languages used by workers in the condensed-matter and field-theory communities. I hope that some of the discussion presented here will lead to attempts to enhance the communication between the two groups and to some substantive progress on what I consider to be the remaining outstanding problem connected with the FQHE.

10.8 Acknowledgments

The author is grateful to S. B. Libby for bringing to his attention the phenomenon of oblique confinement and for emphasizing its possible relevance to the FQHE. The author would also like to express his thanks for useful conversations to numerous others including A. H. MacDonald, G. Semenoff, Z. Tesanovic, P. A. Lee, D. R. Nelson and P. W. Anderson.

10.9 Notes

Note 1. It is interesting that the one other condensed-matter phenomenon yielding extremely accurate quantization of a measurable parameter is the Josephson effect which also depends for its success on a complete absence of dissipation.

Note 2. Choosing the sign of α to depend on the sign of the external B field restores parity conservation.

Summary, Omissions and Unanswered Questions

10.10 Problems

Problem 1. Ordinary insulators and semiconductors also have a gap at the Fermi level but they do not exhibit the QHE (actually they do satisfy the quantization relation but the integer quantum number is zero). In fact they do not conduct at all. What is the difference between the gap in an ordinary insulator and the gap in a QHE device?

APPENDIX

Recent Developments

STEVEN M. GIRVIN

A.1 Introduction

Only a few short years ago a GaAs heterostructure with a mobility of 10^5 cm^2/Vs was considered very high quality. Today commercial MBE machines are producing samples with mobilities on the order of 10^7 cm^2/Vs. This remarkable increase in sample quality is leading to new discoveries at ever lower temperatures and finer energy scales. For the first time, the energy gap for the strong fractions is not dominated by disorder and agrees rather well with theoretical predictions (Willet et al. 1989).

The old odd-denominator rule has been overturned by the discovery of a subtle, nearly gapless, FQHE state at $\nu = 5/2$ which is believed to be associated with a spin unpolarized ground state. Other evidence is beginning to accumulate that even for odd denominators, spin-reversed states may be more prevalant than previously realized.

Experimental evidence for the magnetic-field-induced Wigner crystal has been reported by Andrei et al. (1988). This much sought-after effect is very exciting (and also has generated some controversy (Störmer and Willett 1989, Andrei et al. 1989). The crystallization occurs at filling factors somewhat higher than expected theoretically (Levesque et al. 1984, Lam and Girvin 1984, 1985). Paradoxically however, a Hall state at even lower filling, $\nu = 1/7$ was reported in the same year (Goldman *et al.* 1988, Wakabayashi *et al.* 1988). Willett *et al.* (1988) have also found interesting behavior in the termination of the FQHE series at low fillings. It seems clear that this will continue to be an important direction to pursue.

One other positive development is a new generation of experiments which involve measurements of quantities other than standard transport. Of particular interest are light scattering (Goldberg *et al.* 1988, Pinczuk *et al.* 1988) and luminescence measurements (Kukushkin and Timofeev 1986, 1988a, 1988b, Heiman *et al.* 1988).

These experimental advances are beginning to cause theorists some difficulty, since numerical studies of the competition among possible ground states are limited in the number of particles that can be simulated. This in turn limits the smallest resolvable energy scale. The energy resolution is barely adequate for the strongest members of the hierarchy of fractional states. With the addition of the spin degree of freedom and the interest in going to low filling factors, the dimension of the appropriate state space has become excessively large.

At the same time that progress towards samples with negligible disorder has been made, interesting new developments have occurred in the study of disorder-induced localization of carriers in the integer and fractional effects. Wei *et al.* (1988) have developed a novel method to analyze the scaling behavior of the conductivity associated with the power-law divergence of the localization length at the center of the Landau level (Pruisken 1988). Clark *et al.* (1988) have claimed that the charge of the quasiparticles in the FQHE can be for the first time directly measured from the analog of the Mott minimum metallic conductivity in the presence of disorder.

Finally the integer and fractional effects have been observed in unusual geometries such as ultra-narrow channels (Chang *et al.* 1988b, Smith *et al.* 1988, Roukes *et al.* 1987), and in samples with abrupt changes in density (Haug *et al.* 1988, Washburn *et al.* 1988). The Landauer approach to transport has proved fruitful in analysis of these systems and this experimental and theoretical work has brought together universal conductance fluctuations, the quantum Hall effect and the Aharonov-Bohm effect (Chang *et al.* 1988a, Mankiewich *et al.* 1988, Timp *et al.* 1989, Ford *et al.* 1989). Another development indirectly related to all this is the observation of universal two-point resistance in narrow channel devices (van Wees *et al.* 1988, Wharam *et al.* 1988).

Aside from the great progress within the field itself, one of the most rewarding developments for theorists is the oppurtunity to see the beginnings of cross-fertilization between ideas developed in the QHE and those in other sub-fields such as high-temperature superconductivity and in separate areas such as quantum field theory.

The remainder of this Appendix is organized into sections each of which gives a brief discussion of recent developments in the areas of: off-diagonal long-range order in the FQHE, ring-exchange theories, chiral spin states and high-temperature superconductivity, even-denominator and spin-reversed FQHE states, and the spin-reversed hierarchy. The last section presents a short summary of the items covered.

A.2 Off-Diagonal Long-Range Order

In Chap. 10 the author speculated on the possible form of a Landau-Ginsburg theory of the FQHE. The basic idea is that the system is frustrated if the density deviates from the quantized value $\nu = 1/m$. This frustration can be represented by a gauge field which acquires a curl proportional to the density deviation. This in turn costs gradient (kinetic) energy which is supposed to represent the purely potential energy cost in the underlying microscopic model. This simple picture produces all the correct phenomenology—an incompressible state with vortex excitations carrying fractional charge. The theory contains an adjustable parameter θ (the 'vacuum angle') which controls the vortex charge.

A great deal of progress has been made following the original suggestion (Girvin and MacDonald 1987, Rezayi and Haldane 1988, Read 1987, 1989, Zhang et al. 1989, Jain 1989). Girvin and MacDonald demonstrated (1987) the existence of a peculiar type of off-diagonal long-range order (ODLRO) in the FQHE ground state. This is the type of order which is used to characterize the condensate states of superconductors and superfluids. In Chap. 9 it was shown that the superfluid analogy is very useful for describing the collective excitations of the system. Yet we understand that the nature of the FQHE ground state must be rather different from that of a superfluid. The superfluid ground state spontaneously breaks gauge symmetry. Associated with its definite phase is an indefinite particle number. The FQHE is in some sense dual to this phenomenon because the particle number is better and better defined at larger and larger length scales (i.e., the filling factor is quantized). Hence the phase is wildly fluctuating and we cannot have (ordinary) ODLRO. It turns out that the peculiar type of ODLRO in the FQHE signals not the condensation of electrons (which after all are fermions) but rather the condensation of a composite object consisting of an electron pierced by a tube of (pseudo) magnetic flux [i.e., an 'anyon' (Wilczek 1982, Arovas et al. 1985, Girvin and MacDonald 1987, Girvin 1988)]. This is why the Landau-Ginsburg theory contains a vector potential whose curl is proportional to $\psi^*\psi$—the elementary objects are particles which carry flux. Let us now see how this strange state of affairs comes about.

What is the fundamental ordering which occurs when a system goes into the FQHE state? Our understanding of this starts with Halperin's early interpretation (Halperin 1983) of the significance of Laughlin's wave function. Consider a single-particle wave function in the lowest Landau level. It has the form of an arbitrary polynomial times a gaussian:

$$\phi(z) = \prod_{j=1}^{L}(z - Z_j)\exp(-\tfrac{1}{4}|z|^2). \tag{A.2.1}$$

Note that without loss of generality the polynomial can be parameterized in terms of the locations of its L zeros $\{Z_j\}$. Using the Laughlin plasma analogy (Chap. 7), the (unnormalized) probability density for this state is

$$\rho(z) = e^{-\beta\Phi(z)} \tag{A.2.2}$$

where $\beta = 2/m$ is the (arbitrarily chosen) plasma temperature and

$$\Phi(z) = \tfrac{m}{4}|z|^2 - m\sum_{j=1}^{L}\ln|z - Z_j| \tag{A.2.3}$$

is the electrostatic potential of the plasma at the point z. (Note that my convention for β and for the 2D Coulomb interaction each differ by a factor of two from that of Laughlin.) We see that the particle distribution in the quantum state ϕ is identical to that of a classical 2D Coulomb charge (with charge m) being repelled by a set of unit Coulomb charges located at the positions of the zeros of ϕ. In addition the particle sees a constant background charge of density $\rho_b = -1/2\pi$. Now if the particle density is to be approximately uniformly spread out, the zeros must be approximately evenly distributed with density $1/2\pi$ to neutralize the background. That is, *there must be a finite density of zeros* to balance the effect of the gaussian term which is trying to pull the particle into the origin.

With this picture in mind we can now consider Laughlin's many-particle wave function:

$$\phi(z_1, z_2, ..., z_N) = \prod_{i<j}(z_i - z_j)^m \exp\left(-\tfrac{1}{4}\sum_{k=1}^{N}|z_k|^2\right). \quad (\text{A}.2.4)$$

Following Halperin, we freeze the positions of particles $2, ..., N$ and view ϕ as a wave function for particle one, parametrized by the positions of the other particles. We find immediately that particle one indeed *does* see a finite density of zeros consistent with the above discussion. In fact, each particle sees m zeros located at the positions of the *other* particles. We may say that there are m zeros *bound* to the positions of the particles. This is a slightly confusing point since, if particles are repelled by zeros, how can the zeros be bound to the particles? The answer is that the zeros seen by any given particle are bound only to the *other* particles, not to itself.

The central implication of this binding of the zeros is that, since particles see zeros as repulsive Coulomb charges, the particles see each other as Coulomb charges. Neutrality of the analog plasma pins the density at a filling factor $\nu = 1/m$. More importantly for our present purposes is the fact that long wavelength density fluctuations are severely suppressed, as is nicely illustrated in Fig. 7.7. As usual for a Coulomb problem, this means that the static structure factor (the mean square density fluctuation) vanishes at small wave vectors:

$$s(k) \sim \tfrac{1}{2}k^2. \quad (\text{A}.2.5)$$

We saw in Chap. 9 that it follows in the single-mode-approximation (SMA) that there is an excitation gap.

To summarize, we have the following chain of implications:

Particles see zeros as repulsive 2D Coulomb charges and the binding of zeros to the particles

$\Longrightarrow$ particles see particles as charges
$\Longrightarrow$ plasma screening and charge neutrality
$\Longrightarrow$ vanishing of the $k = 0$ static structure factor
$\Longrightarrow$ existence of an excitation gap (in the SMA)

The existence of an excitation gap is the defining macroscopic feature of the FQHE. We see that the microscopic ordering which produces this gap is the binding of the zeros of the analytic wavefunction (as seen by any given particle) to the positions of the (other) particles.

We are thus led to seek a mathematical object capable of measuring this special type of ordering in the FQHE ground state. Let us first examine the ordinary one-body density matrix used in superfluidity

$$\rho(z,z') = N \int d^2 z_2 \ldots d^2 z_N \, \phi^*(z, z_2, \ldots, z_N) \phi(z', z_2, \ldots, z_N). \quad (A.2.6)$$

When $z = z'$ this is just equal to the particle density. For z far from z', uncorrelated phase fluctuations in ϕ^* and ϕ cause $\rho(z,z')$ to fall rapidly to zero *unless the system has condensed*. In a boson system in zero magnetic field, the ground state wave function has the same sign throughout the configuration space (Feynman 1972). This means that there are no phase fluctuations to kill off the integral in Eq. (A.2.6) and $\rho(z,z')$ goes to a constant for large separations of its argument (at least in three-dimensional systems). This even holds true for the finite temperature generalization of $\rho(z,z')$ as long as the temperature is below the lambda point. This property is called ODLRO (Yang 1962) and is characteristic of bose condensation. The second quantization formulation of the density matrix is

$$\rho(z,z') = \langle \psi^\dagger(z) \psi(z') \rangle. \quad (A.2.7)$$

If the system condenses the field operators take on a vacuum expectation value and we may, to a good approximation, factor the density matrix:

$$\rho(z,z') \sim \langle \psi^\dagger(z) \rangle \langle \psi(z') \rangle. \quad (A.2.8)$$

The vacuum expectation values are nonzero despite the fact that the field operators change the number of particles precisely because the system is in a coherent state of definite phase and indefinite particle number. That is, there is a definite phase relationship between the states $|N\rangle$ and $|N+1\rangle$. It turns out that in two dimensions infrared divergences cause $\rho(z,z')$ to decay algebraically with separation rather than going to a constant. However, I will loosely refer to any state for which the decay is slower than exponential as having ODLRO.

What happens when we evaluate the density matrix for a FQHE system? Two things drastically change the results. The fermion nature of the problem introduces lots of sign changes in the wave function as the configuration is varied. Likewise the magnetic field introduces complex phases into the wave function. The net effect of this is to cause the density matrix to decay even faster than exponential. In fact, it can be shown (Chap. 7, MacDonald and Girvin 1988) (using various analyticity properties of the wave function) that, for any state in the lowest Landau level which corresponds to a uniform liquid, the density matrix is a universal gaussian:

$$\rho(z, z') = \frac{\nu}{2\pi} \exp\left[|z - z'|^2/4 + (z^*z' - zz'^*)/4\right]. \quad (A.2.9)$$

The rapid falloff with distance is due to the huge phase fluctuations of the integrand in Eq. (A.2.6).

The short-range nature of the density matrix does not, however, imply the absence of order in the system. In fact, there is a sense in which the phase factors associated with the Fermi statistics and with the magnetic field combine to cancel each other—not in the sense of eliminating phase fluctuations but in the sense that there is a certain phase predictability to the system. The meaning of this phase predictability should become clear in the following discussion.

The hidden order in the FQHE ground state can be revealed (Girvin and MacDonald 1987, Girvin 1988) by making a singular gauge transformation (Wilczek 1982, Arovas *et al.* 1985) in which one attaches flux tubes containing m flux quanta to each of the particles. This is not a true gauge transform since the curl of the vector potential is nonzero; however, the flux exists only in places that the particles cannot reach (on the other particles). The Hamiltonian becomes

$$H = \sum_{j=1}^{N} \frac{1}{2}\left(p_j + \frac{e}{c}(A + \mathcal{A})_j\right)^2 + \sum_{i<j} V(z_i - z_j), \quad (A.2.10)$$

where A is the vector potential associated with the physical external field and the ($\hat{z}$ component) of the curl of the anyon vector potential $\mathcal{A}$ obeys

$$\nabla_j \times \mathcal{A}_j = m\Phi_0 \sum_{k \neq j}^{N} \delta^2(z_k - z_j) \quad (A.2.11)$$

and $\Phi_0 \equiv hc/e$ is the flux quantum.

Now imagine exchanging two particles by slowly moving them around each other. The quantum amplitude for the system to return to its original state will contain the usual statistics phase plus an additional term, the Berry phase (Berry 1984) which is just what would occur in the Aharanov-Bohm effect when a charge circles a flux tube. This extra phase is just $\theta = m\pi$.

Because the particles never see any actual flux, their classical dynamics remain unaffected and the quantum amplitude associated with any Feynman path integral for the system is unchanged except for this extra phase. Thus the effect of adding the flux tubes is to change the effective statistics of the particles. If m is continuous, the statistics will interpolate continuously between Fermi and Bose! If m is an odd integer and, to compensate for the flux tubes, we change the underlying ('bare') statistics of the particles (by fiat) from Fermi to Bose, *the physics of the system remains unchanged*. This ambiguity of statistics in two dimensions (Wilczek 1982, Arovas et al. 1985, Girvin and MacDonald 1987, Girvin 1988) allows us to treat the FQHE as a problem of hard-core bosons seeing a physical vector potential (from the external field) and a 'fake' gauge field from the solenoids carried by the particles. This is a strong hint that we are on the correct route towards the Landau-Ginsburg theory.

The density matrix in the singular gauge becomes:

$$\tilde{\rho}(z,z') = N \int d^2 z_2 \ldots d^2 z_N \, \phi^*(z, z_2, \ldots, z_N) \\ \times \exp\left(i \int_z^{z'} \mathcal{A} \cdot d\mathbf{l}\right) \phi(z', z_2, \ldots, z_N). \quad (A.2.12)$$

(Note that the line integral is not single-valued, since $\mathcal{A}$ has a curl. However, the phase uncertainty is a multiple of 2π, so the exponential of the integral is well-defined.) Ordinarily a gauge transformation is equivalent to a simple unitary transformation on the density matrix which leaves its eigenvalues invariant. However, a peculiar feature of this 'gauge' change is that the vector potential seen by any particle depends not only on the position of the particle, but on the positions of all the other particles as well. It is this fact which is responsible for the dramatic effect of the singular gauge transformation— it converts $\tilde{\rho}(z, z')$ from gaussian to algebraic. One can show (Girvin and MacDonald 1987) that for Laughlin's state, the exact asymptotic behavior is

$$\tilde{\rho}(z, z') \sim |z - z'|^{-\eta}, \qquad (A.2.13)$$

where the exponent is given by

$$\eta = \frac{m}{2} = \frac{\Gamma}{4} = \beta^{-1}, \qquad (A.2.14)$$

where β is the inverse of the fake temperature of Laughlin's plasma and $\Gamma \equiv \beta m^2$ is the plasma coupling constant. (Note the similarity to the XY model below the Kosterlitz-Thouless point.) In an ordinary superfluid, the density matrix exhibits ODLRO in two dimensions only at zero temperature. The present result is rather curious in that we have algebraic ODLRO even at zero temperature and hence no 'true' condensation. There is a closely related precedent for this behavior however, since a charged boson system in 2D has only algebraic ODLRO. The long-range part of the Coulomb interaction causes the ground-state structure factor to vanish at long wavelengths in just the same was as it does for the Laughlin state. Since the charged system is incompressible, the zero-point fluctuations of the density are too small to keep the phase (which is conjugate to the particle number) well-enough defined to give true ODLRO. The same thing occurs in the FQHE even though the quantum Hamiltonian contains only short-range forces. Thus there is a curious connection between the properties of Laughlin's fake classical Coulomb plasma and the properties of the true *quantum* Coulomb plasma of a charged boson system.

To see how it is that the singular-gauge density matrix can develop algebraic order, imagine starting with $z' = z$ and then dragging z' off to the distant point in which we are interested. We can compute the final phase of the integrand in Eq. (A.2.12) by integrating the gradient of the phase of the integrand along some arbitrary path P from z to z':

$$\delta\theta = \int_z^{z'} d\mathbf{l} \cdot (\nabla \operatorname{Im} \log \phi - \mathcal{A}). \qquad (A.2.15)$$

A little thought shows that the phase fluctuations due to the analytic zeros of ϕ are canceled by the line integral of the gauge field if and only if precisely m zeros are bound to each particle. To see this, consider the special case of P being a large closed circle. The line integral of the gradient of the phase of ψ

is equal to 2π times the number of analytic zeros inside the loop. The line integral of the gauge field is proportional to the amount of (fake) flux inside, and yields $2\pi m$ times the number of particles inside. These two contributions cancel.

This cancellation is perfect for the case of Laughlin's wave function, where the zeros are bound directly on top of the particles. Laughlin's wave function is the exact ground state for short-range repulsive potentials (Chap. 8). As the range of the potential increases, one zero remains on the particles to preserve the antisymmetry, but the locations of the $m-1$ remaining zeros begin to fluctuate about the positions of the particles (Yoshioka 1984a). If we view the zeros as vortices and the flux tubes as antivortices, then the system looks something like the XY model below the Kosterlitz-Thouless temperature, where vortices and antivortices bind together. It was predicted (Girvin and MacDonald 1987) that as the effective 'temperature' (range of interaction) is increased, there will be a vortex unbinding transition in which the singular gauge density matrix suddenly becomes short-ranged and ODLRO would be lost. This transition is roughly analogous to the Kosterlitz-Thouless transition in the sense that it corresponds to a loss of topological order, but it is probably a first-order transition to a Wigner crystal state (Girvin et al. 1986b, Girvin and MacDonald 1987). It was already known that as the range of the interaction increases the FQHE state eventually collapses into a gapless state (Haldane Chap. 8). Rezayi and Haldane (1988) confirmed numerically that at the transition the predicted ODLRO is lost just as the gap collapses.

Once one has identified a correlation function which is long-ranged one can begin to think about a field-theoretic approach to the problem which would formally justify the phenomenological Landau-Ginsburg theory. The infrared properties of $\tilde{\rho}$ allow one to make the usual gradient expansion of the effective action. Note that if the filling factor is $\nu = 1/m$, then $\tilde{\rho}$ is long-ranged only for the choice of m flux quanta per particle in the singular gauge transformation. This then correctly fixes the vacuum angle θ, which in the previous phenomenological theory was an adjustable parameter.

These arguments show us how one arrives at the physics of the FQHE being described by hard-core bosons which carry (fake) magnetic flux. The mean-field theory of these objects replaces the actual local flux density by the mean density. Since at filling factor $\nu = 1/m$, there are m physical flux quanta per particle, we have perfect cancellation between the physical flux and the mean statistical flux:

$$\langle \nabla \times (\boldsymbol{A} + \mathcal{A})\rangle = 2\pi m \langle |\psi|^2 - 1\rangle = 0. \qquad (A.2.16)$$

Thus at the mean-field level of the Landau-Ginsburg theory in Chap. 10, we have turned the original electrons in an external field into bosons in no field. It is only for bosons in zero field that there can be ODLRO, which is the physical reason why only the singular-gauge density matrix is long-ranged. This idea of a mean-field theory of the statistical flux was recently extended (to great effect) by Laughlin (1988a,b) and Fetter *et al.* (1989) to the case of half-statistics particles and will be discussed further below.

The history of the development of our understanding of the FQHE is a curious one. Recall that superfluidity and superconductivity started out with the phenomenological London equations and the Landau-Ginsburg theory. This was followed by the microscopic BCS mean-field theory and then a deeper understanding of the role of broken gauge symmetry in the problem and the relationship of the microscopic theory to the macroscopic. The FQHE started out with not merely a microscopic theory but a deceptively simple looking explicit wave function which turned out to be remarkably accurate and robust (in the sense of being correct more or less independent of the details of the Hamiltonian). The Laughlin wave function is in fact so simple looking and 'obvious' (after the fact, of course!) that it lulls one into thinking that its deeper significance is understood.

However, I would maintain that we do not really understand a phenomenon until we can make a simple mean field theory of the effect which explicitly displays the underlying order. Suppose that we had been given in 1957 a computer printout of the amplitude of the BCS wave function for various configurations of the particles and shown it produced a good variational bound on the ground state energy. Would we then have 'understood' superconductivity? Of course not. We would have missed all the beauty of spontaneous gauge symmetry breaking, the Anderson-Higgs phenomenon, flux quantization and so forth. With these ideas in mind let me return to the discussion above to try to summarize it.

We have found a connection between the FQHE and the problem of bosons in a fluctuating magnetic field (which on the average is zero). If we were dealing with an ordinary superfluid, what would be the simplest mean-field-like ground state that we could write down? If we were being very simplistic we might write

$$\tilde{\phi}(z_1, z_2, ..., z_N) = 1. \qquad (A.2.17)$$

This looks very naive, but actually it is not so bad. It correctly contains the fact that the boson ground state in zero field has ODLRO because it has the same sign everywhere. All that is missing is some Jastrow-like factor to build in the correlations which make ϕ small when the hard cores of the bosons overlap (Feynman 1972). If we *project out* [in the sense of Gutzwiller (Gutzwiller 1963, Brinkman and Rice 1970)] the hard-core overlaps, this qualitatively correct state actually becomes quantitatively quite accurate. Thus this crude wave function is in fact *not* orthogonal to the true ground state (for any finite size) and it correctly captures some of the essential physics of superfluidity, namely the ODLRO.

Suppose we now take this naive trial state as the ground state of the FQHE. This seems preposterous until we recall that to use the boson representation we are working in the singular gauge. It is quite easy to transform this state back to the original regular gauge (where for m odd it describes fermions):

$$\phi(z_1, z_2, ..., z_N) = \prod_{i<j} e^{i \operatorname{Im} \log \phi_m}, \qquad (A.2.18)$$

where ϕ_m is the mth Laughlin function. In other words, our naive guess just yields the phase part of Laughlin's wave function! For any uniform state with an excitation gap, this simple state is in fact not orthogonal to the true ground state precisely because it captures the essence of the physics—the binding of the zeros to the particles which yields ODLRO (back in the singular gauge). Now notice that if we *project* this state onto the lowest Landau level (Girvin and Jach 1984) to minimize its kinetic energy, it becomes *exactly* Laughlin's state! Thus the analogy with the above analysis of the superfluid is quite close and one can argue that we therefore now have a good understanding of the 'meaning' of the Laughlin wave function and of the FQHE itself, in the sense that we have a mean-field theory which captures the essential physics.

Recent work by Read (1987, 1989) and by Zhang *et al.* (1989) have formalized the connection between the Landau-Ginsburg theory of the FQHE and the proper field-theoretic description of particles carrying flux. Read in particular has a somewhat different language in which to view the problem. In the BCS description of superconductivity the phase ordering is revealed by extending the Hilbert space to states of indefinite particle number. These coherent states have a definite phase relationship between substates of $2N$ and $2N+2$ particles (i.e., between states of N and $N+1$ Cooper pairs). Read's picture of the FQHE ordering at $\nu = 1/m$ is that one should consider there to be a coherence between the state with N particles and mN flux quanta and the state with $N+1$ particles and $m(N+1)$ flux quanta. The idea is partly rooted in early suggestions of Anderson (1983) and Girvin (1984a) concerning the close relationship between the state with a true hole at some position and a state with m quasiholes at the same point.

Consider the analytic part of the N-particle wave function with m quasiholes at the point Z

$$\phi(z_1, z_2, ..., z_N; Z) = \prod_{i<j}(z_i - z_j)^m \prod_{k=1}^{N}(z_k - Z)^m. \quad (A.2.19)$$

Now consider the $N+1$-particle Laughlin polynomial, which we may rewrite as:

$$\phi(z_1, z_2, ..., z_N, z_{N+1}) = \prod_{i<j<N+1}(z_i - z_j)^m \prod_{k=1}^{N}(z_k - z_{N+1})^m. \quad (A.2.20)$$

Clearly this is indistinguishable from the wave function in Eq. (A.2.19) if we force particle $N+1$ to be at position Z.

In second quantization language, let $b(Z)$ be the operator which creates a Laughlin quasihole (Chap. 7) at the point Z and let $\psi^\dagger(Z)$ create a (true) electron. Then Read observes that the overlap

$$\langle N+1|\psi^\dagger(Z)b^m(Z)|N\rangle \quad (A.2.21)$$

is nonzero in the FQHE state. The essential point is that each quasihole adds one analytic zero at the point Z to the wavefunction. Now each electron has m zeros attached to it. Hence it looks to the particles as if there is another electron at Z. If we actually put a particle there, we obtain a good overlap with the $N+1$-particle state. Thus in the coherent state the composite operator 'condenses'

$$\langle \psi^\dagger(Z)b^m(Z)\rangle = \rho_0^{1/2}e^{i\theta}. \quad (A.2.22)$$

This occurs only if m zeros are bound to each particle. Hence this overlap is a measure of the underlying order. Clearly there is a very close connection between the condensation of this composite operator [which effectively creates a boson (Read 1987, 1989)] and the condensation detected by the first-quantization singular-gauge density matrix described above.

For the Laughlin state, the overlap described above in Eq. (A.2.21) is essentially perfect because the zeros are bound exactly to the particles. One measure of the how good the overlap is for the *true* ground state can be found by examining the spectral density for creating a hole in the system. Rezayi (1987a) finds numerically for small N that 97% of the spectral weight of the one hole state goes into the ground state of the $N-1$ particle system. As mentioned earlier, Rezayi and Haldane (1988) obtained direct numerical confirmation of the existence of ODLRO and its collapse at the point where the gap collapses as the interaction is varied.

We have come a long way in the last few years towards an understanding of the ordering which occurs in the FQHE and towards a field-theoretic description of the problem. The next step which needs to be taken is to gain control of some of the ultraviolet difficulties in the field theory. By ultraviolet difficulties, I mean the following. The fake vector potential which changes the particle statistics is a strongly fluctuating field which represents a nontrivial perturbation. The behavior of the wave function at short distances (close approaches of the particles) differs considerably from the mean field result. As mentioned in the beginning of this section, the microscopic energies are all potential energies, since the lowest Landau level is degenerate in kinetic energy. When the statistical flux is replaced by a mean field, the dynamics of the particles are affected. The potential energy of density fluctuations is replaced by an effective kinetic energy produced by the fake gauge field. The collective density mode (the magnetoroton describe in Chap. 9) appears as a cyclotron mode at the mean-field level (Read 1987, 1989, Zhang *et al.* 1989). So far no one has succeeded in getting the features (other than the existence of a gap) of the mode dispersion correct. These details (such as the roton minimum) depend on the values of the Haldane pseudopotentials (see Chap. 8) in the interaction and on the short-range correlations in the ground state (Chap. 9).

A.3 Ring Exchange Theories

During the time of completion of the first edition of this book, Kivelson *et al.* (1986, 1987) developed a novel ring-exchange theory of the FQHE (see also Baskaran 1986, Lee *et al.* 1987, Thouless and Li 1987). The physical picture behind this can be best understood by recalling Feynman's ring-exchange picture of the lambda transition in helium (Feynman 1972). Consider evaluating the partition function as a path integral in imaginary time

$$Z = \text{Tr}\, e^{-\beta H} = \text{Tr}\left(e^{-\beta H/N}\right)^N. \quad (A.3.1)$$

Let us take the trace in a representation in which the real-space positions of the particles are specified. Trace means that the paths of the particles must bring the system back to its starting state. Hence, since the particles are indistinguishable, they must either return to their starting points, or to a permutation of their starting points. Appropriate minus signs must be supplied for odd permutations of fermions. In the helium problem all exchange paths enter with the same sign, since the particles are bosons. Given that every permutation can be represented as a set of ring exchanges Feynman argued that the amplitude contributed by a particular exchange ring is exponentially small in the size of the ring:

$$Z \sim \sum_P \exp\left(-\sum_j W_j/\lambda_T\right), \quad (A.3.2)$$

where W_j is the perimeter of the jth exchange ring in a particular permutation, P, and λ_T is the thermal wavelength. It would seem that large ring exchanges are exponentially unimportant, until one realizes that the number of distinct ring exchanges diverges exponentially with size. We may say that the 'Boltzmann' energy factor in Eq. (A.3.2) is counteracted by an 'entropy' term proportional to the ring perimeter. At low enough temperatures the free energy of large rings goes negative and the entropy overwhelms the energy factor (since λ_T diverges at low T). Arbitrarily large ring exchanges begin to contribute significantly to the partition sum: ring exchanges 'condense'. This is the onset of ODLRO in the superfluid. To see why this is equivalent to the ordinary picture of ODLRO, consider introducing a particle at some point z and removing a particle at some point z'. In a path integral sense, there are many different paths which contribute to the amplitude for this process. On top of all the ring exchanges which are happening in the 'background', the extra particle can travel from z to z' or it can take place in a series of exchanges in which some other particle ends up at z'. A little thought shows that if arbitrarily large ring-exchanges are able to contribute to the partition sum, then the many exchange paths from z to z' will make the density matrix long-ranged and hence yield ODLRO. One final point to note is that if the series representing the expansion in ever larger ring exchanges did not diverge, there could be no singularities in the free energy and hence no phase transition.

The application of these ideas to the FQHE is interesting. Normally in a Fermi system there is a random phase cancelation among large ring exchanges due to half of them having minus signs. This prevents ODLRO (unless the particles travel around the rings in pairs!). However, Kivelson *et al.* (1986, 1987) noted that there is a correlation between the sign of the permutation associated with a large ring and the sign contributed by the magnetic flux enclosed by the ring. For special values of the filling factor (in particular $\nu = 1/m$) the Fermi minus signs are canceled by the magnetic field minus signs and large ring exchanges add coherently just as they do for bosons in zero field. Clearly there is a nice connection here between this picture and the effective bose statistics discussed above.

There are numerous technical approximations associated with the original version of the ring exchange theory, including the need to put the particles on a triangular lattice. These issues have been addressed in later work (Lee *et al.* 1987, Baskaran 1986). In particular, D. H. Lee *et al.*, in a beautiful paper, have been able to relate the ring exchange picture to the Laughlin wave function picture. However, there remain some significant difficulties with the proper computation of the sign of various contributions (Thouless and Li 1987, Kivelson *et al.* 1987, 1988, Su *et al.* 1987). So far it has proved impossible to develop a good picture of the collective mode gap within the ring exchange picture, although this should be possible. In my opinion, the acid test of success would be the ability to relate the amplitude for ring exchanges explicitly to Haldane's pseudopotentials and show the collapse of ODLRO and the gap as V_1 is softened.

Despite these annoying technical difficulties, the ring exchange picture has deepened our understanding of the FQHE and is an interesting and novel perspective on the problem. I find particularly appealing the notion that it is only possible to have nonanalytic behavior in the free energy as a function of filling factor (downward cusps at rational fractions) if and only if the ring-exchange series diverges.

A.4 Chiral Spin Liquids, High-T_c, and the FQHE

Recently there as been a remarkable flurry of activity in the field of high-temperature superconductivity (HTSC) (see Geballe and Hulm 1988, Little 1988). Theoretical suggestions for the mechanism have been many and varied, but my purpose here is to examine the suggestion (Kalmeyer and Laughlin 1987, 1989, Laughlin 1988b, Fetter *et al.* 1989) that there is a deep and fundamental connection between HTSC and the FQHE. This discussion is not intended to be a comprehensive or authoritative review, but rather some remarks on recent developments purely from my own personal perspective. Anderson (1987) has proposed that HTSC materials are not Fermi liquids but rather are more like Mott insulators, better pictured as strongly-correlated spin liquids described by a 'resonating valence bond' (RVB) state (Anderson 1973). Prior to the discovery of HTSC, D. H. Lee and J. D. Joannopolous had made an unpublished suggestion to Laughlin that the RVB state might be a second realization in nature of the physics of the FQHE. After the discovery of HTSC this suggestion was investigated in a interesting series of papers by Laughlin and his coworkers (Kalmeyer and Laughlin 1987, 1989, Laughlin 1988a, 1988b, Fetter *et al.* 1989).

Magnetism and spin fluctuations seem to play a significant role in (many) HTSC materials, as would be expected for Mott insulators. The central thread of the argument that there should be a connection with the quantum Hall effect is the following. In a spin-1/2 system we expect there to exist spin wave excitations which can be excited (in neutron scattering, for example) by flipping over one of the spins. Naively this is a spin-1 bose-like excitation. However, it is known that the one-dimensional Heisenberg antiferromagnet exhibits fractionalization of its spin-1 spin waves into pairs of spin-1/2 fermionic excitations (Fadeev and Takhtajan 1981). There are quite general arguments that such a phenomenon should occur for any spin 'liquid' (to be defined below). One way (the only way?) to have a two-dimensional system with an excitation carrying half the expected quantum number is to have some analog of the $\nu = 1/2$ FQHE (which occurs for bosons).

Let us now explore this argument in more detail. Quantum ferromagnets are rather like superfluids—they have a simple broken-symmetry ground state and well-understood low-lying excitations. The spontaneous magnetism of the ferromagnet is the analog of the broken gauge symmetry of the superfluid (Girvin and Arovas 1989). Low-lying spin waves correspond to the Goldstone mode of the superfluid. For the Heisenberg model, for instance, we know the exact ground state (all spins aligned, $S_{\text{tot}} = N/2$). The first excited state is a spin-1 excitation consisting of a translationally invariant superposition of single spin-flips. Quantum antiferromagnets are rather more complex. The Néel order parameter (the staggered magnetisation) does not commute with the Hamiltonian and is not conserved—it fluctuates. These quantum fluctuations make it impossible to write down the exact ground state and low-lying excitations. On general grounds, however, one expects the total spin (which is conserved) to be as small as possible (zero for an even number of spin-1/2 objects); that is, the ground state should be a rotationally invariant singlet. There is no rigorous proof of this for frustrated systems however (see Kalmeyer and Laughlin 1989).

It seems reasonable that there should in principle exist spin Hamiltonians which contain terms which *frustrate* (in the sense used in Chap. 10) the Néel order and which therefore have spin 'liquid' (disordered) ground states. One of the remarkable features of the FQHE is the fact that it has a disordered liquid ground state which is nondegenerate (has an excitation gap). We saw in Chap. 10 and in the discussion of ODLRO in this Appendix that the external magnetic field frustrates (adds kinetic energy to) the electrons in a Landau level. The essence of the FQHE is that it represents a new phenomenon in which the system develops a highly correlated state which allows nature to beat this frustration (at least for certain special filling factors) in a very subtle manner (by cancelling this frustration against the frustration of Fermi statistics). A similar phenomenon ought to exist for frustrated spin systems.

To make this connection more explicit let us explore the concept of a singlet spin liquid. It is perfectly legitimate to view the state of all spins down as the 'vacuum'. A site with an up spin can be viewed as being occupied by a hard core boson. The particle is a boson because spin operators commute on different sites. Equivalently, it takes a pair of Fermi operators to create the object—one to destroy the down electron and one to create the up electron. The boson has a hard-core interaction because there cannot be more than one boson per site (the spin cannot be further raised beyond $S^z = 1/2$). In this language, the Heisenberg Hamiltonian

$$H = J \sum_{\langle ij \rangle} S_i \cdot S_j \tag{A.4.1}$$

becomes (neglecting chemical potential terms and constants)

$$H = \frac{J}{2} \sum_{\langle ij \rangle} \left(b_i^\dagger b_j + b_j^\dagger b_i \right) + J \sum_{\langle ij \rangle} n_i n_j. \tag{A.4.2}$$

We see that the XY piece of the Hamiltonian (which flips spins) corresponds to a hopping kinetic energy term for the bosons and that the Ising piece of the Hamiltonian corresponds to a near-neighbor repulsion among the bosons. The Néel state can be viewed as a crystallization of these bosons into a state with every other site occupied. This state tries to minimize the near-neighbor repulsion of the bosons. However, the kinetic energy causes quantum fluctuations around the classical (Ising) ground state. This means (among other things) that the true ground state will contain complicated correlations and furthermore will actually be a rotationally and translationally invariant singlet (for an even number of spins); that is, the true ground state will be a linear combination of a large number of different Néel states with the order parameter vector pointing in all possible directions. This zero-point-motion restoration of a broken symmetry is analogous to the ground state of a BCS superconductor actually having definite particle number. It may be true, but if focussed on too closely, one misses the proper way to view the physics of the state—as one with definite phase and indefinite particle number. The latter is captured in the mean field theory which ignores the zero-point fluctuations. Correspondingly, the best way to view the Néel state is as a state with a staggered magnetization in a fixed direction, ignoring the zero-wave-vector fluctuations.

One can imagine that it is possible to introduce terms in the Hamiltonian such as second-neighbor repulsions which will frustrate the Néel state and prevent it from ordering. The crystalline lattice will 'melt' into a singlet liquid. At least in the RVB picture, one expects this state to be *locally* singlet. To see what I mean by this, consider dividing the system into two halves and coupling all the spins in a given half ferromagnetically into a state of total spin $S = N/4$. These two halves could then be coupled into a rotationally invariant global singlet. Clearly however if we look *locally* anywhere in the system (except right at the boundary between the two halves) we will see the spins all pointing in the same (randomly fluctuating) direction. What I mean by a local singlet liquid is a state in which the spins in the immediate vicinity of any given point are coupled to a singlet (or to as small a total spin as possible). Consider an operator which is a local coarse-grained spin

$$s_i = \sum_{j=1}^{N} w_{ij} S_j, \quad (A.4.3)$$

where the positive semidefinite weighting factor w_{ij} falls off smoothly with the separation of sites i and j and obeys the sum rule

$$\sum_{j=1}^{N} w_{ij} = 1. \quad (A.4.4)$$

The idea is that the effective spin represented by s should be very small and get smaller as the range of coarse-graining gets larger. But the magnitude of the coarse-grained spin is measured by

$$\langle s_i \cdot s_i \rangle = \sum_{j,k=1}^{N} w_{ij} w_{ik} G_{jk}, \quad (A.4.5)$$

where G is the spin correlation function

$$G_{jk} = \langle S_j \cdot S_k \rangle. \quad (A.4.6)$$

Clearly the magnitude of the coarse-grained spin depends on the properties of the spin correlation function. Now it is easy to show from the definition of the total spin that *any* liquid (i.e., translationally invariant) singlet state satisfies

$$\sum_j G_{ij} = 0. \quad (A.4.7)$$

However, local singlets obey the stronger condition that the coarse-grained spin decays to zero with a finite length scale. Hence the moment sum rules must be finite:

$$\sum_j G_{ij} |R_i - R_j|^{2n} < \infty. \quad (A.4.8)$$

This may hold only for small n or perhaps for all n. The latter possibility occurs in plasmas and has nontrivial consequences for the n-point correlation functions (Blum et al. 1982). The alert reader will begin to see here a connection with the charge neutrality ($n = 0$), perfect screening ($n = 1$), and compressibility ($n = 2$) sum rules of the Coulomb plasma (Baus and Hansen 1980, Caillol et al. 1982). In a plasma, there is local charge neutrality, so that the long range of the Coulomb force of a particle constrained to sit at some point is quickly screened out as one moves away from that point. The charge is 'compensated' by the screening cloud. The analog in the spin problem is to hold one of the fluctuating spins fixed at some site. This spin is compensated within a short distance by the spin screening cloud around it. Indeed we can make the analogy quite specific by returning to our hard-core boson model. For any singlet state, rotational invariance guarantees that

$$G_{ij} = 3\langle S_i^z S_j^z \rangle. \qquad (A.4.9)$$

But this can be rewritten in terms of the boson occupation numbers

$$G_{ij} = 3\langle (n_i - \tfrac{1}{2})(n_j - \tfrac{1}{2}) \rangle \qquad (A.4.10)$$

showing that this is precisely the (connected part of the) density-density correlation function whose Fourier transform $s(q)$ is the static structure factor of the boson 'plasma'. If perfect screening is obeyed, we expect $s(q) \sim q^2$ at small q, as discussed in Chap. 9. As mentioned earlier, the absence of density fluctuations at long length scales is beautifully illustrated in Fig. VII.7.

We can easily generalize the single-mode-approximation (SMA) discussed in Chap. 9 to the spin problem. The ($S^z = 0$ component of the) spin-wave is just a density wave in the bosons:

$$\psi_k = \rho_k |0\rangle, \qquad (A.4.11)$$

where $|0\rangle$ is the ground state and ρ_k is, as usual, the Fourier transform of the boson density operator. On general grounds one expects this to be a good approximation at long wavelengths. Again, if the ground state is a local singlet so that the structure factor vanishes at small wave vector, one expects the spin waves to have a finite gap analogous to the FQHE gap. The boson system is incompressible.

At shorter wavelengths the simple spin-wave picture breaks down. In the FQHE the excitation crosses over from being a density wave (bound particle–hole pair) to being a quasiexciton (bound quasiparticle quasihole pair). This will only happen in the spin problem if there exist fractional quantum number objects. This is a question which we must now explore.

In the FQHE the analyticity of the wave functions seems to play a central role in the fractionalization process. The fact that the (polynomial part of the) wave function depends only on the coordinate z rather than z^* is a reflection of the lack of parity and time reversal symmetry in the problem due to the external magnetic field which dominates the physics. One obvious manifestation of this (besides the handedness implicit in the existence of a Hall conductivity) is the fact that the charge of a quasiparticle depends on the sign of the winding number. Left- and right-handed vortices (quasiparticles) have opposite charge.

A less obvious manifestation of the lack of P- and T-symmetry is the fractional statistics of the quasiparticles. Arovas *et al.* (1984) have a beautiful demonstration of this using Berry's phase (Berry 1984). The reader is referred to the original papers for details, but the main idea is this. A quasiparticle is a vortex. This means that if we drag any particle around the vortex, the phase of the wave function winds by $\pm 2\pi$. This phase winding is what produces the currents circulating around the vortex. Less widely appreciated is the fact that if we adiabatically drag a vortex around a closed loop, the system will acquire an extra, nondynamical Berry's phase $\delta\theta = \pm 2\pi n$, where n is the expectation value of the number of particles inside the loop. For a uniform state, n is simply proportional to the area of the loop. A phase winding proportional to the area is precisely what one would have for a charged particle moving in a uniform magnetic field. Hence vortices see particles the way particles see flux—a fact which was pointed out very early in a cryptic remark in Haldane's seminal paper on the hierarchy (Haldane 1983). Arovas *et al.* note that for a uniform state with filling factor ν

$$\delta\theta = 2\pi\nu N_\Phi, \tag{A.4.12}$$

where N_Φ is the number of flux quanta from the external field enclosed by the path. We deduce immediately from this that a vortex appears to act like a particle with fractional charge νe in the external field. This of course agrees precisely with Laughlin's plasma analogy computation of the fractional charge (Chap. 7). The connection between these two computations is the analyticity of the wave function. We know that associated with the phase winding for particle motion around the vortex, there must also be an analytic zero which makes the modulus of the wave function vanish linearly at the position of the vortex. In the plasma analogy this translates into the vortex carrying a fake plasma charge of $\pm\nu$ times the fake plasma charge of the electrons. Perfect screening then implies a deficit or surplus νe of physical charge. The analyticity connects the radial and angular parts of the wavefunction. The Berry's phase argument concentrates on the angular part, Laughlin's argument concentrates on the radial part, but each yields the same answer.

Arovas et al. also compute the phase change when two quasiparticles are adiabatically exchanged. The calculation proceeds just as before, but the net charge of the vortices modifies the phase winding and causes the vortices to appear to have fractional statistics ν. That is, the excess phase winding (over and above that due to the fractional charge in the uniform field) is

$$\delta\theta_s = \pm\pi\nu, \qquad (A.4.13)$$

the sign depending on the handedness of the exchange path. Clearly for any statistics angle $\delta\theta_s$ not equal to a multiple of π we *must* have broken P- and T-symmetry.

How could this occur for a spin liquid? In the boson representation, there is no gauge field in Hamiltonian of the bosons which would correspond to the external magnetic field in the FQHE. (Actually, Kalmeyer and Laughlin (1987, 1989) find such a field for the frustrated triangular lattice, but even in this case there is still T-symmetry.) A likely possibility is that quantum number fractionalization can only occur if the ground state of the system *spontaneously* breaks T- and P-symmetry (Kivelson and Rokhsar 1988, Laughlin and Kalmeyer 1988) (see, however, Wen et al. 1989). This brings us to the notion of a chiral spin state, which we will discuss shortly. First, however, let us examine why we expect fractional quantum number objects must exist.

There is a general argument why one might expect excitations in a local singlet which carry half-integral spin. The clearest version of this argument that I have heard is that given by Laughlin (1988b). As mentioned previously, this quantum number fractionalization is known to occur in one dimension (Fadeev and Takhtajan 1981), but one might worry that this is a special case. Laughlin's argument for higher dimensions is qualitative, but quite general. Imagine that we have a square lattice of $2N \times 2N$ sites ($N \gg 1$) with a Hamiltonian such that the ground state is a local singlet in the sense defined above.

Now imagine making the lattice larger by one row and one column so that the size is $(2N+1)\times(2N+1)$. Because this is a local singlet with perfect screening of spin fluctuations, we expect the system to be insensitive to boundary effects. The ground state energy should be scarcely affected. This seems like a trivial observation until one realizes that the system now contains an *odd* number of spin-1/2 objects and its ground state cannot possibly be a singlet. Presumably the ground state will take on the lowest possible spin, namely one-half. Because of the special perfect screening properties of the local-singlet ground state we expect the coarse-grained local spin (defined above) to vanish everywhere in the system except for some small region which contains the 'excess' spin-1/2. We can associate the excess energy and spin of this state with a well-defined, localizable quantum 'particle'. Surely then neutral pairs of such particles can exist as finite-energy excitations above the ground state of the neutral $2N\times 2N$ system. Thus in addition to long-wavelength spin-one excitations (spin waves) we expect on quite general grounds that local singlets possess excitations with fractional quantum numbers and furthermore we expect the fraction to be one-half. Note that the above argument fails for systems which are not *local* singlets because the lack of perfect screening prevents us from assigning a well-defined location (and small spatial extent) to the excess spin-1/2.

If one evaluates the partition function of the Heisenberg antiferromagnet (AFM) as a path integral, the problem (with some daring) can be mapped onto a field theory—the nonlinear sigma model. It has been suggested (Dzyaloshinskii *et al.* 1988, Polyakov 1988, Wiegmann 1988, 1989) that this field theory may contain unexpected topological terms which would allow topologically nontrivial field configurations to carry the fractional quantum numbers described above. Recent work (Haldane 1988, Wen and Zee 1988, Fradkin and Stone 1988) indicates that this is *not* actually realized for the Heisenberg antiferromagnet. We know, however, that the Heisenberg AFM on the 2D square lattice is *not* a local singlet. Its ground state almost certainly has Néel order (that is, it is a global singlet with infinitely long-range staggered correlations) (Huse 1988 and references therein). This, then, does not contradict the rather general argument that such fractional quantum number objects should exist for local singlets. (In connection with the HTSC problem, Wiegmann (1988, 1989) argues that the Heisenberg Hamiltonian is only an approximation to the more complete Hubbard Hamiltonian. His analysis of the latter indicates that it contains non-trivial terms which can lead to topological effects.)

Having seen that there is good reason to believe that spin-1/2 excitations ought to exist in local singlets, and having seen that at least in the FQHE there is an intimate connection between fractional charge and fractional statistics, we are lead to the possibility that the ground state of a local singlet must break P- and T- symmetry. Therefore we are led to the notion of a chiral spin liquid (Wen et al. 1989).

Let us approach the idea of chiral order by examining what happens in the FQHE. Each particle in the Laughlin state has m vortices attached to it which cause currents to circulate around it in a preferred direction. Indeed, it is straightforward to show that within the subspace of the lowest Landau level the current operator obeys the identity (Girvin et al. 1986b)

$$\boldsymbol{J} = -\tfrac{1}{2} \nabla \times \rho \hat{\boldsymbol{z}}. \qquad (A.4.14)$$

If we freeze a particle at some spot, there will be a density gradient around it due to the screening cloud and correlation hole. Hence there will necessarily be a circulating current around each particle (although it will be screened out a large distances).

We may now ask what the analog property of a chiral spin state might be. We expect that if the spin at some site is fixed to be up, that there will be a net spin current circulating in a preferred direction around this site. We will take this to be the defining property of a chiral spin state. The precise definition of the current operator depends on the details of the Hamiltonian. For the boson representation of the Heisenberg model the kinetic energy allows the particles to hop along bonds connecting near-neighbor sites. We define a spin current to be a particle current of the bosons (representing the spin-up particles) along the bonds. The corresponding current operator for the bond connecting sites j and k is quite simple and natural:

$$J_{jk} = i(b_k^\dagger b_j - b_j^\dagger b_k). \qquad (A.4.15)$$

In terms of the original spin operators,

$$\begin{aligned} J_{jk} &= i(S_k^+ S_j^- - S_j^+ S_k^-) \\ &= 2(S_k^x S_j^y - S_k^y S_j^x). \end{aligned} \qquad (A.4.16)$$

In general, the actual current operator for the local-singlet Hamiltonian will be more complex, but for simplicity, let us ignore this distinction. In addition, let us for the moment consider a triangular lattice. An operator which measures the correlation between the occupation of a site and the circulating current around that site is

$$\mathcal{O} = S_i^z(S_k^x S_j^y - S_k^y S_j^x), \tag{A.4.17}$$

where sites (i, j, k) lie on the same plaquette (in counter-clockwise order). The rotationally symmetric equivalent of this operator is (to within a sign)

$$\mathcal{O} = \boldsymbol{S}_i \cdot (\boldsymbol{S}_j \times \boldsymbol{S}_k). \tag{A.4.18}$$

Clearly this triple product has a handedness and is odd under time-reversal. A vacuum expectation value

$$\langle \mathcal{O} \rangle \neq 0 \tag{A.4.19}$$

implies spontaneous P- and T-symmetry breaking. This operator expectation value then defines an order parameter for the chiral state and indeed directly measures the chirality.

We are now in a position to draw some hints of possible very deep analogies with the FQHE and Landau-level physics. Consider the one-body density matrix

$$\rho(z, z') = \langle \psi^\dagger(z) \psi(z') \rangle. \tag{A.4.20}$$

It follows from the analyticity of the lowest Landau level wave functions that knowledge of the *diagonal* part of the one-body density matrix (i.e., the density itself) is sufficient to determine (by analytic continuation) the full off-diagonal density matrix (Laughlin Chap. 7, MacDonald and Girvin 1988)

$$\rho(z, z) = \langle \psi^\dagger(z) \psi(z) \rangle \Longrightarrow \rho(z, z'). \tag{A.4.21}$$

In particular, for *any* uniform liquid state with filling factor ν in the lowest Landau level, the density matrix is universal

$$\rho(z, z') = \frac{\nu}{2\pi} \exp\left[(z^* z' - z z'^*)/4 e^{-|z-z'|^2/4}\right]. \tag{A.4.22}$$

This connection between the density and density matrix proves to be rather useful (Laughlin Chap. 7, MacDonald and Girvin 1988).

There exists a rough analog to this relation in spin liquids. From rotational symmetry the spin correlation function in Eq. (A.4.6) satisfies

$$G_{jk} = 3\langle S_j^z S_k^z \rangle$$
$$= \tfrac{3}{4}\langle S_j^+ S_k^- + S_j^- S_k^+ \rangle. \quad (A.4.23)$$

But this can be re-expressed in terms of the boson language as

$$\langle (n_k - \tfrac{1}{2})(n_j - \tfrac{1}{2}) \rangle = \tfrac{1}{4}[\rho(j,k) + \rho(k,j)], \quad (A.4.24)$$

where $\rho(j,k)$ is the one-body off-diagonal density matrix, for the bosons

$$\rho(j,k) = \langle b_j^\dagger b_k \rangle. \quad (A.4.25)$$

Thus we see that singlets have a connection between the diagonal part of the *two*-body density matrix and the off-diagonal one-body density matrix which seems to be analogous to the result we just discussed for the FQHE. It is interesting to speculate on how strong a constraint this is in determining the nature of the allowed ground state for *local* singlets. Does this relationship, combined with the perfect screening (and possibly with the requirement of chiral order) uniquely determine the ground state or at least the class of allowed ground states? In particular is one forced into an FQHE type of state? These interesting questions remain unanswered. In connection with these mysteries it should be noted that Wen et al. (1989) have pointed out that fractional statistics excitations can occur above a P- and T-symmetric ground state provided that there is a doubling of the spectrum with each particle being accompanied by a partner of the opposite statistics. Some recent work by Mele (1989) is also relevant to this point.

Laughlin and Kalmeyer (1987, 1989) considered the possibility that the ground state of the Heisenberg AFM on the 2D triangular lattice might be, in the boson language, the $m = 2$ FQHE state (with the particles restricted to lattice sites). The AFM is frustrated on the triangular lattice (although not fully) and the ground state, rather than being the desired spin liquid, probably has three-sublattice Néel order (Huse 1988). Nevertheless, this paper has attracted a lot of interest to the study of possible spin liquid phases. In particular, the Laughlin-Kalmeyer state has been found to be a singlet (Kalmeyer and Laughlin 1989, Zou et al. 1989) (and it then follows from the plasma argument that it is a local singlet). Laughlin has posed the following question: What spin Hamiltonian has the $m = 2$ FQHE state as its exact ground state? The author, in an unpublished communication to Laughlin, suggested that the answer to this question should involve the fact that the essence of the FQHE is the correlation between charge and vorticity, and that therefore the desired spin state contains chiral order. The Hamiltonian should therefore contain three-spin interactions related to the chirality operator. Following this suggestion, Laughlin (1989) was able to find such a Hamiltonian

$$H = \sum_j D_j^\dagger D_j, \qquad (A.4.26)$$

where

$$D_j = \sum_{k \neq j} \frac{1}{z_j - z_k} \boldsymbol{\sigma}_j \cdot \boldsymbol{\sigma}_k. \qquad (A.4.27)$$

Using the identity (Merzbacher 1970) (valid for vectors which commute with $\boldsymbol{\sigma}$)

$$(\boldsymbol{\sigma} \cdot \boldsymbol{A})(\boldsymbol{\sigma} \cdot \boldsymbol{B}) = \boldsymbol{A} \cdot \boldsymbol{B} + i\boldsymbol{\sigma} \cdot (\boldsymbol{A} \times \boldsymbol{B}), \qquad (A.4.28)$$

it is easy to show that Laughlin's Hamiltonian contains (among other terms) the three-spin chiral interaction

$$\sum_{j \neq k \neq l} \frac{\boldsymbol{r}_{jk} \times \boldsymbol{r}_{jl} \cdot \hat{\boldsymbol{z}}}{|\boldsymbol{r}_{jl}|^2 |\boldsymbol{r}_{jk}|^2} (\boldsymbol{\sigma}_j \cdot \boldsymbol{\sigma}_k \times \boldsymbol{\sigma}_l). \qquad (A.4.29)$$

Clearly this is the key term which guarantees the chiral order of the ground state.

It is somewhat unsatisfactory that this Hamiltonian contains (relatively) long-range $1/r$ interactions because it clouds the issue of whether or not it is possible to obtain a *local* singlet using only short-range interactions. In the FQHE, the analyticity of the wave functions allows one to have a short-range interaction in the Hamiltonian and end up with long-range (pseudo) interactions (and hence perfect screening) in the plasma analogy for the wave function.

A second unsatisfactory aspect of this Hamiltonian is that it is neither P- nor T-symmetric. Ideally, one would like to find a P- and T-symmetric Hamiltonian with local interactions whose ground state spontaneously breaks these symmetries. One could imagine however that even with a symmetric Hamiltonian, the mean-field theory of such a system might yield an *effective* Hamiltonian for the broken symmetry phase which is similar to Laughlin's.

Presumably something along the lines of the following Hamiltonian, which is both local and symmetric, would be interesting to investigate:

$$H = \sum_{\langle ijk \rangle} \left(\mathcal{O}_{\langle ijk \rangle} \right)^2, \qquad (A.4.30)$$

where the sum is over plaquettes of the triangular lattice and $\mathcal{O}_{\langle ijk \rangle}$ is the chirality operator defined in Eq. (A.4.18). If one could find a short-range interaction for which a chiral spin state is the exact zero-energy ground state, one would have the analog of the Haldane pseudopotential interactions (Chap. 8). The reader is directed to the paper by Wen et al. (1989) for a detailed discussion of these issues in chiral spin states.

In connection with HTSC there has been a lot of interest in the question of the nature of holes injected into strongly correlated spin liquids. Laughlin (1988a) proposes to construct the 'holon' of the RVB theory (Kivelson, Rokhsar and Sethna 1987) in the following manner. First take the spin liquid to be described by the FQHE $m = 2$ state. The justification for this is that the particles (up spins) are bosons, the (lattice) filling is one-half, and the resulting state is correctly a local singlet. Next, form a 'spinon' (Anderson 1987) as a fractional-spin quasihole analogous to the FQHE quasihole. This is done in the usual way (Chap. 7) by creating a vortex in the boson (up-spin) fluid which repels from the origin, one half of a boson. This localizes a net spin-1/2 (with $S^z = -1/2$) near the site of the vortex. The spinon is a charge-zero, spin-1/2 object. Now remove a spin-down electron from the vicinity of the vortex core. This forms the holon, which is a charge-one, spin-zero object and is the current-carrying particle involved in the transport. Based on the FQHE analogy, Laughlin finds the holon to obey half-statistics. Zee uses the term 'semions' to describe these objects. Because pairs of semions are not fermions but bosons (Tao and Wu 1985), Laughlin (1988a) has proposed that holons pair to form a superconducting state. Laughlin's analysis of this state is based on a mean-field theory of statistics similar in spirit (but more complex and detailed in practice) to that first used in the fractional Hall effect (Chap. 10, Girvin and MacDonald 1987, Canright and Girvin 1989a) and discussed in the previous section on ODLRO. Superconducting pairing of semions has been confirmed in an RPA calculation

(Fetter et al. 1989) and by direct numerical studies of fractional statistics particles on a lattice (Canright and Girvin 1989b). A useful review may be found in a recent paper by Chen et al. (1989).

These ideas represent fascinating developments, but their relevance to real HTSC materials must still be considered highly speculative. Two problems which need to be further addressed are the following. First, the band width of the hole is some very large number related to the hopping matrix element t^* which is considerably larger than the spin energy J. It is hard to see how this will not overwhelm the spin energy at a reasonable level of hole doping. A counter argument to this might be that it is known that for a single hole, the band width for *coherent* motion is rather small (Kane et al. 1989, Gros and Johnson, 1989).

The second problem is that studies of semion superconductivity to date have not dealt with the fact that fractional statistics objects are vortices in some background fluid. As such they see a strong effective magnetic field from the background (Canright and Girvin 1989b). It may be more appropriate to view the ground state as some sort of FQHE hierarchy state rather than as the pairing of semions in the absence of a background field. Anderson has made some speculations along these lines. For the case of finite hole doping it may be appropriate to drop the boson representation of the hole liquid and use a fermion (with spin) representation. The problem of finding singlet states of electrons in the lowest Landau level has arisen in the FQHE with the discovery of the $\nu = 5/2$ state. There may well be useful connections with singlet states for doped antiferromagnets.

The experimental and theoretical status of the HTSC field is in such a rapid state of flux that it is impossible to give it a proper review. In any case I have focussed here just on current speculations regarding interesting connections with the fractional Hall effect.

A.5 Even-Denominator and Spin-Reversed States

The long-standing odd-denominator rule for the FQHE filling fraction follows in the Laughlin picture (Chap. 7) [and its hierarchical generalization (Chap. 8)] from the Fermi statistics. Initial indications (Clark et al. 1986) of a $\nu = 5/2$ state seen in a drop in ρ_{xx} have been confirmed by the observation of a Hall plateau (Willett et al. 1987). This state is quite subtle, appearing only in the very highest-mobility samples and has an extremely small activation energy on the order of tens of milliKelvin (Gammel et al. 1988). The peculiar properties of the state and the fact that it is visible only in the second Landau level suggest that it is of a different nature than previously discovered states. Second Landau-level states are reached at low magnetic fields, where the Zeeman energy is reduced. This suggests the possibility that the 5/2 state circumvents the odd-denominator rule by reversing half of its spins, perhaps to form a singlet. This suspicion has received strong support from experiments (Eisenstein et al. 1988) using magnetic fields tilted away from the normal to the plane of the electron gas. The idea is that the Landau-level 2D orbital motion of the electrons is sensitive only to the *normal* component of the field, while the spins, which are free to orient themselves in any direction, see a Zeeman energy proportional to the *total* field. As one increases the total field, keeping the normal component fixed, one finds that the 5/2 state rapidly disappears indicating that it is a spin-unpolarized state unable to take advantage of the Zeeman energy. Some other spin-polarized state becomes more favorable at high fields.

An early paper by Halperin (1983) proposed a class of states with half the spins reversed. Their polynomial part has the 'Greek-Roman' form (so-called because of the notation used in some papers on the subject)

$$\Psi_{mmn}(z_1, ..., z_{N/2}; z_{[1]}, ..., z_{[N/2]}) = \prod_{i<j}^{N/2} \prod_{k,l}^{N/2} (z_i - z_j)^m (z_{[i]} - z_{[j]})^m (z_k - z_{[l]})^n, \quad (A.5.1)$$

where $[j] \equiv j + N/2$, coordinates $\{1, 2, ..., N/2\}$ refer to up-spin electrons, and coordinates $\{[1], [2], ..., [N/2]\}$ refer to down-spin electrons. The filling factor for such states is easily computed by the following argument. The average number of zeros that an electron sees on the other electrons is $(m+n)/2$. In analogy with the Laughlin plasma, the filling factor is the inverse of this quantity:

$$\nu_{mmn} = \frac{2}{m+n}. \quad (A.5.2)$$

Thus, for example, Halperin proposed the state Ψ_{332} as a spin-reversed alternative to the 2/5 hierarchical state. There is some numerical evidence that this state would in fact be the lowest energy state at $\nu = 2/5$ if it were not for the Zeeman energy cost (Chakraborty and Zhang 1984, Zhang and Chakraborty 1984, Chakraborty and Pietiläinen 1989).

The even-denominator state at 5/2 consists of a filled lowest Landau level (both spin states), which we can ignore, plus a half-filled second Landau level. It is possible to show that we can entirely ignore the fact that we are dealing with the second Landau level by simply pretending that we have the usual lowest Landau-level orbitals, but modifying the Haldane pseudo-potentials describing the interaction (MacDonald 1984c). Hence we are interested in finding a state with filling factor $\nu = 1/2$. The state Ψ_{331} seems to be just what we need. It allows us to have an even denominator by reversing half the spins. However, a technical problem arises which the alert reader may already have spotted.

The problem is this. Up-spin and down-spin electrons are not really distinguishable, and we cannot get away with simply using different coordinates z_j and $z_{[j]}$ to describe them. What we really need is a wave function which is properly antisymmetric under exchange of combined space and spin coordinates. Nevertheless there is something partially correct about using the form described above, for the case of a Hamiltonian which is spin-independent. Let us now see how this works.

Let α and β represent the spin-up and spin-down states, respectively. Then the correct way to write down the wave function is

$$\Phi_{mmn} = \sum_P \frac{(-1)^P}{\sqrt{N!}} \Psi_{mmn}(z_{P_1}, ..., z_{P_{N/2}}; z_{P_{[1]}}, ..., z_{P_{[N/2]}})\phi_s, \quad (A.5.3)$$

where the sum is over all permutations P of N objects, P_j is the image of j under that permutation, and ϕ_s is the spin function

$$\phi_s = (\alpha_{P_1}, ..., \alpha_{P_{N/2}}; \beta_{P_{[1]}}, ..., \beta_{P_{[N/2]}}). \quad (A.5.4)$$

It is clear from the construction that this state satisfies the antisymmetry requirements. Furthermore, we must have m odd to prevent the wave function from vanishing under antisymmetrization.

Now suppose that $\mathcal{O}$ is some operator (such as the Hamiltonian or some other observable). Then the matrix element

$$\langle \Phi_{mmn} | \mathcal{O} | \Phi_{mmn} \rangle \quad (A.5.5)$$

can be reduced to a matrix element of the Greek-Roman part alone:

$$\langle \Psi_{mmn} | \mathcal{O} | \Psi_{mmn} \rangle. \quad (A.5.6)$$

This follows from the fact that any components in the bra and ket having different permutations which mix up and down spinor ordering will have vanishing matrix elements for any spin-independent operator. This is why the Greek-Roman form is so useful and appears in the literature without the bother of the full antisymmetrization. However, there is one technical difficulty which we have not faced. For a spin-independent Hamiltonian, the total spin is a good quantum number which can be used to classify the eigenfunctions. It is not at all clear that Φ_{mmn} is an eigenfunction of total spin, and in general it turns out *not* to be. We are particularly interested in spin-singlet states (essentially because they are the most different from the fully spin-polarized states that form the odd-denominator hierarchy). Clearly the state already satisfies $S^z = 0$ by construction. The requirement that

$$S^+ S^- |\Phi_{mmn}\rangle = 0 \quad (A.5.7)$$

can be shown to reduce to the Fock cyclic condition (Hammermesh 1962) on the Greek-Roman wave function

$$\sum_{j=1}^{N/2} \left(e(i,[j]) - \delta_{ij} \right) \Psi_{mmn}(z_1, ..., z_{N/2}; z_{[1]}, ..., z_{[N/2]}) = 0, \quad (A.5.8)$$

where $e(a,b)$ is the permutation which interchanges indices a and b.

It is straightforward to show that the Greek-Roman form generates a legal singlet if and only if $m = n + 1$. This follows from the fact that the $\nu = 2$ state for the Landau level with both spin states completely full is a single Slater determinant which is necessarily a singlet. This state has the form

$$\Psi_{\text{Slater}} = \Psi_{110}. \quad (A.5.9)$$

Hence any state of the form

$$\Psi = \Psi_{2p,2p,2p} \Psi_{110} \quad (A.5.10)$$

is a legal singlet.

From this discussion it is clear that Halperin's 2/5 state (Ψ_{332}) is an eigenfunction of total spin but that Ψ_{331} is *not* and hence cannot be used to describe the desired $\nu = 1/2$ state. This state has a good energy however and might be relevant to situations in which the Hamiltonian is effectively spin-dependent (Yoshioka et al. 1989, MacDonald et al. 1989). There are very small spin effects originating in the lack of inversion symmetry in GaAS, but these are probably too small to be significant (MacDonald et al. 1989). A recent study of the FQHE in two-layer structures treated the layer number as a kind of pseudospin and found a spin-dependent Hamiltonian which had as its ground state Ψ_{331} in certain parameter regions (Yoshioka et al. 1989).

The difficulty of finding a state with $\nu = 1/2$ and which is a legal singlet was overcome in an important paper by Haldane and Rezayi (1988a,b) which developed the so-called 'hollow-core' model. The physical idea behind this state is to imagine pairing up- and down-spin electrons into effective bosons. The Greek-Roman wave function for this state is

$$\Psi_{\text{HC}} = \Psi_{331} \, \text{per}|M|, \qquad (A.5.11)$$

where the $N/2 \times N/2$ matrix M is given by

$$M_{ij} = \left(z_i - z_{[j]}\right)^{-1}. \qquad (A.5.12)$$

The permanent of the matrix is like the determinant, except that all the permutations enter with the same sign. The effect of this permanent on the wave function is rather subtle. It causes the wave function to satisfy the Fock condition without changing the filling factor in the thermodynamic limit. Note that an important additional effect is that it cancels one factor of $(z_i - z_{[j]})$ in ψ_{331}, which means that up- and down-spin particles fail to avoid close approaches—their relative angular momentum can be as small as zero. It turns out that this state is the exact nondegenerate ground state for the Hamiltonian in which the Haldane pseudo-potentials are: $V_m = \{0, 1, 0, 0, 0, ...\}$. Because of the lack of repulsion at the shortest distances, this is known as the 'hollow-core model.' It is the hollow core which allows the up- and down-spin particles to pair into bosons.

A curious mathematical identity (Haldane and Rezayi 1988b) allows one to rewrite the ground state exactly in the suggestive form

$$\Psi_{\text{HC}} = \Psi_{222} \, \det|\widetilde{M}|, \qquad (A.5.13)$$

where
$$\widetilde{M}_{ij} = \left(z_i - z_{[j]}\right)^{-2}. \qquad (A.5.14)$$

Note that Ψ_{222} is just the Laughlin $m = 2$ state (for spinless bosons) and that the determinant restores the antisymmetry for exchange of two particles of the same spin.

It would appear that the hollow-core Hamiltonian is a rather poor representation of the actual Coulomb interaction, which is a maximum at zero separation even if it is softened by the effect of the finite thickness of the electron inversion layer (Yoshioka 1984b 1985b, MacDonald and Aers 1984, Girvin *et al.* 1986b, Zhang and Das Sarma 1986). There are several effects which tend to ameliorate this difficulty, however. Besides the finite-thickness correction to the pseudo-potential, it turns out that the corrections needed to simulate the second Landau level soften V_0 and strengthen V_2 while barely changing V_1. In addition, Rezayi and Haldane (1989) have found that previously neglected Landau-level mixing effects tend to soften V_0. On top of all this, numerical studies show that the hollow-core state remains the ground state up to fairly significant values of V_0. The message of the 5/2 state may be that previous tests of our understanding of the size of the pseudo-potential parameters (such as agreement between predicted and measured excitation gaps) do not apply to even angular-momentum channels such as V_0 due to the Pauli principle. Various corrections (such as Landau level mixing) are expected to be strongest for V_0 since the particles approach each other closely in that channel.

Still the correlations built into the hollow-core state are rather extreme. The two-point function for like spins obeys
$$g_{\uparrow\uparrow} \sim |z_\uparrow - z_\uparrow|^6, \qquad (A.5.15)$$
while for unlike spins
$$g_{\uparrow\downarrow} \sim |z_\uparrow - z_\downarrow|^0. \qquad (A.5.16)$$

The state Ψ_{331} has much more 'reasonable' correlations, and, as noted above, numerical evidence shows that it is the true ground state for weakly spin-dependent forces (Yoshioka *et al.* 1989, MacDonald *et al.* 1989). One wonders if there is some other state between these two extremes which is the physically observed state. There are no clues at present, and the hollow-core model certainly remains the leading contender to explain the 5/2 even-denominator state.

A.6 Spin-Reversed Hierarchy

In addition to the body of theoretical work which has considered the possibility of spin-unpolarized ground states (Halperin 1983, Chakravarty and Zhang 1984, Zhang and Chakravarty 1984, Yoshioka 1986, Rasolt and MacDonald 1986, Haldane Chap. 8, Yoshioka et al. 1988, Maksym 1988) studies have been made of the nature of spin-reversed quasiparticles (Chakraborty 1986, Chakraborty et al. 1986, Rezayi 1987b) in the FQHE. A great deal of interest has appeared in the last year in the possibility that some states *already observed* in the FQHE hierarchy may be only partially spin-polarized or perhaps unpolarized. The same method of tilting the magnetic field which was useful in the study of the 5/2 state is giving indications of spin-reversed states at some odd-denominators (Clark 1989, Eisenstein et al. 1989, Syphers and Furneaux 1988a,b). The experimental situation is a bit unclear at present because of various conflicting claims regarding the interpretation of these experiments. One problem is that the tilted field affects the confinement of the electrons in the finite-thickness inversion layer potential well. These effects have been considered by Syphers and Furneaux (1988a,b) and by Chakraborty and Pietiläinen (1989). However, there may be additional effects of the tilted field since it breaks the rotational invariance in the plane of the electron gas (Nickila and MacDonald 1989, Chakraborty and Pietiläinen 1989). Connected with this are Landau level mixing effects, which can destroy the particle-hole symmetry within the lowest Landau level (Haug et al. 1987).

Because of the present uncertainties in the experimental and theoretical situation, I will limit my discussion here to a brief overview of some of the basic issues. One situation which gives a tilted-field effect is to have a spin-polarized ground state, but have the lowest-lying quasiparticle state be spin-reversed (Chakraborty et al. 1986, Chakraborty 1986, Rezayi 1987b). This would give an excitation gap which would increase with total B until the Zeeman energy exceeds some threshold. At that point, the spin-aligned quasiparticle would be the lowest excitation. This situation can be summarized by saying that there is a level-crossing among the low-lying excited states.

A second possibility is that the ground state for some fillings is partially spin-polarized or perhaps a spin singlet. In that case the excited quasiparticle energies would be spin-symmetric except for the Zeeman energy, which would favor the spin-aligned quasiparticle. As a function of total B, there would not in this case be a level crossing among the excited states (as was the case above), but rather a level crossing in the *ground-state* energy (in other words, a phase transition). Eisenstein *et al.* (1989) believe that this may be the case for the $\nu = 8/5$ state. They observe that activation energy for dissipation decreases linearly with increasing total B out to some critical field, and then, rather than seeing the state collapse, they see the activation energy begin to *increase* linearly with total B.

This change of behavior may be associated with a transition from a spin-singlet state to a spin-polarized state. The scenario they suggest is the following. Suppose that at zero tilt, the ground state is a spin-singlet related to Halperin's $\nu = 2/5$ state (Halperin 1989) by particle-hole symmetry. (Note that when spin is included the particle–hole symmetry connects ν with $\nu' = 2 - \nu$, *not* $\nu' = 1 - \nu$.) There is evidence that this state may be preferred at least in the absence of Zeeman energy (Chakraborty and Zhang 1985, Zhang and Chakraborty 1984, Morf and Halperin 1986). The quasielectron and quasihole above this state can lower their energy using the Zeeman splitting since they see no preferred orientation associated with the background fluid. This would explain the linear *decrease* of the activation energy with total B. At higher fields, the ground state switches to the spin-polarized 8/5 state, which consists of the lowest spin state fully occupied and the next spin state 3/5 occupied. If it happens that the spin asymmetry of the background fluid causes the quasiparticles to prefer to be spin-reversed (Chakraborty *et al.* 1986, Chakraborty 1986, Rezayi 1987b), then their activation energy will *rise* linearly with total B. At very high fields (presumably beyond the reach of the Eisenstein experiment), there would be a level crossing among the excited states and the spin-aligned quasiparticles would become the lowest lying. Their energy relative to the spin polarized parent fluid would then be independent of field, so the activation energy would saturate. This scenario would appear to explain the data, but additional puzzling features remain to be cleared up (Eisenstein *et al.* 1989).

It is safe to say that our understanding of the 'simple' fractional states is probably fairly good, but our understanding of the hierarchical states is poorer. Adding spin to the problem yields entirely new possibilities for hierarchical sequences (Rezayi 1989), which make for many subtleties and complications.

A.7 Summary and Conclusions

The field of the quantum Hall effect has been remarkably active in both experiment and theory in the last few years. The theoretical ideas associated with the FQHE are now beginning to find their way into other areas of condensed matter physics and continue to generate great interest. One outstanding experiment yet to be done is to directly observe the collective magneto-roton mode discussed in Chap. 9. Further experiments need to be done in connection with the Wigner crystal and the spin-reversed hierarchy. As the quality of samples continues to improve I expect that we will be able to go to lower and lower temperatures to pursue ever more subtle and surprising states of highly-correlated two-dimensional electron systems. As Feynman used to say, "There's room at the bottom."

References

Abrahams E, Anderson P W, Licciardello D C and
    Ramakrishnan T V 1979 Phys Rev Lett **42** 673
Aers G C and MadDonald A H 1984 J Phys C **17** 5491
Affleck I 1983 J Phys C **16** 5839
    1985 : 1986 Nuclear Physics B **265** 409
Aharanov Y and Bohm D 1959 Phys Rev B **115** 485
Anderson P W 1958 Phys Rev B **112** 1900
    1963 Phys Rev B **130** 439
    1973 Mat Res Bull **8** 153
    1983 Phys Rev B **28** 2264
    1987 Science **235** 1196
Ando T 1974 J Phys Soc Japan **37** 622
    1977 J Phys Soc Japan **43** 1616
    1983a In: *Recent Topics in Semiconductor Physics* eds H Kamimura and Y Toyozawa (World Scientific, Singapore) 72
    1983b J Phys Soc Japan **52** 1740
    1984a J Phys Soc Japan **53** 3101
    1984b J Phys Soc Japan **53** 3126
    1985 *Proceedings of Int Conf on Electronic Properties of Two-Dimensional Systems* **VI** [1986 Surface Science **170** 243]
Ando T, Fowler A B and Stern F
    1982 Rev Mod Phys **54** 437

Ando T, Matsumoto Y and Uemura Y
    1975 J Phys Soc Japan **39** 279
Ando T and Uemura Y 1974 J Phys Soc Japan **36** 959
Andrei E Y, Deville G, Glattli D C, Williams F I B,
    Paris E and Etienne B
    1988 Phys Rev Lett **60** 2765
    1989 Phys Rev Lett **62** 973; 1926(E)
Anzin V B, Veselago V G, and Zavaritskiĭ V N and
    Prokhorov A M
    1984 Pis'ma Zh Eksp Teor Fiz **39** 231
    [1984 JETP Lett **40** 1002]
Aoki H 1977 J Phys C **10** 2583
    1978 J Phys C **11** 3823
    1983 J Phys C **16** 1893
Aoki H and Ando T 1981 Solid State Commun **38** 1079
    1985 Phys Rev Lett **54** 831
Arnol'd V I and Avez A 1968 *Ergodic Problems of*
    Classical Mechanics (Benjamin New York)
Arovas D P, Schrieffer J R and Wilczek F
    1984 Phys Rev Lett **53** 722
    1985 Nuclear Physics B **251** 117
Ashcroft N W and Mermin N D 1976 *Solid State Physics*
    (Holt, Rhinehart and Winston New York) 574
Aubry S and André G 1980 Ann Israel Phys Soc **3** 133
Avron J and Seiler R 1985 Phys Rev Lett **54** 259
Avron J, Seiler R and Simon B
    1983 Phys Rev Lett **51** 51
Azbel' M Ya 1964 Zh Eksp Teor Fiz **46** 929
    [1964 Sov Phys JETP **19** 634]
Baraff G A and Tsui D C 1981 Phys Rev B **24** 2274
Baskaran G 1986 Phys Rev Lett **56** 2716
Baus M and Hansen J-P 1980 Phys Rep **59** 1
Berry M V 1984 Proc R Soc Lond A **392** 45
BIPM 1983 Comptes Rendus in *17th Conf Gén des*
    Poids et Mesures 97
Bliek L, Braun E, Melchert F, Schlapp W, Warnecke P
    and Weimann G 1983 Metrologia **19** 83
Blum L, Gruber C, Lebowitz J L and Martin P
    1982 Phys Rev Lett **48** 1769
Boebinger G S, Chang A M, Störmer H L and Tsui D C
    1985a Phys Rev B **32** 4268
    1985b Phys Rev Lett **55** 1606
    1985c unpublished
Boebinger G S, Chang A M, Störmer H L, Tsui D C,
    Hwang J C M, Cho A and Weimann G

1985d *Proceedings of Int Conf on Electronic Properties of Two-Dimensional Systems* **VI** [1986 Surface Science **170** 129]
Brézin E, Zinn-Justin J and Le Guillou S C
    1976 Phys Rev B **14** 3110; 4976
    Brézin E, Hikami S and Zinn-Justin J
    1980 Nuclear Physics B **165** 528
Brézin E, Gross D J and Itzykson C
    1984 Nuclear Physics B **235** 24
Brézin E, Nelson D R and Thiaville A
    1985 Phys Rev B **31** 7124
Briggs A, Guldner Y, Vieren J P, Voos M, Hirtz J P and Razeghi M 1983 Phys Rev B **27** 6549
Brinkman W F and Rice T M 1970 Phys Rev B **2** 4302
Bychkov Yu A, Iordanskii S V and Eliashberg G M
    1981 Pis'ma Zh Eksp Teor Fiz **33** 152
    [1981 JETP Lett **33** 143]
Byers N and Yang C N 1961 Phys Rev Lett **7** 46
Cage M E and Girvin S M
    1983a,b Comments Sol State Phys **11** 1; 47
Cage M E, Dziuba R F, Field B F, Williams E R, Girvin S M, Gossard A C, Tsui D C and Wagner R J 1983 Phys Rev Lett **51** 1374
Cage M E, Field B F, Dziuba R F, Girvin S M, Gossard A C and Tsui D C 1984 Phys Rev B **30** 2286
Cage M E, Dziuba R F and Field B F 1985 IEEE Trans Instrum Meas **IM-34** 301
Caillol J M, Levesque D, Weis J J and Hansen J P
    1982 J Stat Phys **28** 325
Canright G S and Girvin S M 1989a Phys Rev Lett (submitted)
    1989b Phys Rev Lett (submitted)
Cardy J L and Rabinovici E 1982 Nuclear Physics B **205** 1
Cardy J L 1982 Nuclear Physics B **205** 17
Chakraborty T 1985 Phys Rev B **31** 4016
    1986 Phys Rev B **34** 2926
Chakraborty T and Pietiläinen P
    1988 *The Fractional Quantum Hall Effect: Properties of an incompressible quantum fluid* Springer Series in Solid State Sciences **85** (Springer-Verlag Berlin Heidelberg New York Tokyo)
    1989 Phys Rev B **39** 7971
Chakraborty T, Pietiläinen P and Zhang F C
    1986 Phys Rev Lett **57** 130
Chakraborty T and Zhang F C 1984 Phys Rev B **29** 7032
Chalker J T 1984 Surface Science **142** 182

Chang A M 1985 in *Advances in Solid State Physics* **25**
 ed P Grosse (Vieweg, Braunschweig) 405–411
Chang A M, Berglund P, Tsui D C, Störmer H L
 and Hwang J C M 1984a Phys Rev Lett **53** 997
Chang A M, Paalanen M A, Störmer H L, Hwang J C M and
 Tsui D C 1984b Surface Science **142** 173
Chang A M, Paalanen M A, Tsui D C, Störmer H L and
 Hwang J C M 1983 Phys Rev B **28** 6113
Chang A M, Störmer H L and Tsui D C 1985 unpublished
Chang A M, Timp G, Chang T Y, Cunningham J E,
 Chelluri B, Mankiewich P M, Behringer R E
 and Howard R E 1988a Surface Science **196** 46
Chang A M, Timp G, Chang T Y, Cunningham J E,
 Mankiewich P M, Behringer R E and Howard R E
 1988b Solid State Commun **67** 769
Chang L L and Esaki L 1980 Surface Science **98** 70
Chang L L and Ploog K 1985 eds *Molecular Beam Epitaxy
 and Heterostructures* (Martinus Nijhoff
 The Hague)
Chiao R Y, Hansen A and Moulthrop A A
 1985 Phys Rev Lett **54** 603
Chiu K W and Quinn J J 1974 Phys Rev B **9** 4724
Cho A Y and Arthur J R 1975
 Prog Solid State Chem **10** 157
Chui S T 1985a Phys Rev B **32** 8438
 1985b Phys Rev B **32** 1436
Chui S T, Hakim T M and Ma K B
 1985 (preprint) published as
 [1986 Phys Rev B **33** 7110]
Clark R G, Haynes S R, Suckling A M, Mallett J R,
 Wright P A, Harris J J and Foxon C T
 1989 Phys Rev Lett **62** 1536
Clark R G, Mallett J R, Haynes S R, Harris J J
 and Foxon C T 1988 Phys Rev Lett **60** 1747
Clark R G, Nicholas R J, Usher A, Foxon C T and
 Harris J J 1985 *Proceedings of Int Conf on
 Electronic Properties of Two-Dimensional
 Systems* **VI** [1986 Surface Science **170** 141]
Clark R G, Nicholas R J and Usher A
 1986 Surface Science **170** 141
Clothier W K 1965 Metrologia **1** 36
Coleman S 1973 Commun Math Phys **31** 259
Cutkosky R D 1974 IEEE Trans Instrum Meas **IM-23** 305
Dana I, Avron Y and Zak J 1985 J Phys C **18** L679
Das Sarma S and Stern F 1985 Phys Rev B **32** 8442

Davidson J S, Robinson G Y and Dahlberg E D
    1985 Bull Am Phys Soc **30** 254
Davidson J S, Dahlberg E D, Valois A J
    and Robinson G Y 1986 Phys Rev B **33** 2941; 8238
Deaver B S and Fairbank W N 1961 Phys Rev Lett **7** 43
Delahaye F and Reymann D
    1985 IEEE Trans Instrum Meas **IM-34** 316
de Vega H J and Schaposnik F A
    1986 Phys Rev Lett **56** 2564
Deser S, Jackiw R and Templeton S
    1982a Phys Rev Lett **48** 975
    1982b Ann Phys (NY) **140** 372
Dingle R, Störmer H L, Gossard A C and Wiegmann W
    1978 Appl Phys Lett **33** 655
Doll R and Näbauer M 1961 Phys Rev Lett **7** 51
Dzyaloshinskii I E, Polyakov A M and Wiegmann P B
    1988 Phys Lett **A127** 112
Ebert G. von Klitzing K, Probst C and Ploog K
    1982 Solid State Commun **44** 95
Ebert G, von Klitzing K, Ploog K and Weimann G
    1983a J Phys C **16** 5441
Ebert G, von Klitzing K, Probst C, Schuberth E,
    Ploog K and Weimann G
    1983b Solid State Commun **45** 625
Ebert G, von Klitzing K, Maan J C, Remenyi G,
    Probst C, Weimann G and Schlapp W
    1984 J Phys C **17** L775
Ebert G, von Klitzing K and Weimann G
    1985 J Phys C **18** L257
Edwards J T and Thouless D J 1972 J Phys C **5** 807
Edwards S F and Anderson P W 1975 J Phys F **5** 965
Eisenstein J P 1985 Appl Phys Lett **46** 695
Eisenstein J P, Störmer H L, Narayanamurti V and
    Gossard A C 1985a *Proceedings of the 17th Int
    Conf on the Physics of Semiconductors* ed D J
    Chadi and W A Harrison (Springer-Verlag Berlin
    Heidelberg New York Tokyo) 309
Eisenstein J P, Störmer H L, Narayanamurti V,
    Cho A Y, Gossard A C and Tu T W
    1985b Phys Rev Lett **55** 875
Eisenstein J P, Willett R L, Störmer H L, Tsui D C,
    Gossard A C and English J H
    1988 Phys Rev Lett **61** 997
Eisenstein J P, Störmer H L, Pfeiffer L, West K W
    1989 Phys Rev Lett **62** 1540

Emery V J 1985 Mol Crystals **121** 384
Englert Th and von Klitzing K
    1978 Surface Science **73** 70
Englert Th, Tsui D C, Gossard A C and Uihlein Ch
    1982 Surface Science **113** 295
Essam J W 1980 Rep Prog Physics **43** 833
Fadeev L D and Takhtajan L A 1981 Phys Lett **85A** 375
Fang F F and Stiles P J 1983 Phys Rev B **27** 6487
                               1984 Phys Rev B **29** 3749
Fano G, Ortolani F and Colombo E
    1986 Phys Rev B **34** 2670
Feenberg E 1969 *Theory of Quantum Fluids* Pure and
    Applied Physics Series **31**
    (Academic London) 166
Fetter A L, Hanna C B, Laughlin R B
    1989 Phys Rev B **39** 9679
Feynman R P 1972 *Statistical Mechanics* (Benjamin
    Reading Mass) (and references therein)
Field B F 1985 IEEE Trans Instrum Meas **IM−34** 320
Fisher D S 1982 Phys Rev B **26** 5009
Ford C J B, Thornton T J, Newbury R, Pepper M,
    Ahmed H, Peacock D C, Ritchie D A, Frost J E F
    and Jones G A C 1989 Appl Phys Lett **54** 21
Fowler A B, Fang F F, Howard W E and Stiles P J
    1966 Phys Rev Lett **16** 901
Foxon C T, Harris J J, Wheeler R G and Lacklison D E
    1985 : 1986 J Vac Sci Tech **B4** 511
Fradkin E and Stone M 1988 Phys Rev B **38** 7215
Friedman M H, Sokoloff J B, Widom A and Srivastava Y N
    1984a Phys Rev Lett **52** 1587
    1984b Phys Rev Lett **53** 2592
Fukuyama H and Lee P A 1978 Phys Rev B **18** 6245
Furneaux J E and Reinecke T L
    1984a Surface Science **142** 186
    1984b Phys Rev B **29** 4792
Furneaux J E, Syphers D A, Brooks J S, Schmiedeshoff
    G M, Wheeler R G and Stiles P J 1985
    Proceedings of Int Conf on Electronic
    *Properties of Two-Dimensional Systems***VI**
    [1986 Surface Science **179** 154]
Gammel P L, Bishop D J, Eisenstein J P, English J H,
    Gossard A C, Ruel R, Störmer H L
    1988 Phys Rev B **38** 10128
Gavrilov M G, Kvon Z D, Kukushkin I V and Timofeev V B
    1984 Pis'ma Zh Eksp Teor Fiz **39** 420

[1984 JETP Lett **39** 170]
Gavrilov M G, Kukushkin I V, Kvon Z D and Timofeev V B
    1985 *Proceedings of the 17th Int Conf on the Physics of Semiconductors* ed D J Chadi and W A Harrison (Springer-Verlag Berlin Heidelberg New York Tokyo) 287
Geballe T H and Hulm J K 1988 Science **239** 367
Girvin S M   1984a Phys Rev B **29** 6012
            1984b Phys Rev B **30** 558
            1988   *Interfaces, Quantum Wells, and Superlattices* eds C R Leavens and R Taylor, NATO ASI Series B **179** (Plenum New York London) 333
Girvin S M and Arovas D P 1989 Physica Scripta **T27** 156
Girvin S M and Jach T 1984 Phys Rev B **29** 5617
Girvin S M and Jonson M 1982 J Phys C **15** L1147
Girvin S M and MacDonald A H 1987 Phys Rev Lett **58** 1252
Girvin S M, MacDonald A H and Platzman P M
    1985 Phys Rev Lett **54** 581
    1986a J Mag and Mag Mat **54–57** 1428
    1986b Phys Rev B **33** 2481
Giuliani G F and Quinn J J
    1985a Phys Rev B **31** 6228
    1985b Solid State Commun **54** 1013
    1985c *Proceedings of Int Conf on Electronic Properties of Two-Dimensional Systems* **VI**
    [1986 Surface Science **170** 316]
Giuliani G F, Quinn J J and Ying S C
    1983 Phys Rev B **28** 2969
Goldberg B B, Heiman D, Graf M J, Broido D A, Pinczuk A, Tu C W, English J H and Gossard A C
    1988 Phys Rev B **38** 10131
Goldin G A, Menikoff R and Sharp D H
    1985 Phys Rev Lett **54** 603
Goldman V J, Shayegan M and Tsui D C
    1988 Phys Rev Lett **61** 881
Goodall R K, Higgins R J and Harrang J P
    1985 Phys Rev B **31** 6597
Gornik E, Lassnig R, Störmer H L, Seidenbusch W, Gossard A C, Wiegmann W and von Ortenberg M
    1985a *Proceedings of the 17th Int Conf on the Physics of Semiconductors* ed D J Chadi and W A Harrison (Springer-Verlag Berlin Heidelberg New York Tokyo) 303
Gornik E, Lassnig R, Strasser G, Störmer H L,

Gossard A C and Wiegmann W
   1985b Phys Rev Lett **54** 1820
Götze W and Hajdu 1978 J Phys C **11** 3993
   1979 Solid State Commun **29** 89
Gros C and Johnson M D 1989 (unpublished)
Guldner Y, Hirtz J P, Vieren J P, Voisin P, Voos M and
   Razeghi M 1982 J Physique **43** L613
Guldner Y, Hirtz J P, Briggs A, Vieren J P and Voos M
   1984 Surface Science **142** 179
Gurewich A V and Mints R G
   1984 Pis'ma Zh Eksp Teor Fiz **39** 318
   [1984 JETP Lett **39** 381]
Gusev G M, Kvon Z D, Neizvestnyi I G, Ovsyuk V N and
   Cheremnykh P A
   1984 Pis'ma Zh Eksp Teor Fiz **39** 446
   [1984 JETP Lett **39** 541]
Gutzwiller M C 1963 Phys Rev Lett **10** 159
Haavasoja T, Störmer H L, Bishop D J, Narayanamurti V,
   Gossard A C and Wiegmann W
   1984 Surface Science **142** 294
Hagen C R 1984 Ann Phys (NY) **157** 342
   1985a Phys Rev D **31** 331
   1985b *ibid.* 848
   1985c *ibid.* 2135
Hadju J 1983 in: *Lecture Notes in Physics* **177**
   ed G Landwehr (Springer-Verlag Berlin
   Heidelberg New York Tokyo) 327
Haldane F D M 1983 Phys Rev Lett **51** 605
   1985 Phys Rev Lett **55** 2095
   1988 Phys Rev Lett **61** 1029
Haldane F D M and Chen L 1984 Phys Rev Lett **53** 2591
Haldane F D M and Rezayi E H
   1985a Phys Rev Lett **54** 237
   1985b Phys Rev B **31** 2529
   1988a Phys Rev Lett **60** 956
   1988b *ibid.* 1886
Haldane F D M and Wu Y–S 1985 Phys Rev Lett **55** 2887
Halperin B I 1982 Phys Rev B **25** 2185
   1983 Helv Phys Acta **56** 75
   1984 Phys Rev Lett **52** 1583; **52** 2390(E)
   1985 *Proceedings of Int Conf on*
   Electronic Properties of Two-Dimensional
   Systems **VI** [1986 Surface Science **170** 115]
Halsey T C 1985 Phys Rev B **31** 5728 (and references
   therein)

Hammermesh M 1962 *Group Theory and its Application to Physical Problems* (Addison-Wesley Reading, Mass) 249
Hamon B V 1954 J Science Instrum **31** 450
Hansen J P and Levesque D 1981 J Phys C **14** L603
Harper P G 1955 Proc Phys Soc (London) **A265** 317
Hartland A 1984 in *Precision Measurement and Fundamental Constants II*, NBS Spec Publ **617** ed B N Taylor and W D Phillips 543
Hartland A, Davies G J and Wood D R 1985 IEEE Trans Instrum Meas **IM-34** 309
Haug R J, von Klitzing K, Nicholas R J, Maan J C and Weimann G 1987 Phys Rev B **36** 4528
Haug R J, MacDonald A H, Streda P and von Klitzing K 1988 Phys Rev Lett **61** 2797; **62** 608(E)
Hawrylak P and Quinn J J 1985 Phys Rev B **31** 6592
Heiman D, Goldberg B B, Pinczuk A, Tu C W, Gossard A C and English J H 1988 Phys Rev Lett **61** 605
Heinonen O and Taylor P L 1983 Phys Rev B **28** 6119
Heinonen O, Taylor P L and Girvin S M 1984 Phys Rev B **30** 3016
Higgs P W 1964a Phys Lett **12** 132
    1964b Phys Rev Lett **13** 508
    1966 Phys Rev **145** 1156
Hikami S 1981 Phys Lett **98 B** 208
    1984 Phys Rev B **29** 3726
Hikami S and Brézin E 1986 *Proceedings of Int Conf on Electronic Properties of Two-Dimensional Systems* VI [1986 Surface Science **170** 262]
Hofstadter D 1976 Phys Rev B **14** 2239
Horing N and Yildiz M 1976 Ann Phys NY **97** 216
Huse D A 1988 Phys Rev B **37** 2380
Ichimaru S 1973 *Basic Principles of Plasma Physics: A Statistical Approach* (Benjamin Reading Mass) 1
Ihm J and Phillips J C 1985 J Phys Soc Japan **54** 506
Igarashi T, Wakabayashi J and Kawaji S 1975 J Phys Soc Japan **38** 1549
Imry Y 1982 In: *Proceedings of the Taniguchi International Conference on Anderson Localization* eds Y Nagaoka and H Fukuyama, Springer Series in Solid State Science (Springer-Verlag Berlin Heidelberg New York Tokyo) 198
Inoue M and Nakajima S 1984 Solid State Commun **50** 1023

Iordansky S V 1982 Solid State Commun **43** 1
Ishikawa K 1984 Phys Rev Lett **53** 1615
Isihara A 1983a Surface Science **133** L475
        1983b Solid State Commun **46** 265
Itzykson C 1985 *Recent Developments in Quantum Field Theory* ed J Ambjorn *et al.* (Elsevier Amsterdam Oxford New York)
        1986 To appear in: *Quantum Field Theory and Quantum Statistics, Essays in Honor of the 60th Birthday of E S Fradkin* (Adam Hilger Bristol)
Iwasa Y, Kido G and Miura N
        1983 Solid State Commun **46** 239
Jackiw R and Rebbi C 1976 Phys Rev D **13** 3398
Jackiw R and Templeton S 1981 Phys Rev D **23** 2291
Jackson J D 1975 *Classical Electrodynamics* (Wiley New York)
Jain J K 1989 Phys Rev Lett **63** 199
Jancovici B 1981 Phys Rev Lett **46** 386
Jones R G and Kibble B P
        1985 IEEE Trans Instrum Meas **IM-34** 181
Jonson M and Girvin S M 1984 Phys Rev B **29** 1939
José J V, Kadanoff L P, Kirkpatrick S and Nelson D R
        1977 Phys Rev B **16** 1217
Joynt R 1982 *PhD Thesis*, University of Maryland (College Park Maryland)
        1984 J Phys C **27** 4807
        1985 J Phys C **18** L331
Joynt R and Prange R E 1984 Phys Rev B **29** 3303
Kallin C and Halperin B I 1984 Phys Rev B **30** 5655
        1985 Phys Rev B **31** 3635
Kalmeyer V and Laughlin R B 1987 Phys Rev Lett **59** 2095
        1989 Phys Rev B **39** 11879
Kane C L, Lee P A and Read N 1989 Phys Rev B **39** 6880
Kaplit M and Zemel J N 1968 Phys Rev Lett **21** 212
Kastalski A, Dingle R, Cheng K Y and Cho A Y
        1982 Appl Phys Lett **41** 274
Kawaji S 1978 Surface Science **73** 46
        1983 *Proceedings of the International Colloquium on the Foundations of Quantum Mechanics* eds S Kamefuchi *et al* (Phys Soc of Japan Tokyo) 327
Kawaji S, Igarashi T and Wakabayashi J
        1975 Prog Theor Phys Suppl **57** 176
Kawaji S, Wakabayashi J, Yoshino J and H Sakaki

References

  1984 J Phys Soc Japan **53** 1915
Kazarinov R F and Luryi S 1982 Phys Rev B **25** 7626
Kennedy T A, Wagner R J, McCombe B D and Tsui D C
  1977 Solid State Commun **22** 459
Khmel'nitzkii D E 1983 Piz'ma Zh Eksp Teor Fiz **82** 454
Kinoshita T and Lindquist W B
  1981 Phys Rev Lett **47** 1573
Kinoshita J, Inagaki K, Yamanouchi C, Yoshihiro K,
  Endo T, Murayama Y, Koyanagi M, Moriyama J,
  Wakabayashi J and Kawaji S
  1984 J Phys Soc Japan Suppl **53** 339
Kirk W P, Kobiela P S, Schiebel R A and Reed M A
  1985 (preprint) published as
  [1986 J Vac Sci Tech **A4** 2132]
Kivelson S A, Kallin C, Arovas D P and Schrieffer J R
  1986 Phys Rev Lett **56** 873
  1987 Phys Rev B **36** 1620
  1988 Phys Rev B **37** 9085
Kivelson S A and Rokhsar D S 1988 Phys Rev Lett **61** 2630
Kivelson S A, Rokhsar D S and Sethna J P
  1987 Phys Rev B **35** 8865
Kohmoto M 1985 Ann Phys (NY) **160** 343
Kohn W 1961 Phys Rev **123** 1242
Komiyama S, Takamasu T, Hiyamizu S and Sasa S
  1985 Solid State Commun **54** 479
Kosterlitz J M and Thouless D J 1973 J Phys C **6** 1181
Kubo R 1957 J Phys Soc Japan **12** 570
Kuchar F, Bauer G, Weimann G and Burkhard H
  1984 Surface Science **142** 196
Kuchar F, Meisels R, Weimann G and Burkhard H
  1985 *Proceedings of the 17th Int Conf on the*
  Physics of Semiconductors ed D J Chadi and
  W A Harrison (Springer-Verlag Berlin
  Heidelberg New York Tokyo) 275
Kukushkin I V and Timofeev V B
  1985 *Proceedings of Int Conf on Electronic*
  Properties of Two-Dimensional Systems **VI**
  [1986 Surface Science **170** 148]
  1986 Pis'ma Zh Eksp Teor Fiz **44** 179
  [1986 JETP Lett **44** 228]
  1988a Sov Phys JETP **67** 594
  1988b Surface Science **196** 196
Lam P K and Girvin S M 1984 Phys Rev B **30** 473
        1985 Phys Rev B **31** 613 (E)
Landau L D and Lifshitz E M 1960 *Electrodynamics of*

Continuous Media (Pergamon Press Oxford London New York Paris)
Laughlin R B 1981 Phys Rev B **23** 5632
　　　　　　1983a Phys Rev Lett **50** 1395
　　　　　　1983b Phys Rev B **27** 3383
　　　　　　1984a Surface Science **141** 11
　　　　　　1984b Phys Rev Lett **52** 2304
　　　　　　1984c Surface Science **142** 163
　　　　　　1984d In: Springer Series in Solid State Sciences **53** eds G Bauer, F Kuchar and H Heinrich (Springer-Verlag Berlin Heidelberg New York Tokyo) 272
　　　　　　1985a Physica **126B** 254
　　　　　　1985b *Energy and Technology Review* (National Technical Information Service Springfield Virginia June 1985) 11
　　　　　　1985c *Proceedings of the 17th Int Conf on the Physics of Semiconductors* ed D J Chadi and W A Harrison (Springer-Verlag Berlin Heidelberg New York Tokyo) 255
　　　　　　1988a Phys Rev Lett **60** 2677
　　　　　　1988b Science **242** 525
　　　　　　1989   Ann Phys **191** 163
Laughlin R B, Cohen M L, Kosterlitz J M, Levine H, Libby S B and Pruisken A M M
　　1985 Phys Rev B **32** 1311
Laughlin R B and Kalmeyer V 1988 Phys Rev Lett **61** 2631
Lavine C F, Wagner R J and Tsui D C
　　1982 Surface Science **113** 112
Lee D H, Baskaran G and Kivelson S
　　1987 Phys Rev Lett **59** 2467
Lerner I V and Lozovik Yu E
　　1978 Zh Eksp Teor Fiz **78** 1167
　　[1980 Sov Phys JETP **51** 588]
Levesque D, Weis J J and Hansen J P
　　1983 in *Applications of the Monte Carlo Method in Statistical Physics*, Topics in Current Physics **36** ed K Binder (Springer-Verlag Berlin Heidelberg New York Tokyo) 37
Levesque D, Weis J J and MacDonald A H
　　1984 Phys Rev B **30** 1056
Levine H, Libby S B and Pruisken A M M
　　1983 Phys Rev Lett **51** 1915
　　1984a,b,c Nuclear Physics B **240** FS[12] 30, 49, 71
Licciardello D and Thouless D J 1978 J Phys C **11** 925

References

Lin B J F, Tsui D C, Paalanen M A and Gossard A C
  1984 Appl Phys Lett **45** 695
Little W A 1988 Science **242** 1390
London F 1950 *Superfluids* **Vol 1**: *Macroscopic Theory
  of Superconductivity* (Wiley New York) 151
Long A P, Myron H W and Pepper M
  1984 J Phys C **17** L433
Luryi S and Kazarinov R F 1983 Phys Rev B **27** 1386
                          1985 unpublished
MacDonald A H 1984a Phys Rev B **29** 6563
              1984b Phys Rev B **30** 4392
              1984c Phys Rev B **30** 3550
MacDonald A H and Aers G C 1984 Phys Rev B **29** 5976
MacDonald A H, Rice T M and Brinkman W F
  1983 Phys Rev B **28** 3648
MacDonald A H and Girvin S M
  1986 Phys Rev B **33** 4414 *ibid* **34** 5639
  1988 Phys Rev B **38** 6295
MacDonald A H and Murray D B
  1985 Phys Rev B **32** 2707
MacDonald A H and Středa P 1984 Phys Rev B **29** 1616
MacDonald A H, Aers G C and Dharma-wardana N W C
  1985 **31** 5529
MacDonald A H, Liu K L, Girvin S M and Platzman P M
  1986 Phys Rev B **33** 4014
MacDonald A H, Oji H C A and Girvin S M
  1985 Phys Rev Lett **55** 2208
MacDonald A H, Yoshioka D and Girvin S M
  1989 Phys Rev B **39** 8044
Mahan G D
  1981 *Many Particle Physics* (Plenum New York)
Maksym P A 1988 *Proceedings of the International
  Conference on the Application of High Magnetic Fields
  to Semiconductor Physics*, Wurzburg
  (Springer-Verlag Berlin Heidelberg New York Tokyo)
Mankiewich P M, Behringer R E, Howard R E, Chang A M,
  Chang T Y, Chelluri B, Cunningham J and Timp G
  1988 J Vac Sci Tech **B6** 131
Marra W C, Fuoss P H and Eisenberger P E
  1982 Phys Rev Lett **49** 1169
Mele E J 1989 Physica Scripta **T27** 82
Mendez E E, Heiblum M, Chang L L and L Esaki
  1983 Phys Rev B **28** 4886
Mendez E E, Wang W I, Chang L L and L Esaki
  1984a Phys Rev B **30** 1087

Mendez E E, Chang L L, Heiblum M, Esaki L, Naughton M, Martin K and Brooks J
1984b Phys Rev B **30** 7310
Mendez E E, Washburn S, Esaki L, Chang L L and Webb R A 1985 *Proceedings of the 17th Int Conf on the Physics of Semiconductors* ed D J Chadi and W A Harrison (Springer-Verlag Berlin Heidelberg New York Tokyo) 397
Mermin N and Wagner H 1966 Phys Rev Lett **17** 1133
Merzbacher E 1970 *Quantum Mechanics* 2nd ed (Wiley New York London Sydney Toronto) 271
Morf R and Halperin B I 1986 Phys Rev B **33** 2221
Moriyasu K 1983 *An Elementary Primer for Gauge Theory* (World Scientific Singapore) 96
Mott N F and Davis E A 1979 *Electronic Processes in Non-Crystalline Materials*, 2nd ed (Clarendon Press Oxford)
Nagaoka Y and Fukuyama H 1982 eds *Anderson Localization* Springer Series in Solid State Sciences **39** (Springer-Verlag Berlin Heidelberg New York Tokyo)
Nakajima K, Tanahashi T and Akita K 1982 Appl Phys Lett **41** 194
Nee T W and Prange R E 1967 Physics Letters **25A** 8
Nee T W, Koch J F and Prange R E 1968 Phys Rev **168** 779
Nelson D R and Halperin B I 1979 Phys Rev B **19** 2457
Ng K C 1974 J Chem Phys **61** 2680
Nicholas R J, Stradling R A and Tidey R J 1977 Solid State Commun **23** 341
Nickila J D and MacDonald A H 1989 Bull Am Phys Soc **34** 914
Niemi A J and Semenoff G W 1983 Phys Rev Lett **51** 2077
Niu Q and Thouless D J 1984a Phys Rev A **17** 2453
1984b Phys Rev B **30** 3561
Niu Q, Thouless D J and Wu Y-S 1985 Phys Rev B **31** 3372
Nixon C T, Harris J J, Wheeler R G and Lacklison D E 1985 (see Foxon *et al.* 1985)
Northrup T G 1963 *The Adiabatic Motion of Charged Particles* (Interscience New York)
Obloh H, von Klitzing K and Ploog K 1984 Surface Science **142** 236
Oji H 1984 J Phys C **17** 3059

References

Oji H and Středa P 1985 Phys Rev B **31** 7291
Ono Y 1982 J Phys Soc Japan **51** 237
Onsager L 1949 Nuovo Cim **6** Suppl **2** 249
          1952 Phil Mag **43** 1006
Paalanen M A, Tsui D C and Gossard A C
     1982 Phys Rev B **25** 5566
Paalanen M A, Tsui D C, Lin B J F and Gossard A C
     1984 Surface Science **142** 29
Pepper M and Wakabayashi J 1982 J Phys C **15** L861
                            1983 J Phys C **16** L113
Pinczuk A, Valladares J P, Heiman D, Gossard A C,
     English J H, Tu C W, Pfeiffer L and West K
     1988 Phys Rev Lett **61** 2701
Pines D 1964 *Elementary Excitations in Solids*
     (Benjamin New York) 149
Pippard A B
     1964 Phil Trans Royal Soc (London) **A265** 317
Platzman P M, Girvin S M and MacDonald A H
     1985 Phys Rev B **32** 8458
Platzman P M and Wolff P A 1973 *Waves and Interactions
     in Solid State Plasmas [Solid State Physics
     Supplement 13]* (Academic New York)
Pokrovsky V L and Talapov A L 1985 J Phys C **18** L691
Polyakov A M 1975a Phys Lett B **59** 79
              1975b *ibid.* 82
              1988   Mod Phys Lett A **3** 325
Portal J C, Nicholas R J, Brummel M A, Razeghi M and
     Poisson M A 1983 Appl Phys Lett **43** 293
Powell T G, Dean C C and Pepper M
     1984 J Phys C **17** L359
Prange R E 1981 Phys Rev B **23** 4802
            1982 Phys Rev B **26** 991
Prange R E and Joynt R 1982 Phys Rev B **25** 2943
Pruisken A M M 1984   Nuclear Physics B **235** FS[11] 277
                1985a Phys Rev B **32** 2636
                1985b in *Localization, Interaction and
     Transport Phenomena* Springer Series in Solid
     State Sciences **61** eds B Kramer, G Bergmann
     and Y Bruynseraede (Springer-Verlag Berlin
     Heidelberg New York Tokyo)
                1985c Phys Rev B **31** 416
                1985d Columbia University preprint
                1988   Phys Rev Lett **61** 1297
Pruisken A M M and Schaefer L
     1981 Phys Rev Lett **46** 490

1982 Nuclear Physics B **200** 20
Pudalov V M and Semenchinskii S G
  1983 Pis'ma Zh Eksp Teor Fiz **38** 173
  [1983 JETP Lett **38** 202]
  1984a Pis'ma Zh Eksp Teor Fiz **39** 143
  [1984a JETP Lett **39** 170]
  1984b Zh Eksp Teor Fiz **86** 1431
  [1984b Sov Phys JETP **59** 838]
  1984c Solid State Commun **51** 19
Pudalov V M, Semenchinskii S G and Edelman V S
  1984d Pis'ma Zh Eksp Teor Fiz **39** 474
  [1984d JETP Lett **39** 576]
Rajaraman R 1982 *Solitons and Instantons* (North Holland Amsterdam)
Rammal R, Toulouse G, Jaeckel M T and Halperin B I
  1983 Phys Rev B **27** 5142
Rasolt M and MacDonald A H 1986 Phys Rev B **34** 5530
Rasolt M, Perrot F and MacDonald A H
  1985 Phys Rev Lett **55** 433
Rauh A, Wannier G H and Obermair G
  1974 Phys Status Solidi(b) **63** 215
Razeghi M, Hirtz P, Larivain J P, Blondeau R, de Cremoux B and Duchemin J P
  1981 Electron Lett **17** 643
Razeghi M, Duchemin J P and Portal J C
  1985 Appl Phys Lett **46** 46
Read N 1987 Bull Am Phys Soc **32** 923
  1989 Phys Rev Lett **62** 86
Rendell R W and Girvin S M 1981 Phys Rev B **23** 6610
Rezayi E H 1987a Phys Rev B **35** 3032
       1987b Phys Rev B **36** 5454
       1989   Phys Rev B **39** 13541
Rezayi E H and Haldane F D M 1985 Phys Rev B **32** 6924
                1988 Phys Rev Lett **61** 1985
                1989 Bull Am Phys Soc **34** 913
Ricketts B W 1985 J Phys D **18** 885
Riedel E K 1981 Physica **106A** 110
Rikken G L J A, Myron H W, Wyder P, Weimann G, Schlapp W, Horstman R E and Wolter J
  1985a J Phys C **18** L175
Rikken G L J A, Myron H W, Langerak C J G M and Sigg H
  1985b *Proceedings of Int Conf on Electronic Properties of Two-Dimensional Systems* **VI**
  [1986 Surface Science **170** 160]
Ringwood G A and Woodward L M

References

    1984 Phys Rev Lett **53** 1980
Rogers F J 1980 J Chem Phys **73** 6272
Roukes M L, Scherer A, Allen S J Jr, Craighead H G,
    Ruthen R M, Beebe E D and Harbison J P
    1987 Phys Rev Lett **59** 3011
Sakaki H, Hirakawa K, Yoshino J, Svensson S P,
    Sekiguchi Y, Hotta T and Nishii S
    1984 Surface Science **142** 306
Schiebel R A 1983 *Proc IEEE Int Conf Electron
    Devices* 711
Schlesinger Z, Allen S J, Hwang J C M and Platzman P M
    1984 Phys Rev B **30** 435
Schonfeld J F 1981 Nuclear Physics B **185** 157
Schulman L S 1981 *Techniques and Applications of Path
    Integration* (Wiley and Sons New York)
    (and references therein)
Seager C H and Pike G E 1974 Phys Rev B **10** 1421
Shah J, Pinczuk A, Gossard A C and Wiegmann W
    1985 Phys Rev Lett **54** 2045
Shiraki Y 1977 J Phys C **10** 4539
Sichel E K, Sample H H and Salerno J P
    1985 Phys Rev B **32** 6975
Simon Ch, Goldberg B B, Fang F F, Thomas M K and
    Wright S 1985 Phys Rev B **33** 1190
Singh R and Chakravarti S 1985 SUNY preprint published as
    1986 Phys Rev Lett **57** 245
Small G W 1983 IEEE Trans Instrum Meas **IM-34** 446
Smith R P, Closs H and Stiles P J
    1984 Surface Science **142** 246
Smith T P and Stiles P J
    1984 Solid State Commun **52** 941
Smith T P, Goldberg B B, Stiles P J and Heiblum M
    1985a Phys Rev B **32** 2696
Smith T P, Heiblum M and Stiles P J 1985b *Proceedings
    of the 17th Int Conf on the Physics of Semi-
    conductors* ed D J Chadi and W A Harrison
    (Springer-Verlag Berlin Heidelberg New York
    Tokyo) 393
Smith T P III, Lee K Y, Hong J M, Knoedler C M,
    Arnot H and Kern D P 1988 Phys Rev B **38** 1558
Smrčka L 1984 J Phys C **17** L63
Smrčka L and Středa P 1977 J Phys C **10** 2153
Stauffer D 1979 Phys Reports **54** 2
Stein D, von Klitzing K and Weimann G
    1983 Phys Rev Lett **51** 130

Stell G 1964 *Equilibrium Theory of Classical Fluids*
    ed H L Frisch and J L Lebowitz (Benjamin
    New York) II-171
Störmer H L and Tsang W T
    1980 Appl Phys Lett **36** 685
Störmer H L and Tsui D C 1983 Science **220** 1241
Störmer H L, Dingle R, Gossard A C, Wiegmann W and
    Sturge M D 1979 Solid State Commun **29** 705
Störmer H L, Gossard A C and Wiegmann W
    1981 Appl Phys Lett **39** 712
Störmer H L, Tsui D C and Gossard A C
    1982 Surface Science **113** 32
Störmer H L, Chang A M, Tsui D C, Hwang J C M,
    Gossard A C and Wiegmann W
    1983a Phys Rev Lett **50** 1953
Störmer H L, Schlesinger Z, Chang A, Tsui D C,
    Gossard A C and Wiegmann W
    1983b Phys Rev Lett **51** 126
Störmer H L, Tsui D C, Gossard A C and Hwang J C M
    1983c Physica **117B** & **118B** 688
Störmer H L, Haavasoja T, Narayanamurti V, Gossard A C
    and Wiegmann W
    1983d J Vacuum Sci Technol B **1** 423
Störmer H L, Chang A M, Tsui D C and Hwang J C M 1985
    *Proceedings of the 17th Int Conf on the
    Physics of Semiconductors* ed D J Chadi and W A
    Harrison (Springer-Verlag Berlin Heidelberg
    New York Tokyo) 267
Störmer H L and Willett R L 1989 Phys Rev Lett **62** 972
Středa P 1982 J Phys C **15** L717
    1983 J Phys C **16** L369
Středa P and Oji H 1984 Phys Lett A **102** 201
Středa P and Smrčka L 1983 J Phys C **16** L195
Středa P and von Klitzing K 1984 J Phys C **17** L483
Su W P 1984 Phys Rev B **30** 1069
    1985 Phys Rev B **32** 2617
Su W P and Schrieffer J R 1981 Phys Rev Lett **46** 738
Su W P, Schrieffer J R and Heeger R J
    1979 Phys Rev Lett **42** 1698
Su Z-B, Pang H-B, Xie Y-B and Chou K-C
    1987 Phys Lett **A123** 249
Syphers D A, Fang F F and Stiles P J
    1984 Surface Science **142** 208
Syphers D A and Furneaux J E
    1988a Surface Science **196** 252

References

            1988b Solid State Commun **65** 1513
Tao R 1984 Phys Rev B **29** 636
        1985 J Phys C **18** L1003
Tao R and Thouless D J 1983 Phys Rev B **28** 1142
Tao R and Haldane F D M 1986 Phys Rev B **33** 3844
Tao R and Wu Y-S 1984 Phys Rev B **30** 1087
               1985 Phys Rev B **31** 6859
Tausenfreund B and von Klitzing K
    1984 Surface Science **142** 220
Taylor B N 1985 NBS Journ Research **90** 91
Theis T 1980 Surface Science **98** 515 (and references therein)
Thompson A M 1968 Metrologia **4** 1
Thouless D J 1977   Phys Rev Lett **39** 1167
               1981   J Phys C **14** 3475
               1982   In: *Proceedings of the Taniguchi International Conference on Anderson Localization* eds Y Nagaoka and H Fukuyama, Springer Series in Solid State Science (Springer-Verlag Berlin Heidelberg New York Tokyo) 191
               1983a Phys Rev B **27** 6083
               1983b Phys Rev B **28** 4272
               1984a Phys Reports **110** 279
               1984b J Phys C **17** L325
               1984c *XIIIth Int Colloquium on Group Theoretical Methods in Physics* ed W W Zachary (World Scientific Singapore) 319
               1985   J Phys C **18** 5211
Thouless D J, Kohmoto M, Nightingale M P and den Nijs M 1982 Phys Rev Lett **49** 405
Thouless D J and Li Q 1987 Phys Rev B **36** 4581
Timp G, Mankiewich P M, DeVegvar P, Behringer R, Cunningham J E, Howard R E, Baranger H U and Jain J K 1989 Phys Rev B **39** 6227
Tinkham M 1975 *Introduction to Superconductivity* (McGraw-Hill New York)
Tosatti E and Parrinello M
    1983 Lett Nuovo Cimento **36** 289
Trugman S A 1983 Phys Rev B **27** 7539
Trugman S A and Doniach S B 1982 Phys Rev B **31** 5280
Trugman S A and Kivelson S 1985 Phys Rev B **26** 3682
Tsui D C and Allen S J 1981 Phys Rev B **24** 4082
Tsui D C and Gossard A C 1981 Appl Phys Lett **38** 550
Tsui D C and Kaminsky G 1979 Phys Rev Lett **42** 595

Tsui D C, Englert Th, Cho A Y and Gossard A C
    1980 Phys Rev Lett **44** 341
Tsui D C, Gossard A C, Field B F, Cage M E and
    Dziuba R F 1982a Phys Rev Lett **48** 3
Tsui D C, Störmer H L and Gossard A C
    1982b Phys Rev Lett **48** 1559
Tsui D C, Störmer H L and Gossard A C
    1982c Phys Rev B **25** 1405
Tsui D C, Störmer H L, Hwang J C M, Brooks J S and
    Naughton M J 1983a Phys Rev B **28** 2274
Tsui D C, Dolan G and Gossard A C
    1983b Bull Am Phys Soc **28** 365
Tsukada M 1976 J Phys Soc Japan **41** 1466
van der Wel W, Harmans C P J M and Mooij J E
    1985 *Proceedings of Int Conf on Electronic
    Properties of Two-Dimensional Systems* **VI**
    [1986 Surface Science **170** 226]
Van Dyke R S Jr, Schwinberg P B and Dehmelt H G
    1981 *Atomic Physics* **7** ed D Kleppner and
    F M Pipkin (Plenum New York) 340
van Wees B J, van Houten H, Beenakker C W J,
    Williamson J G, Kouwenhoven L P, van der Marel D
    and Foxon C T 1988 Phys Rev Lett **60** 848
Villain J 1977 J Phys C **10** 1717
Vinen W F 1961 Proc Royal Soc London **A260** 218
Vollhardt D and Wölfle P 1980 Phys Rev B **22** 4666
von Klitzing K 1983 In: *Atomic Physics* **8**
    eds I Linden, J Rosen and S Svanberg
    (Plenum New York) 43
        1984 Physica **126B** 242
von Klitzing K, Dorda G and Pepper M
    1980 Phys Rev Lett **45** 494
von Klitzing K, Tausendfreund B, Obloh H and Herzog T
    1983 in *Lecture Notes in Physics* **177** ed G
    Landwehr (Springer-Verlag Berlin Heidelberg
    New York Tokyo) 1
von Klitzing K and Ebert G, 1984 In: Springer Series in
    Solid State Sciences **53** eds G Bauer,
    F Kuchar and H Heinrich, (Springer-Verlag
    Berlin Heidelberg New York Tokyo) 242
von Klitzing K, Ebert G, Kleinmichel N, Obloh H,
    Dorda G and Weimann G 1984 *Proceedings of the
    17th Int Conf on the Physics of Semi-
    conductors* ed D J Chadi and W A Harrison
    (Springer-Verlag Berlin Heidelberg New York

References

Tokyo) 271
Wagner R J, Kennedy T A, McCombe B D and Tsui D C
   1980 Phys Rev B **22** 945
Wagner R J, Lavine C F, Cage M E, Dziuba R F and
   Field B F 1982 Surface Science **113** 10
Wakabayashi J and Kawaji S
   1978 J Phys Soc Japan **44** 1839
   1980 Surface Science **98** 299
Wakabayashi J, Kawaji S, Yoshino J and Sakaki H 1985
   In: *Proc of Electronic Properties of Two-Dimensional Systems* **VI**
   [Surface Science **170** 136 1986]
Wakabayashi J, Kawaji S, Yoshino J and Sakaki H
   1986 J Phys Soc Japan **55** 1319
Wakabayashi J, Fukano A, Kawaji S, Koike Y
   and Fukase T 1988 J Phys Soc Jpn **57** 3678
Wang Z Z, Saint-Lager M C, Monceau P, Renard R,
   Greisser P, Meershaut A, Guemas L and Rouxel J
   1983 Solid State Commun **46** 325 (and references therein)
Wannier G H 1978 Phys Status Solidi(b) **88** 757
Washburn S, Webb R A, Mendez E E, Chang L L and
   Esaki L 1985 Phys Rev B **31** 1198
Washburn S, Fowler A B, Schmid H and Kern D
   1988 Phys Rev Lett **61** 2801
Wegner F 1979a,b Z Physik B **36** 209; B**38** 207
   1983   Z Physik B **51** 279
Wei H P, Tsui D C and Razeghi M
   1984 Appl Phys Lett **45** 666
Wei H P, Chang A M, Tsui D C and Razeghi M
   1985a Phys Rev B **32** 7016
Wei H P, Chang A M, Tsui D C, Pruisken A M M and
   Razeghi M 1985b *Proceedings of Int Conf on Electronic Properties of Two-Dimensional Systems* **VI** [1986 Surface Science **170** 238]
Wei H P, Tsui D C and Pruisken A M M
   1986 Phys Rev B **33** 1488
Wei H P, Tsui D C, Paalanen M A, Pruisken A M M
   1988 Phys Rev Lett **61** 1294
Weiss D, Stahl E, Weimann G, Ploog K and von Klitzing
   1985 In: *Proc of Electronic Properties of Two-Dimensional Systems* **VI**
   [Surface Science **170** 285 1986]
Wen X G, Wilczek F and Zee A 1989 Phys Rev B **39** 11413
Wen X G and Zee A 1988 Phys Rev Lett **61** 1025

Wharam D A, Thornton T J, Newbury R, Pepper M,
    Ahmed H, Frost J E F, Hasko D G, Peacock D C,
    Ritchie D A and Jones G A C 1988 J Phys C **21** L209
Widom A and Clark T D 1982 J Phys D **15** L181
Wiegmann P B 1988 Phys Rev Lett **60** 821
    1989 Physica Scripta **T27** 160
Wilczek F 1982 Phys Rev Lett **49** 957
Wilczek F and Zee A 1983 Phys Rev Lett **51** 2250
Willett R L, Eisenstein J P, Störmer H L,
    Tsui D C, Gossard A C and English J H
    1987 Phys Rev Lett **59** 1776
Willett R L, Störmer H L, Tsui D C, Gossard A C
    and English J H 1989 Phys Rev B **37** 8476
Willett R L, Störmer H L, Tsui D C, Pfeiffer L N,
    West K W, Baldwin K W 1988 Phys Rev B **38** 7881
Wilson B A, Allen, Jr S J and Tsui D C
    1981 Phys Rev B **24** 5887
Witten E 1979 Nuclear Physics B **149** 285
Woltjer R, Mooren J, Wolter J and André J P
    1985 Solid State Commun **53** 331
Wu Y S 1984 Phys Rev Lett **53** 111
Wysokinski K I and Brenig W 1983 Z Phys B **54** 11
Yang C N 1962 Rev Mod Phys **34** 694
Yoshihiro K, Kinoshita J, Inagaki K, Yamanouchi C,
    Moriyama J and Kawaji S 1983 Physica B **117** 706
Yoshihiro K, Kinoshita J, Inagaki K, Yamanouchi C,
    Endo T, Murayama Y, Koyanagi M, Wakabayashi J
    and Kawaji S 1985a (preprint)
Yoshihiro K, Kinoshita J, Inagaki K, Yamanouchi C,
    Murayama Y, Endo T, Koyanagi M, Wakabayashi J
    and Kawaji S 1985b In: *Proc of Electronic
    Properties of Two-Dimensional Systems* **VI**
    [Surface Science **170** 193 1986]
Yoshioka D J 1984a Phys Rev B **29** 6833
    1984b J Phys Soc Japan **53** 3740
    1985a In: *Proc of Electronic Properties
    of Two-Dimensional Systems* **VI**
    [Surface Science **170** 125 1986]
    1985b preprint published as:
        [1986 J Phys Soc Japan **55** 885]
        1986 J Phys Soc Japan **55** 3960
Yoshioka D J and Fukuyama H
    1979 J Phys Soc Japan **47** 394
Yoshioka D J and Lee P A 1983 Phys Rev B **27** 4986
Yoshioka D J, Halperin B I and Lee P A

References

        1983 Phys Rev Lett **50** 1219
Yoshioka D J, MacDonald A H and Girvin S M
        1988 Phys Rev B **38** 3636
        1989 Phys Rev B **39** 1932
Zak J 1964 Phys Rev A **134** 1607
Zavaritskiĭ V N and Anzin V B
        1983 Pis'ma Zh Eksp Teor Fiz **38** 249
        [1983 JETP Lett **38** 294]
Zawadzki W and Lassnig R
        1984a Solid State Commun **50** 537
        1984b Surface Science **142** 225
Zhang F C and Chakraborty T 1984 Phys Rev B **30** 7320
Zhang F C, Vulovic V Z, Guo Y and Das Sarma S
        1985 Phys Rev B **32** 6920
Zhang F C and Das Sarma S
        1986 Phys Rev B **33** 2903
Zhang S C, Hansson T H and Kivelson S
        1989 Phys Rev Lett **62** 82
Zheng H Z, Tsui D C and Chang A M
        1985a Phys Rev B **32** 5506
Zheng H Z, Choi K K, Tsui D C and Weimann G
        1985b *Proceedings of Int Conf on Electronic Properties of Two-Dimensional Systems* **VI**
        [1986 Surface Science **170** 209]
Zil'berman G E 1957a Zh Eksp Teor Fiz **32** 296
        [1957a Sov Phys JETP **5** 208]
        1957b Zh Eksp Teor Fiz **33** 387
        [1957b Sov Phys JETP **6** 299]
Zou Z, Doucot B and Shastry B S 1989 Phys Rev B **39** 11424

Index

Absolute resistance standard, 2, 44-50
  measurement uncertainties of, 45, 49, 50, 385
Activation energy
  in FQHE, 179-180, 184, 377, 201-206, 215, 231, 385, 437; *see also* Excitation energy gap in FQHE
    disorder dependence, 184, 204, 205, 377-378, 401
    as a function of the filling factor, 206, 207, 208, 212, 218
    magnetic field dependence, 184, 201, 203, 206
    sample mobility dependence, 184, 201-202
    theoretical estimate, 204, 377
  in IQHE, 52, 382-383
AFM, see Heisenberg antiferromagnet
Aharonov-Bohm effect, 18, 323-324, 402, 408
Anderson-Higgs mechanism, 372, 390
Anderson localization, see Localization
Angular momentum
  azimuthal, 236
  conservation of, 245
  eigenstates of, 35, 235-236, 240, 245, 246, 271, 272, 277, 320
  relative angular momentum of quasiparticles, 320
  relative angular momentum operator, 306, 312
  total z-component operator of an N-particle system, 306
Anyon, 404
  effective statistics, 408
  mean-field theory, 410
  vector potential, 407
Asymptotic freedom, 161
Axial anomalies in (2+1)-D field theories, 398
Azimuthal symmetry, see Rotational invariance in 2-D

Back-flow, 348
Backscattering, 76
Basis states of magnetophonons (magnetorotons), 386-387
BCS model of superconductivity, 352, 389, 413
Berry's phase, 408, 422-423
Boson representation of Heisenberg model, 419
Boundary conditions
  disk (or ring) geometry, 18, 31, 42, 83, 252-254, 312, 328
  in effective field theory of localization, 153
  periodic boundary conditions (toroidal geometry), 31, 78-79, 104, 109, 112, 326-329
  spherical geometry, 31, 326, 329
Breakdown of dissipationlessness in IQHE devices, 58-62, 88, 92-93

Bulk nature of QHE, 18, 32
Bulk leakage current, 55

Carrier density in two-dimensional devices, 42, 282
Center of mass motion, 236, 245, 328, 359-360
Charge-density operator, 292
Charge-density wave (CDW), see Wigner crystal
Charge neutrality condition in plasma, 247-248, 255, 405, 421
Chern (topological) number, 109, 328, 394
Chiral anomaly in QHE, 111
Chiral spin state (liquid), 403, 423, 425, 429
  order parameter, 426
Chirality operator, 426, 428
Circulation, 102, 368
  quantization of, 102, 368, 393
Classical plasma, see Plasma
Classical probability distribution function, 246, 404
Coherent potential approximation (CPA), 120, 121, 147-148
Coleman's theorem, 160
Collective excitation modes, 353-354
  in Feynman's theory of 4He, 354-357
  in FQHE (neutral elementary excitations), 230, 280, 321, 336-337, 343-344, 347
    density-wave approach to, 357-365, 373-377; see also Single mode approximation/Density-wave excited state
    energy, 298, 344-349, 360-362, 373, 375
    mean-field, 414
    quasiexciton interpretation of, 373-374
Conductance, 2, 7
  Hall conductance as a topological invariant, 108-110, 112-113, 118, 146, 381
  in electronic disorder problem,
    'bare', in effective field theory, 151, 152
    Green's function formulation of, 112-113, 126-127, 135-136
    renormalized, in scaling theory, 165-167
    SCBA, 128, 153-158, 170, 171
  Kubo formula, 80, 107-108
Conductivity tensor, 1, 7, 30
  in FQHE, 3, 183
    temperature dependence, 176, 184
  in IQHE, 1, 2, 38, 383-384
    frequency dependent conductivity, 87, 91
    temperature dependence, 51-55, 170-171
  Kubo formula for, 108
  in ordinary Hall effect, 6
Conformal invariance of the action, 161

Continuity equation, 368
Corbino disk, 18, 42, 83 230
Correlation function,
  one point, 247
  two-point (or pair), 249, 266-268, 270-271, 274-276, 333, 361
  computational methods for, 254-257
Coset space, 158
Coulomb interaction in 2-D, 22, 26, 29, 34, 113, 175, 179, 184, 204, 247, 362
  energy, 31, 358, 388
  in FQHE finite-size studies, see Lowest-Landau-level/Second-Landau-level Coulomb interaction
  modified, 184, 205, 377, 435
  pseudo-Coulomb potential in the 2DOCP, 247, 363
  screening of, 30, 235
CP(N-1) model, 158, 169
Critical current in 2-d QHE devices, 58
  influence of magnetic field, 59
Critical exponents in scaling theory of localization, 159, 160, 168, 173
Crystallization transition, see Wigner-crystal transition
Current density operator, 367
  in a magnetic field, 369
Current distribution in two-dimensional QHE devices, 56-58
Current-voltage characteristics in FQHE, 195
Cusp in ground state energy of FQHE, 291, 316, 329-330, 386, 394-395
Cyclotron energy (or gap), 31, 236, 279, 305, 310, 388
Cyclotron frequency, 7, 178, 235
Cyclotron radius, 13, 156
Cyclotron resonance in Si inversion layer, 375, 378, 388
Cylindrical geometry, 111

Daughter state, 189-190, 233, 288, 320-325
Debye length, 270, 278
Degeneracy of FQHE ground state, 235-238, 260, 280, 313, 375, 418
Density matrix (one body), 406
  of FQHE ground state, 407, 426
    in the singular gauge, 408-409
    second quantised expression of 406
  of spin systems (boson rapresentation), 427
  two-body, 427
Density of carriers in 2-D conducting systems, 39-42, 180
Density of states in disorder-broadened Landau bands, 12, 15-16, 206, 212, 382
  calculation of, 120-122, 129-130, 147-148, 176

Index

SCBA, 122, 147-148
  gap in, 137
  measurements of, in a strong magnetic field, 211-218
Density operator, 355
  projected on the nth Landau level, 358, 387, 396
  commutation relation, 387, 396
Density-wave excited state
  in Feynman's theory of helium, 355-357
  in FQHE, 357-360; *see also* Collective excitation modes
  inter-Landau level (magnetoplasmon), 374-376
  intra-Landau-level (magnetophonon, magnetoroton), 360-364, 373-374
  spin-density wave, 376
  valley-polarization wave, 376
  wave-vector dependence, 359-360
Dielectric constant in 2-D conducting systems, 26, 29, 179, 181, 204
Diffusion constant, 156
Diffusion propagator, 135
Discrete broken symmetry in FQHE, 113, 209, 210, 328
Disk geometry scheme, 246-254, 312
Disorder
  role of disorder in FQHE, 176, 179, 204-205, 223, 231, 234, 236, 377-378, 385, 395-396, 402
  role of disorder in IQHE, 12-17, 63-64, 69, 117, 128-130, 402, 381-182; *see also* Localization theory
Disorder-broadened quasiparticle band, see Quasiparticle band
Dissipation
  flux-flow in type-II superconductors, 372
  lack of, in QHE at zero temperature, 2, 381
  thermal activation of, in QHE, 372, 377, 382-383
Domain walls, 209, 210
Drift velocity, 13, 225, 373

Edge current, 118, 123-127, 135
  connection with topology, 137-139
  diamagnetic, 150, 153
Edge effects in 2-D QHE devices, 56-58
Edge states, 15, 18-21, 32, 123, 135, 139-144, 220
Effective field theory of localization in IQHE, 146-158
  'bare' conductance, 151-158
  critical fluctuations, 149-152
  renormalized conductance, 165-167
  saddle point of, 147-148
  scaling implications, 158-173
  topological term, 139, 144, 150, 161, 166

Effective mass in 2-D conducting systems, 26, 28, 33, 39-42, 181
Effective potential in 2-D conducting systems, 25, 29, 204, 205
Einstein relation, 156
Electrical metrology, 37, 384
Electrochemical potential, 15
Electron density
  for Hartree-Fock Wigner crystal wave function, 251
  for Laughlin ground state wave function, 249, 250-252
  for quasielectron wave function, 264
  for quasihole wave function, 264, 266, 273, 274
Electron-electron interaction, see Coulomb interaction
Electron-phantom interaction, 270
Electronic disorder problem, see Localization theory/Disorder
Elementary excitations in quantum systems, 353-354
Elementary excitations in FQHE
  fractionally charged (quasiparticles), 175, 177, 197, 201, 260; *see also* Quasiparticle
  neutral, 230, 344, 357-360; *see also* Collective excitation modes in FQHE
Energy (per electron) of FQHE ground state
  in finite-size numerical studies, 329-333
  Hartree-Fock Wigner crystal wave function, 253
  Laughlin ground state wave function, 249
    approximate formula, 252, 253, 329
  as a function of the filling factor, 291
Ensemble average over random potentials, 132-133
Even denominator filling factor, 190, 401-402, 430-435; *see also* Spin-unpolarized states
Exact diagonalization of few particle-systems, 204, 304, 325, 328, 385, 386, 402
Exactness of the FQHE, 276-280
Excitation energy gap in FQHE, 175-176, 179-180, 184, 207, 303, 363, 377, 389, 418, 437; *see also* Activation energy
  filling factor dependence, 362-363, 347
  finite-size system numerical results, 344-350
  quasielectron-quasihole, 296, 298, 342-345, 347, 377
  single mode approximation, 362-263, 405
Excitation spectrum in FQHE, 5, 175-177, 184, 197, 230, 303
  in finite-size studies, 343-352
  in single mode approximation, 362

Exciton (electron-hole pair), 290-293, 295
Extended states
  in disorder-broadened Landau bands, 14-17, 20-22, 73, 77-80, 82-83, 88-89, 92, 129, 130, 169-170, 175, 206, 215, 220
    energy shift of, 222-223, 224, 227
  importance of, in QHE, 12, 17, 69, 117, 176, 222
  in Landau levels, 11
  percolation nature of, 84, 87, 220

Feynman-Bijl formula, 356, 361
Feynman-Hellmann theorem, 73, 79
Feynman's theory of ^{4}He, 348, 354-357
Field-theoretic approach to localization theory, 117, 132-173
  averaged density of states, 120-122, 147-148, 176
  averaged single-particle propagators, 119, 125, 132-133, 135
  continuous local gauge symmetry between advanced and retarded fields, 132-133, 144-146, 147, 148, 150
  effective field theory, 146-158; see also Effective field theory
  generating functional, 132
  IQHE in field-theoretic language, 144-146
  replica method, 132-133
  scaling implications of, see Scaling theory
  source-term formalism in, 134-136, 150, 165
    conductance, 135-136, 151, 152, 153-158, 165-167
  topological classification of field configurations and
    topological charges, 135, 137-139, 144, 153, 161-164
Filling factor of Landau levels, 11
  fractional filling in FQHE, 177, 183, 186, 188, 311, 316
    as a continued fraction, 190, 323
    even-denominator, 190, 401-402, 430-435
    observed in FQHE, 183, 185-187, 190, 231, 234
    for spinless boson systems, 314, 316
    for spinless fermion systems, 248, 314, 316
    for spin-1/2 fermion systems, 311, 315-316
  integer filling in IQHE, 11, 38
Fine-structure constant, alfa, 45
  measurement of alfa through a measurement of the
  Quantum Hall resistance, 51, 384
Fixed points in scaling theory of IQHE, 159, 168, 173

Flow diagram in 2-parameter scaling theory of IQHE, 168, 223, 224
Fluctuation-dissipation theorem, 363
Flux quantization condition, 102-103, 392-393
Flux quantum unit, 19
  density of, 379
Fluxoid, 102, 395
Fock cyclic condition, 433, 434
FQHE, see Fractional Quantum Hall Effect
Fractional charge of quasiparticles, 175, 177, 193, 229-230, 233, 260-261, 270, 276-280, 324, 342, 402, 422
  experimental test of, 230, 402
Fractional Quantum Hall Effect (FQHE), 3, 175-177, 183-184, 233-234
  basic experiment, 177
  breakdown of FQHE
    due to disorder, 223, 231, 235
    at low filling factors, 230, 231
  conditions for observation of, 177-183
  Coulomb interaction in FQHE, 22, 175, 179, 184, 209, 235, 236 263
  discrete broken symmetry in, 113, 209, 210,
  integer vs. fractional QHE, 173
Fractional quantum numbers, 192
Fractional statistics, 209, 321, 323, 422
  wave function, 283, 288
Frustration ansatz, 390-391, 403

Gapless excitations, 350-351, 372, 376
Gauge argument for IQHE, 4, 17-22, 73, 103, 131, 192, 201
Gauge transformation
  equivalent to a translation, 358, 380
  singular, 407
Gaussian localization, 91
Goldstone mode, 34, 144, 147-148, 159, 160, 362
Grassmann variables, 133
Greek-Roman wave function, 431-435
Green's function (single particle)
  advanced, 119
  diffusion propagator, 136
  in electronic disorder problem, 119, 125
    field-theory expression of, 132-133
  free, in an electric field, 95
  retarded, 119
  in terms of a self energy, 125
Ground state for FQHE, 3, 175, 176, 237, 277, 278, 386
  degeneracy of, 235-238, 260, 280, 313, 330, 344
  energy, see Energy of FQHE ground state
  Laughlin ground state, see Laughlin ground state for FQHE
  pseudo-crystal ground state, 229, 386

Index

Wigner crystal ground state, see Wigner crystal
X-ray diffraction structure factor of, 229
Group velocity, 13, 373
Guiding center coordinates, 306, 312
Gutzwiller projection, 412

Half-integral spin excitations, 417, 423-425
Hall angle, 54, 384
Hall conductance, 2, 7
  electron-electron interaction effect on, 113
  as a topological invariant, 103, 108-110, 112-113, 118, 146, 381
Hall resistance, 8-9, 38
  absolute resistance standard, 44-50, 384-385
Hall steps, see Plateaus
Hall voltage, 8, 38, 43, 44
  distribution in a strong magnetic field, 212, 217
Halperin states, 311, 316, 431; see also spin-unpolarized states
Hamiltonian
  with chiral spin interaction, 429
  Heisenberg, 419
  for interacting electrons in a magnetic field, 234, 305
  singular gauge transformation of, 407
  for spin systems
Hard-core boson, 408
Harper equation, 106
Hartree-Fock charge-density-wave state, see Wigner crystal ground state
Heat capacity measurements, 382
Heisenberg antiferromagnet (AFM)
  ground state, 418, 419, 424, 427
  hamiltonian, 419
  Néel state, 419
  in one dimension, 417
  partition function, 424
  spin-current operator, 425
  triangular lattice, 426, 427
Heterojunctions, 4, 27
Heterostructures, 27-30, 39-42, 177, 180-183, 186, 401; see also 2-D conducting systems
  advantages of, 27, 28
  Coulomb interaction in, 29, 184, 204
  finite extent of electrons in the third direction, 204
Hierarchy of FQHE sequence (Hierarchical model), 3, 5, 176, 184-192, 201, 231, 289-290, 303, 321-325, 372, 385, 430
  spin-reversed, 402, 436-437
High-temperature superconductivity (HTSC), 403, 417, 429-430
Hollow-core model, see Spin-unpolarized states
Holon, 429

Homotopy properties of QHE, 4
Hopping conduction, see Variable-range hopping conductance
HTSC, see High-temperature superconductivity
Hubbard-Stratonovich transformation, 396
Hypernetted-chain method (HCM)
  one parameter, 254-257
  for radial distribution function of Laughlin ground state, 250-251
  for structure factor of Laughlin ground state, 337
  two-parameter, 266-268

Impurities, 3, 12-17
  density of states in impurity potentials, 12, 16
  effects in QHE, 12-17, 69-99
  potential, 70-74
Incompressibility of Laughlin ground state, 3, 190, 209, 246, 309, 362-363, 371, 394
Independent electron approximation in QHE, 3, 22, 25
Inelastic length scale, see Thouless length
Instantons, 118, 161, 162
  contribution of a single instanton to the free energy, 163-165
  dilute instanton gas approximation, 167
Integer Quantum Hall Effect (IQHE), 1-2, 37-38
  basic experiment, 42-44
  conditions for observation of, 1, 11-12, 69, 70
  at finite temperature, 382-384
  integer vs fractional QHE, 173
  numerical studies, 89-93
  spin effects in, 31
Inter-particle spacing
  in Laughlin ground state, 237, 309, 325
  in two-electron problem, 237
Inversion layers, 4, 22-30; see also 2-D conducting systems
  band gap in, 181
  resonance line shape in Si inversion layers, 375, 389
  tunneling into, 389
Ionic disk
  energy, 253-254
  radius, 270
IQHE, see Integer Quantum Hall Effect

Jastrow product, 244
Josephson effect, 398

Kohn's theorem, 374, 376, 379
Kosterlitz-Thouless phase transition, 210, 369, 371, 410
Kubo formula 107-108
  failure of, 86-87, 99

Kubo formula (*contd.*)
  in periodic potential with magnetic field, 107-108, 112
  in scattering potential, 80

Lambda transition in helium, 406, 415
Landau bands, disorder-broadened, 12
  density of states in, 12, 15, 16, 92, 129-130, 147-148, 206, 211-217, 382
  energy levels, 13
  extended states, 14-17
  localized states, 14-17
Landau diamagnetism, 32
Landau gauge, 9, 10
Landau levels, 10, 119, 236
  degeneracy of, 10, 11, 119, 156, 235, 388
  energy levels, 10, 119, 236
  extended states, 11
  filling factor of, 11
  localized states in, 11,
  mixing of higher, 91, 204, 205, 212, 224, 277, 280, 358, 388, 435, 436
  number of states in, 10
  spacing between, 31
Landau-Ginsburg macroscopic order parameter, 368, 390
Landau-Ginsburg theory of FQHE, 372, 389-398, 403
  fictitious vector potential, 390, 397, 403-404
  flux quantization, 393
  fluxoid, 395
  free-energy functional, 390, 394
  frustration, 390-391,403
  microscopical justification of, 396-398
  phonon gap, 394-395
  in terms of hard-core bosons, 408, 410
  topological contribution to the effective action, 394, 397
  vortex state, 392-393
Laughlin ground state for FQHE, 189-190, 209, 229, 233, 303-304, 310, 314, 351-352, 354, 361, 385, 410-411
  exact wave function, 277-278
  generalized Laughlin states (non-variational derivation)
    for spinless particle systems, 313-314
    for spin-1/2 electron systems, 314-315
  numerical analysis in finite-size systems, 329, 332-333, 336-339, 349
  principal daughter states, 320-325
  variational wave function in disk geometry, 246, 405
    electron density of, 247-248, 261, 277 280
    energy of, 249, 252-253, 291, 329, 332
    equivalent one component plasma to, 247-249, 265-266
    Halperin's interpretation of, 277-278, 404-405
    radial distribution function of, 249-252
    structure factor of, 299, 336-337, 339, 361, 405
  variational wave function in periodic geometry, 328-329
  variational wave function in spherical geometry, 328-329
Laughlin-Halperin theory of IQHE, see Gauge argument for IQHE
Levinson's theorem, 77
Light scattering, see Raman scattering
Localization length, 90, 120, 201
  power-law divergence of, 402
Localization theory, 2, 4, 74, 84, 96, 117, 120-121,
  conventional localization, 158-161, 166
  density of states in, 120-122
  field-theoretic approach to, see Field-theoretic approach to localization theory
  in FQHE, 236, 378, 402
  in IQHE, 128-131, 381, 402
  in a magnetic field, 161-171, 382
  magnetization in, 122-126
  scaling of, see Scaling theory
  strong localization, 159
  transport parameters in, 126-129
  weak localization, 117, 137, 159, 170, 225
Localized states
  in disorder-broadened Landau bands, 14-17, 20-22, 32, 77, 79, 82, 88-90, 92, 129-131, 206-207, 213, 215
    single-particle density of, 15, 16, 206, 382
  in Landau levels, 11
  of quasiparticles in FQHE, 236, 378, 315
  relevance to IQHE, 12-17, 63-64, 69, 117, 128-130, 381
  in rotational invariant gauge, 30
Longitudinal collective mode (upper-hybrid mode), 354
Lorentzian approximation, 12, 31
Lorentzian density of states, 12
Lowest-Landau-level approximation, 15, 31, 90
  validity of, 72
Lowest-Landau-level Coulomb interaction (in FQHE finite-size numerical studies)
  energy-level spectrum, 343-350
  ground state energy, 332
  structure factor, 336-337, 340-341
  truncated pseudopotential method, 348-350
Luminescence measurements, 402

Madelung energy, 252

Index

Magnetic dependence of the transport parameters, 170-171
Magnetic length, 10, 70, 201
  of a quasiexciton, 296, 373
Magnetic moment of a 2D electron gas, 66
Magnetic monopole, 209
Magnetic units, 12, 119
Magnetization in 2D electronic disorder problem
  calculation of the disorder-averaged magnetization, 122-126
  experimental measurement of, 213-216, 382
Magnetocapacitance of a 2D electron gas, 66, 382
Magnetophonon, 359, 360-364, 386-387
Magneto-photoresponse of a 2D electron gas, 66
Magnetoplasmon, 359, 374-376
Magnetoroton minimum, 230, 280, 298-299, 336, 359-360, 363-364, 373, 378, 386-387, 438
  mean-field theory of, 364
  two-magnetoroton bound state, 365, 372, 387
Many-body effects in IQHE, 52, 113, 381
Many-quasiparticle state, 283-285
Mass action, law of, 231, 377
Massless excitations, see Goldstone modes
MBE, see Molecular Beam epitaxy
Mean-field theory of FQHE, see Landau-Ginsburg theory of FQHE
Mermin-Wagner theorem, 160
Metal-insulator transition, 120, 159
Metal Oxide Semiconductor Field Effect Transistor (MOSFET), 23, 39-42, 180, 180, 181
Metrological applications of IQHE, 2, 37, 44-51, 384-385
Microwave conductivity, 364, 389
Minima in the longitudinal (diagonal) conductivity/resistivity
  in FQHE, 3, 183, 185-187
  in IQHE, 2, 9, 43-44, 62-65
Mixing of higher Landau levels, 91, 204, 205, 212, 224, 277, 280, 358, 388, 435
Mobility edge, 159, 207, 215
Mobility gap, 20-21, 69, 73, 79, 88, 137, 201, 381
Mobility in 2-D conducting systems
  in Hall-step region, 63
  temperature dependence of, 182
  in zero magnetic field, 22, 24, 39-41, 180-182, 401
Mobility threshold, 202
Molecular Beam Epitaxy (MBE), 27, 33, 181, 401

Monte Carlo method for,
  charge density of a many-body-system wave function, 257-259
Monte Carlo results for
  electron density of quasiparticle wave function, 274
  radial distribution function of Laughlin ground state, 250
  structure factor of Laughlin's ground state
  two-dimensional one component plasma, 249
Mott insulator, 417
Multiple-connected geometry, 18, 368

Néel order parameter, 418
Néel state, 418, 419
  frustration of, 418, 419
Non-linear sigma model, 159, 161, 167, 222, 424

Oblique confinement, 398
Off-diagonal long-range order (ODLRO)
  in bose condensation, 406, 416
  in FQHE, 403-414
  analogy with BCS theory, 411-413
Order-disorder transition of FQHE
  in three-parameter scaling theory, 223, 378
  with the formation of a Wigner glass, 231, 375, 378, 396
Ordinary or classical Hall Effect, 6, 7
Oscillator strength, 356-357
  projected, 360-361, 363, 365, 379

Parity symmetry in 2-D, 325, 398
  breaking of
    in FQHE, 372, 391, 422
    in spin systems (spontaneously), 423, 425, 426
Partially polarized states, 311
Participation ratio, 160
Particle-hole symmetry, 388, 436-437
Percolation
  models, 57
  orbit, 98
  threshold, 87
  in a smooth potential, 83-88
Periodic geometry (in FQHE finite-size studies), 326-328
  excitation spectrum in, 345-346, 348, 350-351
  ground state energy in, 330-331
  quasiparticle charge analysis, 342
  structure factor in, 337-341
Phantom particle, 265
Phase fluctuations in FQHE ground state, 403-404
Phase shift, 77

Phase transition
  absence of, in FQHE, 184, 209-211
  in tilted field, 431, 437
Plasma
  two-dimensional one component plasma (2DOCP), 247, 248, 265, 278, 314, 363, 404
    charge density of, 248, 314
    charge neutrality condition, 247-248, 255, 421
    correlation functions, 249, 254-257, 266-267
    dimensionless coupling parameter of, 248, 254-257, 266, 314
    liquid nature of, 248-249
    plasma Debye's length, 270, 278
    potential energy, 247, 265
    with short-range forces, 278
    structure factor, 297, 299, 336-337
    sum rules, 276, 277, 314, 421
  two-dimensional two-component plasma, 316
Plateaus in the Hall resistance.
  in FQHE, 183, 185, 233
    widths, 378
  in IQHE, 9, 43-44
    widths and shapes of, 62-65, 382
Point defects in 2D, see Vortices
Pontryagin index, 162; see also topological invariants
Potential
  gaussian white noise random potential, 72
  impurity potential 70-74
  periodic potential with magnetic field, 104-110
  scattering potential, 74-80, 95
  smooth potential, 80-83
  weak potential, 73-74
Projection operator for the nth Landau level, 358
Propagator, see Green's function
Pseudopotential method, 303-316, 321, 429
Pseudopotential parameters, 307-308, 313, 320, 432
  for Coulomb potential, 307, 332
  ground-state energy in terms of, 331-332
  in hollow-core method, 432

Quantization condition, 1, 381, 384
Quantum antiferromagnet, see Heisenberg antiferromagnet
Quantum ferromagnet, 418
Quantum numbers
  fractional, 192
  integer, 101, 192, 276
  topological, see Topological quantum numbers
Quasiexciton, see Quasiparticle pair

Quasiparticle (quasielectron or quasihole), 190, 233, 234, 260, 269, 276, 316, 341, 395; see also Elementary excitations in FQHE
  analogy with particles, 269-276
  creation energy (charged disk approximation), 271
  fractional charge, see Fractional charge of quasiparticles
  identification with vortex, 370-371, 393-394, 422
  operators, 261, 262, 301
  quasielectron
    creation energy, 264, 269
    electron density, 264-268
    wave function, 262-263, 272,
  quasihole
    creation energy, 264, 267
    electron density, 264, 266
    mean-square orbit radius, 273-276
    wave function, 260-261, 272-274, 317
  relative angular momentum of, 320
  size of, 201, 270, 271
  for spin-1/2 fermion systems
  statistics of, 175, 205-209, 320-321, 323, 422
  as a topological entity, 210
Quasiparticle band, 207
  density of (localized and extended) states in, 211-217
  energy shift of extended states due to disorder, 223-227
  total number of states in, 207
Quasiparticle pair
  of like sign, 281-283, 286-289, 319-320
    separation between, 281
  of opposite sign (quasiexciton), 205, 233, 290, 293, 345, 347, 365, 373-374, 422
    contribution to the structure factor, 297, 299
    creation energy, 296, 298, 320, 342-343, 345, 347, 371, 373
    wave function, 293-294, 373
Quenched average over random fields, 173

Radial distribution function
  of classical liquid (computational methods), 254-257
  of Hartree-Fock Wigner crystal ground state wave function, 251
  of Laughlin's ground state wave function, 249-252, 361
  of lowest-Landau-level Coulomb interaction ground state (finite-size numerical studies), 333
Raman scattering, 389, 402
Renormalization group, 117; see also Scaling theory

Index

equations in scaling theory of localization, 159, 168, 224
  real-space, 386
Replica trick method, 131-132, 172
  replica symmetry breaking and restoration, 119, 132-133, 144-146
Residual interaction, 306
Resistance, 2, 7-9
  absolute resistance standard, 2, 44-50, 384-385
  fractional quantization of, 192, 193
  Hall resistance, 8
  integer quantization of, 8-9, 38, 192
Resistivity tensor, 1, 7, 30
  in FQHE, 3, 183, 185-189
    temperature dependence, 176, 184, 205-206, 208, 210, 213
    thermal activation of the dissipation, 197-201, 206-208, 211, 215, 372, 377
  in IQHE, 1, 2, 38, 186
    current dependence and critical current of, 58
    scattering parameter dependence, 7
    temperature dependence, 51-55, 170, 171, 205-206, 213
    thermal activation of dissipation, 52, 206, 215, 217, 382-384
  in ordinary Hall effect, 6
Resonance line shape in Si inversion layers, 375, 378
Resonating valence bond (RVB), 417, 429
Ring-exchange theory of FQHE, 386, 415-417
Rotational invariance, 101-102
  in 2D systems, 236, 305, 309, 325-327, 329, 341
Rotationally invariant symmetric gauge, 9, 35, 235, 239,
Roton minimum
  in FQHE, see Magnetoroton minimun
  in ^{4}He, 363
RVB, see Resonating valence bond

Scaling theory of (magneto-)transport in 2-dimensions, 117-118, 120, 137, 384; *see also* Renormalization group
  experimental verification of, 170, 177, 223-228, 402
  one-parameter scaling of conventional localization, 117, 158-161, 166
  three-parameter scaling theory in FQHE, 223, 378
  two-parameter scaling of localization in a magnetic field (IQHE), 118, 161-173, 222, 223
    flow diagram for the IQHE, 169, 223
    instantons, 161-165, 167-168
    phase transition, 169
    topological term, 139, 144, 150, 161, 166
Scattering potential, 74-80, 95
Scattering time in 2-D conducting systems, 39-41, 178,180, 181,382
SCBA, see Self-consistent Born approximation
Screening in inversion layers, 30
Second-Landau-level Coulomb interaction (FQHE finite-size studies)
  excitation spectrum, 350-351
  structure factor, 338, 340-341
  truncated pseudopotentials method, 350-351
Second-Landau-level states in FQHE, 431, 432
Self-consistent Born approximation (SCBA), 91, 118, 120, 224
  conductance in localization problem, 118, 127-128, 153-158
  density of states in localization problem, 91, 122, 147-148
  self-energy in localization problem, 121-122
Semion, 429-430
Shear stress, 354
Sigma model, see Non-linear sigma model
Single-mode approximation (SMA) ansatz, 297-299, 336-337, 346-348, 357-360, 386
  breakdown of, 365
  generalization to spin systems, 421
  for inter-Landau-level excitations, 358-359, 374-376
  for intra-Landau-level excitations, 358-364
  momentum conservation, 359-360
  multi-magnetophonon (-magnetoroton) basis states, 386-387
  quantitative accuracy of, 297, 299, 337, 364, 365
Single-particle eigenstates
  in gauge-invariant formalism, 312
  in Landau gauge, 10
  in symmetric gauge, 35, 235
Single-particle excitation, 354, 355, 357
Spherical geometry (in FQHE finite-size studies), 326, 328
  curvature effects in, 326, 333
  excitation spectrum, 344-347, 349
  ground state energy in, 329-330, 332
  radial distribution function in, 333-334
  structure factor in, 336-337
Spinon, 429
Spinor variables, 328-329
Spin correlation function, 420-421, 427-428
Spin-current density, 425
Spin-density wave excitation, 376-377, 417, 421, 424

Spin-1/2 fermion systems
  principal incompressible ground states of, 311, 314-316
  quasielectron exitations of, 318
  quasihole exitations of, 317, 318
  two-quasihole state, 320
Spin hamiltonian
  with chiral interaction, 429
  Heisenberg, 419
Spin liquid, singlet, 417-430; see also chiral spin state
  locally singlet, 419-420
    half-integral spin excitations in, 417, 423-425
    hamiltonian for exact ground state, 428
Spin resonance, 389
Spin-reversed ground states, 401, 402, 431, 436; see also Spin unpolarized/Partially polarized states
Spin-reversed quasiparticles, 436-437
Spin-unpolarized ground states, 310-311, 316, 401, 402, 436
  Halperin's wave function 431, 434
  hollow-core model, 434-435
Spin-wave excitation, 417, 421
Stability of Laughlin's incompressible states, 248, 303, 315
Staggered magnetization, see Néel order parameter
Static susceptibility, 364-366
Statistical flux, 407-408; see also Fractional statistics
  mean field, 410-411, 414
Statistics of charged elementary excitations in FQHE, see Quasiparticles/Fractional statistics
Structure factor of FQHE ground state
  direct observation of, 229, 389
  dynamic, 334,
  finite-size numerical studies results, 337-341
  fraction of $S(Q)$ contributed by collective mode, 337
  intrinsic (or guiding center) static, 314, 316
  projected static, 335, 346, 360-362
    peak in, 336, 363, 365
  static, 293, 295, 297-299, 335-336, 356, 361, 405
    Hartree-Fock, 375
Sum rule
  in classical plasma, 276-277, 421
  oscillator strength, 356-357, 363 379
Superlattice, 180
Symmetric polynomial, 272
Symmetry (continuous) between
  replica-fields, 132, 133, 144-146, 150, 158-173

Tao-Thouless theory of FQHE, 209
Temperature dependence of the transport parameters
  for FQHE, 176, 184, 197-201, 205-208, 210-211, 213, 215, 377
  for IQHE, 51-55, 170-171, 205-206, 213, 217, 382-384
Thermodynamic limit (in FQHE finite-size studies), 304, 325-326, 329, 335
Thermoelectric power, 389
Thermomagnetic transport in 2-d QHE devices, 65-66
Theta-vacuum angle, 118, 162, 394, 398, 403, 410
Thouless length, 171, 224, 225, 384
Thouless number, 90, 92
Three-electron problem, 239-242
Tilted magnetic field, 431, 436
Time reversal symmetry, 76, 325, 383
  breaking of
    in FQHE, 372, 391, 422-423
    in spin systems (spontaneously), 423, 425, 426
Topological aspects of IQHE, 101-116, 201, 381
Topological quantum number (or invariant), 101-103
  Hall conductance as a, 103, 108-110, 112-113, 146, 381
  quasiparticle in FQHE as, 210
  topological term in field-theory of localization problem, 118, 138-139, 143-144, 146, 153, 161
    identification with magnetic flux quanta, 143
  unperturbed topological excitation, see Instanton
  vortex state, see Vortex
Translational invariance, 11, 12, 320, 325, 329
  breakdown of, 12, 29, 69, 309
Translationally invariant geometries for finite-size studies, 325-329
Transverse collective mode (lower-hybrid mode), 354, 359, 362
Truncated Hamiltonians (or pseudopotentials), 303, 308; see also Pseudopotential parameters
  for Coulomb interaction, 348-352
Tunneling into inversion layers, 389
Two-dimensional conducting systems (inversion layers), 4, 22-30, 39, 42; see also Heterostructure
  carrier density in, 42
  Coulomb interaction in, 29, 204-205

Index

critical current for dissipation breakdown in, 58
current distribution in, 56
effective potential, 25, 29
material parameters, 40, 41, 180-183, 401
relevant to FQHE, 180-183
Two-electron problem, 236-238
inter-particle spacing, 237

Ultra-narrow channel, 402

Valley index, 33, 52, 303, 376
Valley-polarization wave, 376-377
Vandermonde polynomial, 248
Variable-range hopping conductivity, 52, 55, 198, 201, 383-384
Vector potential
Landau gauge, 10
singular gauge, 407
symmetry gauge, 34, 235
Vortex
in FQHE, 369-372
finite energy, 369, 371
fractional charge of, 370
fractional statistics of, 422
identification with Laughlin's quasiparticles, 370-371, 420
Landau-Ginsburg theory, 392-394
in superfluid helium, 209, 367-369

in type-II superconductors, 372

Ward identities, 133
Weak potential, 71, 73-74
Wigner crystal, 3, 176, 184, 195-196, 229-231, 237-238, 310, 343, 378, 386, 388, 401, 438
ground state wave function (Hartree Fock), 250
energy of, 237, 253, 329
radial distribution function, 251
instability, see Crystallization transition
phonon, 298-301
energy, 298
transition in FQHE, 5
upon increase of the range of the interaction, 348-352, 410
at low density, 230-231, 252, 319, 363-365, 389, 395, 401
Wigner glass, 231, 375, 378, 396

Young graph, 310, 315-316
Young symmetry, 315, 320

Zeeman energy, 31, 310-311, 343, 376, 431, 432, 436
Zero-point motion, 419
Zeros of FQHE ground state wave function, 277-278, 404-405, 409-410